STEPHEN JAY GOULD

EIGHT LITTLE PIGGIES

REFLECTIONS IN NATURAL HISTORY

PENGUIN BOOKS

PENGUIN BOOKS

Published by the Penguin Group
Penguin Books Ltd, 27 Wrights Lane, London W8 5TZ, England
Penguin Books USA Inc., 375 Hudson Street, New York, New York 10014, USA
Penguin Books Australia Ltd, Ringwood, Victoria, Australia
Penguin Books Canada Ltd, 10 Alcorn Avenue, Toronto, Ontario, Canada M4V 3B2
Penguin Books (NZ) Ltd, 182–190 Wairau Road, Auckland 10, New Zealand

Penguin Books Ltd, Registered Offices: Harmondsworth, Middlesex, England

First published in Great Britain by Jonathan Cape 1993
Published in Penguin Books 1994
5 7 9 10 8 6 4

Printed in England by Clays Ltd, St Ives plc

For Agnes Pilot

for her unfailing intelligence,
loyalty, and integrity

Contents

A Reflective Prologue   *11*

1 | THE SCALE OF EXTINCTION

    1   Unenchanted Evening   23
    2   The Golden Rule: A Proper Scale for Our
             Environmental Crisis   41
    3   Losing a Limpet   52

2 | ODD BITS OF VERTEBRATE ANATOMY

    4   Eight Little Piggies   63
    5   Bent Out of Shape   79
    6   An Earful of Jaw   95
    7   Full of Hot Air   109

3 | VOX POPULI

Evolving Visions

    8   Men of the Thirty-Third Division: An Essay on
             Integrity   124
    9   Darwin and Paley Meet the Invisible
             Hand   138
    10   More Light on Leaves   153

Time in Newton's Century

11   On Rereading Edmund Halley   168
12   Fall in the House of Ussher   181

4 | MUSINGS

Clouds of Memory

13   Muller Bros. Moving & Storage   198
14   Shoemaker and Morning Star   206

Authenticity

15   In Touch with Walcott   220
16   Counters and Cable Cars   238

5 | HUMAN NATURE

17   Mozart and Modularity   249
18   The Moral State of Tahiti—and of
        Darwin   262
19   Ten Thousand Acts of Kindness   275
20   The Declining Empire of Apes   284

6 | GRAND PATTERNS OF EVOLUTION

*Two Steps towards a General Theory of Life's
        Complexity*

21   The Wheel of Fortune and the Wedge of
        Progress   300
22   Tires to Sandals   313

*New Discoveries in the Earliest History of
        Multicellular Life*

23   Defending the Heretical and the
        Superfluous   326
24   The Reversal of *Hallucigenia*   342

7 | REVISING AND EXTENDING DARWIN

25   What the Immaculate Pigeon Teaches the
        Burdened Mind   355

26   The Great Seal Principle   371
27   A Dog's Life in Galton's Polyhedron   382
28   Betting on Chance—and No Fair
         Peeking   396

8 | REVERSALS—FRAGMENTS OF A BOOK
     NOT WRITTEN

29   Shields of Expectation—and Actuality   409
30   A Tale of Three Pictures   427
31   A Foot Soldier for Evolution   439

Bibliography   *457*
Index   *467*

A Reflective Prologue

THESE ESSAYS, volume six in a continuing series, confront history on the broadcast scale of life's evolution during 3.5 billion years. Since macrocosms are fractals of microcosms, the series also records a personal history. In the sweetest introduction I have ever received (for a talk at the Academy of Natural Sciences in San Francisco) former major league ballplayer and current ecoactivist Bruce Bochte recounted my article on why Joe DiMaggio's 56 game hitting streak is the greatest accomplishment in the history of baseball (see Essay 31 in *Bully for Brontosaurus*). He then pointed out that I was working on a 208 monthly essay streak, also unbroken since its inception in January 1974. Over such a long stretch of adulthood, ranging from relative youth (in my early thirties) to distinct middle age (I just passed the half-century mark), many passages must be noted as the ineluctable changes of life unfold. Two aspects of ontogeny seem especially relevant to this continuing series.

First, *Eight Little Piggies* is a book of middle life, and it does contrast, entirely favorably I think (but I am no longer talking to my thirtysomething self), with my youthful *Ever Since Darwin*. I suppose that the major sign of this particular passage lies in my exploration of a traditional essay genre that I had previously shunned—the contemplative and highly personal ruminations in Section 4, "Musings." These essays, on memory, persistence, and authenticity, talk about the importance of unbroken connections within our own lives and to our ancestral generations—a theme of supreme importance to evolutionists who study a world in which extinction is the ultimate fate of all and prolonged per-

11

sistence the only meaningful measure of success.

These essays may treat familiar themes, but at least they follow my idiosyncratic procedure of building, via oddly tangential connections, from a small and concrete item or incident to a broad generality—from sitting with my grandfather on some warehouse steps to characteristic pathways of false memory in our favorite stories (Essay 11); from a graveyard and the invisibility of a large factory in Amana, Iowa, to our need for bucolic myths and the false concept of past golden ages (Essay 12); from calling cards and a visit with a 97-year-old paleontologist who knew C. D. Walcott to the importance we place on continuity and nonvicarious experience (Essay 13); and from a breakfast in San Francisco to a taxonomy of authenticity and the role of vernacular customs and architectures in the preservation of regional diversity (Essay 14). All these essays feature our treatment and distortion of historical records—a kind of ultimate subject for any paleontologist!

Second, the six volumes form a sensible series, each with a different central focus appropriate to its time in three ways: stage in my own life, reaction to current events, and position in the developing logic of an extended discourse on evolution and history. The first volume, *Ever Since Darwin* (1977), centers upon the basic explication of Darwinian principles (where else would one start?). *The Panda's Thumb* (1980) develops the largely unrecognized extensions and corrections of Darwinism that run so counter to many sociocultural hopes and expectations (as in the principle of imperfection embodied in the title example). *Hen's Teeth and Horse's Toes* (1983) had an immediate focus that now, and happily, seems a bit outdated (but by no means dead)—the attack of "creation science" (biblical literalism) upon teaching evolution, and our victories both in courtrooms and in cogent and decent argument. *The Flamingo's Smile* (1985) emphasizes the importance of randomness and unpredictability in the history of life. This theme had a double and immediate origin at two levels—my own bout with cancer at the most personal, and the proposal and successful development of the asteroidal impact theory of mass extinction at the broadest and most general. *Bully for Brontosaurus* (1991), following a longer gap for rumination and synthesis, then put the two central themes together—the mechanics of Darwinism with the unpredictability of complex temporal sequences—to form, finally, a full scale disquisition on the nature of history and its primary theme of contingency (also explored in my intervening *Wonderful Life,* 1989).

I like to think of these volumes as building rather than replacing. The old foci carry over and weave together, tightening up thereby and leaving room for new extensions. Yet one theme of transcendent (and growing) importance has been almost absent (and shamefully so) from my writings heretofore. How can any naturalist, any self-professed lover of diversity, ignore the subject of anthropogenic environmental deterioration and massive extinction of species on our present earth? Oh, I have not entirely bypassed this central concern of my profession. Side comments and paragraphs abound, and even a full essay or two (Essay 29 on nuclear winter in *The Flamingo's Smile,* for example). But I have never addressed this theme centrally and head on.

My reluctance reflects no failure of strong feelings. Quite the contrary. If anything, I have desisted because my feelings are too powerful—lying in the domain that Wordsworth described as "thoughts that do often lie too deep for tears" (and perhaps for words as well). I had never found any distinctive words to convey these common emotions. I could not bear merely to write the shibboleths of the movement—or, even worse, to emote for show, catharsis, or accountability, but with nothing different to add to the hyperabundance of current expostulations.

Perhaps I have finally found something to say that might be helpful, rather than only repetitive. *Eight Little Piggies* includes a section—placed *primus inter pares*—on the sadness of anthropogenic destruction. But if I have finally found a voice, I came to it in my usual way—entirely unanticipated and via a quirky item arising from a personal experience then adumbrated along a forest of tangents. I went to French Polynesia with my son in the summer of 1991 and learned that the island of Moorea had served as the model for Rogers and Hammerstein's Bali Ha'i of *South Pacific.* I also knew that Moorea, and other adjacent islands, had recently experienced a tragically unnecessary extinction of a large, beautiful, and historically important fauna of land snails (the genus *Partula*) that had been the life's work of Henry Crampton, a great scholar of land snails who was revered within the profession, if unknown outside. (My technical research is on land snails, so this is my community.)

Finally, I had a place to stand. By focusing on the cruel and ironic erasure of Crampton's lifelong struggle, rather than only on the animals, I could construct a humanistic reversal for the usual focus on animal victims (primary and perfectly appropriate of course, but so often said by people with a far better sense of the necessary poetry). The Bali Ha'i setting was irresistible, especially with the titular pun of "unenchanted evening." So I rented *South Pacific* (bless the VCR as a new research tool) and wove a nexus of humanistic references into a piece that tries to construct a reversed, people-centered perspective on the tragedy of extinction. But I could also use the emotionalism of Ezio Pinza and his great song* to end with a rereversal and acknowledge—as must be done—the primal (and primary) rights of nature and her beauty. The package worked at least for some people. Of two close colleagues who had studied *Partula* before its extinction on Moorea, the American wrote to tell me that he had been moved to tears, while the Englishman, from the world of stiff upper lips, wrote to inform me that his wife had cried.

I then round out this first section of essays with a most general perspective (though arising from a few squirrels) based on the geologist's primary theme of time scales and their limited domains of application (Essay 2), followed by a small story about the first death of a marine invertebrate species in our times—just a tale on one level, but also a powerful symbol of impending trouble (Essay 3).

If these two sections on musings and extinctions represent ad-

---

*This reference, at the very end of the essay, elicited a firestorm of correspondence. I love nothing more than a brouhaha about a really tiny and tangential item in my essays. I knew this one was coming! Here I am, writing about the deepest issues of life and death, and what do I get? A horde of letters on Pinza's final note, most telling me that he ended on the third above the tonic, and some accusing me of gross musical stupidity (the only form of ignorance that I probably do not possess). But, as I said, I knew it was coming and should have been on my guard, for all parties in this controversy are right. The main rendition does end on the third, but the line reappears several times and often ends on the tonic. I never liked that ending on the third (awkward, and even the great Pinza didn't always make it right on pitch). Moreover, *tonic*, with its multiple meanings, works so much better in a literary sense for the last line of the essay. In any case, I did ponder the issue, and I made a deliberate decision, perhaps not for the best. But thanks to everyone for writing. Details are all that matters: God dwells there, and you never get to see Him if you don't struggle to get them right.

ditions based on changing times and personal growth (aging might be a more honest description), the other six sections follow the traditions of previous volumes and tell new stories about new themes (mixed with some golden oldies) in the domains of evolutionary theory and the history of life.

Section 2, "Odd Bits of Vertebrate Anatomy," takes four central principles for explaining evolutionary legacies and transitions and illustrates them with intriguing peculiarities of restricted parts in vertebrate anatomy: the fact that earliest land vertebrates had up to eight digits per limb and that five is therefore not canonical (the eponym for this entire volume); the tail bend of ichthyosaurs; the evolution of mammalian ear bones from reptilian jaw bones; and why the evolution of swim bladders from lungs (and not vice versa as Darwin assumed and as many texts still proclaim) is neither paradoxical nor contrary to our usual view of evolutionary sequences in vertebrates.

"Vox Populi," title of the next section, is a double entendre. My essays on the central works of individual scientists have always been my personal favorites. They are, literally, *vox populi* as essays about individual scientists. But the title also recalls a trenchant line from Darwin, where he argues that the old motto of *vox populi, vox dei* (the voice of the people is the voice of God) cannot apply to science. This is not a plea for elitism, but a recognition that traditional ways of thought often block understanding. All my essays on individuals try to rescue interesting and honorable (not necessarily correct) effort from the opposite dangers of historical legacy—scorn for enemies and hagiography for heroes (opposite in content perhaps, but eerily similar in debarring our sympathetic understanding).

Section 5, "Human Nature" treats the subject that, at least in a legitimately parochial sense, is the most pressing and important at the interface of evolutionary biology and human life: how to avoid the pitfalls of biological determinism (and its unfortunate legacy in social use) and the simple silliness of sociobiology when done in the strictly adaptationist and speculative mode (still the canonical form in pop culture and not at all rare in professional literature)—and to discover what our biological heritage, and the principles of evolutionary change, truly have to tell us about the nature of our mentality and behavior.

The next two sections treat the main body of evolutionary

theory, but in a personal and iconoclastic spirit—Section 6 on broad patterns in time and Section 7 on basic Darwinian theory and some important revisions. Section 6 is divided into two parts. The first (Essays 21 and 22) stand together as a duet and try to present a coherent case for the central role of sensible unpredictability as the major feature of life's history. Essay 21 sets out the necessary and sufficient major condition, and Essay 22 forges a double whammy by adding an equally powerful theme that makes the result even "worse" (for those committed to conventions of progress and predictability), but ever so much more fascinating for those who wish to grasp the richness of sensible history. The second pair of essays continues the wonderful and rapidly developing story of surprises in documenting the vast anatomical explosion that heralded the first flowering of multicellular life (see my book *Wonderful Life* for the basic story).

The essays of Section 7, "Revising and Extending Darwin," may be the most challenging in the book, but this subject, too often ignored in popular presentations, cannot be bypassed by anyone who wishes to grasp the depth of evolutionary theory. (The low road, too often taken, merely speaks of the power of natural selection and leaves readers with the cardboard view that evolution may be equated with the building of nicely working organic machinery). This section treats principles additional to (and, in important ways, restrictive of) natural selection—internal constraints and historical legacies (Essays 25–27) and randomness as a force for change, not merely as a source of raw material for natural selection (Essay 28). Evolution is much more than a story of matching form to local environments, with increments of general progress slowly accumulating through time—the usual view of pure Darwinian functionalism. Any genealogy is a complex tale of interplay between these Darwinian themes and a set of forces, based on the internal genetic and developmental architecture of organisms, that produce different historical patterns and conceptual meanings. The material may be more difficult, but I have tried to approach this vital subject by concrete example: the coloration of pigeons, fish tails and frog calls, different ranges of morphological variety in domesticated dogs and cats, and the eye tissue of completely blind mole rats.

Perspectives that are reversed or orthogonal to traditional modes form my stock in trade. (I have long argued that conceptual locks are far more powerful than factual lacks as barriers to scientific understanding.) About twenty years ago, I got an idea for such a reversed book on the history of evolutionary thought: I would describe this science via the organisms studied, rather than the folks so engaged. Instead of chapters on Lamarck, Cuvier, Darwin, and Mendel, I would write about trilobites, ammonites, and *Tyrannosaurus.* For a variety of reasons, I never pursued the project to full fruition, but I still like the idea, and putative chapters can stand as full essays. The last section (Section 8 on reversals) offers two such chapters. Essays 29 and 30 are a miniature *Rashomon*— two different perspectives (one based on specimens, the other on iconography) on three major worldviews applied sequentially to the study of *Cephalaspis,* most famous of the earliest vertebrates. (I generally try to expunge redundancy when I gather these several years of monthly essays into a single collection, though I have allowed two pieces to finish with the same wonderful last lines of Darwin's *Origin of Species.* But in the duet of Essays 29 and 30, the redundancy is studied and intentional, for a pair of such different looks at the same story carries its own power, just as the great Japanese film taught us). The last essay, "A Foot Soldier for Evolution," tells the story of a clam studied by nearly every important evolutionist from Lamarck on. The title is a double entendre, as organisms are the rank and file of this grand history (my main reason for planning such a book, for history by organisms must be more democratic than chapters on people, where we are driven to rank from Darwin down); and clams, in particular, are Pelecypoda in technical parlance (meaning "hatchet foot") and therefore truly foot soldiers. The epilogue to this essay provides, at my expense, the most important of all lessons in science—and a fitting end to the book.

All these essays first appeared as columns in my monthly series, "This View of Life," in *Natural History* magazine. I regard this volume as a true natural selection, since many have been dumped and the others improved and then organized into a sensible and coherent sequence, more organic than linear in its webs of cross referencing. The many that I now dislike or regard as substandard are on the scrap heap, so I will stand by all items in the present winnowing and reshaping. But some, inevitably, please me more than others, or at least serve as better exemplars of my chosen style. I think that I am condemned to like best the essays that are most difficult or most focused on particulars of little public knowledge or approbation. I am not hopelessly rarefied or ethereal, and I do feel quite warmly towards some of the most evident (if vital) themes and homely illustrations—as in musings on distortions of memory and myths of past golden ages (Essays 13 and 14). But I do so wish that some of the more complex pieces could receive some share of attention. I especially like Essay 29 because focusing on specimens rather than scientists so well highlights the crucial duality of all scientific activity—tension between the necessary social embeddedness of all scientific thinking and progress towards more adequate factual knowledge of an external reality (by pathways often tortuous and circuitous). If everyone knew the beauty and oddness of actual *Cephalaspis* fossils, this essay would be a sure winner, but knowledge is a prerequisite for this kind of love. For this reason, we need more "hands-on" science education, and we must resist the terrible current trend to confuse museums with theme parks (wonderful things in their proper domain), and to replace real specimens with large, throbbing, blinking glitz in order, ultimately, to pack more bodies into the gift shop. Also, please study Essay 25, even if you revile pigeons. C. O. Whitman was one of our greatest biologists and the key theme of his pigeon work really does speak to the nature of mind as a product of evolution.

I adore all the odd animals and anatomical bits that I pack into these essays, but I have always liked my "people pieces" best (my favorite essays in the last three volumes are "The Titular Bishop of Titiopolis," on Nicolaus Steno in *Hen's Teeth and Horse's Toes,* "Adam's Navel," on Gosse's bizarrely magnificent treatise *Omphalos* in *The Flamingo's Smile,* and "In a Jumbled Drawer," on the conventional career of N. S. Shaler and the iconoclasm of William James in *Bully for Brontosaurus*).

People ask how I keep finding honorable intent in reviled characters by the simple expedient of rediscovering either the full logic of their argument or the social context of their claims. Are all thinkers so worthy of respectful resurrection? Of course not; many are reviled in their own time for good reasons that remain equally cogent today. For the most part, I don't choose to write about these people (though see my book *The Mismeasure of Man* for proof that I do not only seek virtue in historical figures). I praise Archbishop Ussher (Essay 12), so falsely labeled as a unique reactionary shoving his finger into the crumbling dike of revealed religion in order to hold back the flood of science, when he actually represented a large research tradition, humanistically motivated and successful in its own terms (though wrong about the age of the earth due to a false premise at the core of the argument). I warm to a paranoid dyspeptic like Eugène DuBois (Essay 8) when he fairly sticks to his guns in a noble (if losing) argument, but when posterity, not even trying to grasp the subtlety, invariably reports that he finally labeled his precious "apeman" a giant gibbon, although he actually, by this trope of argument (and when you grasp his full system rather than picking at isolated straws), tried to affirm the immediately ancestral status of his *Homo erectus*.

But the saints often need intellectual resurrection as well (though I'd rather be misunderstood in clouds of celestial music than bubbles of boiling magma). Halley is so tied to his eponymous comet that his work on the earth's age is largely ignored and usually interpreted ass-backwards when mentioned at all (Essay 11). Goethe's oracular reduction of all plant form to a leaf archetype needs to be read for its unconventional form of *scientific* excellence (Essay 10).

I also like to find unusual entrées to important subjects generally treated under conventional formats. Thus I approach Darwin's personal views on race and sex not by analyzing the *Descent of Man* (though I do not ignore this primary document), but by studying his very first publication, an article with Captain FitzRoy "On the Moral State of Tahiti" (Essay 18). As for the hottest topic of "modularity" in cognitive science, I never found a distinctive way in until I learned that an English dilettante named Daines Barrington had published an article in England's leading scientific journal on Mozart's musical abilities as a child of eight. (Barrington wrote when young Wolfgang represented a generic prodigy and had not yet become, so to speak, Mozart. Thus, he could be presented as a type, a general puzzle in why musical ability could be so hypertrophied in an otherwise ordinary boy). All this, in the propitious time of Mozart's bicentennial year, led me to Darwin on mind, Tinbergen on the evolution of behavior and cognition in general, my own puzzling over Michelangelo's stunning effect in using modularity to carve his statue of Moses, even, via Monty Python and back to Mr. Barrington, on the liter-

ary convention (and odd name) of litotes to mask wonderment—
thus giving me a string to tie my beginning to a stylistically gentle
end. Not bad.

I often wonder what I am doing every month. I can't be just a
dilettante or poseur, or the streak wouldn't be continuing with
such undiminished commitment and ardor, such love of each little new thing learned. I guess I see myself in the guise of
Papageno, the bird catcher of Mozart's *Magic Flute*. He captures
(but does not harm) the most beautiful objects of nature. He is
capable of naive wonder, counting as originality in breaking
through conventional prejudice—as when frightened by the
Moor Monastatos, he stops and upbraids himself: How foolish;
black birds exist, so why not black men? He wins, at the end, the
greatest Darwinian prize of continuity, as he finally gets his
Papagena and the promise of many little Papagenos and Papage-
nas to follow (I think of them as essays). But do not doubt his
overwhelming modesty amidst all the showy confidence. I sense
how rich and complex it is out there, and what a tiny, tiny part any
of us have been able to understand. And I feel much like
Papageno as he struggles to sing but can produce no words
(though he hums a lovely melody) because his mouth is padlocked.

Nonetheless, if I could have but one Mozartian wish—although it be as unrealistic as the original claim itself, and although this series will stop, *deo volente*, in January 2001—I would
request the time to write as many monthly essays as the Don had
women in Spain. *Ma in Ispagna . . .*

1 | The Scale of Extinction

1 | Unenchanted Evening

TAHITI IS THE STEREOTYPE, virtually the synonym, of enchantment in our legends. Nurse Forbush from Little Rock might have been a charmer (especially when played by Mary Martin), but the South Sea locale made a strong contribution to "some enchanted evening." (And Ezio Pinza, that greatest of operatic Don Giovannis, slumming on Broadway, didn't hurt the scene either.)

Some aspects of the legend need correction. Point Venus, for example, still the landing spot for many tourists, is not named for the beauty of Tahiti's women, but for astronomy and Captain Cook, who set up his instruments at the site to measure the transit of the planet Venus across the disk of the sun in 1769. Charles Darwin himself was beguiled both by the name and the place when he arrived on the *Beagle* in November 1835:

> . . . We landed to enjoy all the delights of the first impressions produced by a new country, and that country the charming Tahiti. A crowd of men, women and children, was collected on the memorable point Venus, ready to receive us with laughing, merry faces.

Darwin, however, broke ranks with male convention in expressing a lack of enthusiasm for Tahiti's women: "I was much disappointed in the personal appearance of the women; they are far inferior in every respect to the men." He objected most of all to the current fad in coiffure:

An unbecoming fashion in one respect is now almost universal: it is the cutting of hair, or rather shaving it, from the upper part of the head, in a circular form, so as to leave only an outer ring of hair. The missionaries have tried to persuade the people to change this habit: but it is the fashion, and that is sufficient answer at Tahiti as well as at Paris.

Darwin's heterodox judgment had not been widely shared. Captain Bligh, who got such a bum rap from Charles Laughton, may not win any medals for grasping human psychology, but he was a great seaman and no more dictatorial than the normal run of British shipmasters. The celebrated mutiny on his *Bounty* owed as much to Fletcher Christian's longing for Tahiti and the woman he left behind as to any of Bligh's shipboard policies.

Tahiti may be beautiful, but the title for "picture perfect paradise" has usually been awarded—and rightly so in my judgment (for I have just returned from my first visit to French Polynesia)—to the neighboring island of Moorea. Located just twelve miles northwest of Tahiti, Moorea is an extinct volcano, with a soaring crater rim, deeply dissected by later erosion into jagged peaks and draperies. Seen from Tahiti, especially when enshrouded by its usual entourage of seemingly personal clouds, Moorea becomes a most fitting symbol of beauty combined with mystery. One day on Tahiti, Charles Darwin scaled a local peak and received his dose of Moorea's spell:

> From the point which I attained, there was a good view of the distant island of Eimeo [the old name for Moorea]. . . . On the lofty and broken pinnacles, white massive clouds were piled up, which formed an island in the blue sky, as Eimeo itself did in the blue ocean.

This impression of beauty and mystery has certainly persisted. Oscar Hammerstein used Moorea as his model for Bali Ha'i, the off-limits paradise of delight in *South Pacific*:

> Bali Ha'i will whisper
> On the wind of the sea:
> "Here am I, your special Island
> Come to me, come to me!"

Bali Ha'i. A photograph of Moorea from Crampton's monograph on *Partula*. *Carnegie Institution of Washington.*

Who could resist these enticements, especially for a few francs and a forty-minute ferry ride? So my son Ethan and I visited Moorea on our recent trip. We were not disappointed. Unfortunately, the lure of Bali Ha'i has attracted other guests, some not so harmless. This essay is a story of genocide in paradise, a preventable wholesale slaughter, just completed in one human generation. You do not know the tale only because it pitted snail against snail, rather than man against man. But do not breathe a sigh of relief for moral exculpation of our species. Snails killed snails, but humans imported the agent of death—consciously and for decent motives, but with tragic and easily avoidable misperception.

Oceanic islands are our great natural laboratories of evolution, the source of so many ideas about organic change and of so many classical examples from finches on the Galápagos to flies on Hawaii. The combination of geographic isolation and difficult access, with frequent absence of predators or competitors, provides explosive possibilities for creatures who manage to reach these bounteous havens. (On the Galápagos, for example, finches radiated into a series of ecological roles usually filled by several families of birds on continents. Some species eat seeds of varying sizes; others act like woodpeckers; one species uses cactus needles to pry insects out of crevices. Darwin was bamboozled during his celebrated visit and classified these birds into several groups. He only learned the true story and significance when a professional ornithologist surveyed his collection in London and recognized the anatomical signature of finches beneath all the diversity.)

Land snails provide some of our finest and most intensely studied examples—and for obvious reasons. Few manage to make the long and fortuitous ocean voyage (by such odd means of transport as natural rafts, mud on birds' feet, or hurricanes if the distances are not too great). The lucky immigrants often find an open world divided into numerous separate pieces (the islands of a chain), each available for colonization and each the eventual source of an evolutionary radiation. Moreover, with their legendary lack of range and their hermaphroditic nature, small founding populations of snails (right down to the absolute minimum of one) can readily become the source of isolated colonies and, eventually, new species. One rat on an island is a transient memory (unless she is a pregnant female); one snail, any snail, may be the progenitor of a vast and changing population.

The high islands of the Pacific are the most promising places of all, for they combine maximal isolation with ecological diversity (the full range from seashore to volcanic mountain top). Many of these islands are formed by single and fairly symmetrical volcanoes. The volcano sides are often dissected into a series of radiating valleys from crater rim to sea. Since most land snails prefer moisture, they often live on the valley floors, but not on the intervening ridges. This common geography adds yet another ingredient to the evolutionary caldron—a source of isolation within islands, as each valley becomes its own separate pocket. On the most diverse of oceanic islands, almost every valley may house a separate species of a particularly prolific snail.

The great radiations of Pacific island land snails are a glory of evolution and a source of joy and knowledge to those of us who have followed Darwin's footsteps into a profession. (I must confess to some jealousy here, for I have devoted my career to the less diverse land snails of low Atlantic islands—to *Poecilozonites* on

Bermuda, and *Cerion* on the Bahamas). Darwin's own Galápagos house a classic example—more than sixty endemic species of the family Bulimulidae. Even more famous (to cognoscenti, for I do not expect a general murmur of recognition here) are two great radiations on more isolated central Pacific islands—the several hundred species of Achatinellidae in the Hawaiian islands and the one hundred or so species of the genus *Partula* on Tahiti, Moorea, and their far-flung neighbors.

These high-island Pacific snails occupy an honored place in the history of evolutionary thought as foci for one of our great and extended debates. No animals seemed better suited for resolving a major issue about causes of organic change: What is the role of environment in evolution? In particular, do organisms change their form to fit altered conditions? And, if so, does environment work its influence directly by Lamarckian inheritance of characters acquired during life, or does form map environment through the indirect route of Darwinian adaptation by natural selection of the most fit in a random spectrum of variation?

Against these two different versions of adaptation—Lamarckian and Darwinian—other evolutionists asserted that form would not match environment in any clear way. The grossly maladapted will die, of course, but if variation arises only rarely and in definite directions, and if most alternatives are well enough suited to local environments, then adaptation will not shape the differences among populations. "Internal causes" (direction of rare mutations), rather than external shaping (by natural selection), will then predominate in the production of evolutionary change.

What better test of these issues than the high-island Pacific snails? For if every valley housed a different population, think of the natural experiment thus provided. Many valleys would have nearly identical environments, but be colonized by only distantly related snails. If adaptation ruled, and climate shaped evolution in predictable ways, then the different snails of separate but similar valleys should evolve strong likenesses as adaptations to common conditions. But if "internal factors" predominated, then no correlation of form and environment should be found among populations.

John T. Gulick (1832–1923), son of an American missionary who worked in Hawaii, fired the first important salvos in a series of works published between 1872 and 1905. Although Gulick

A map of Moorea taken from Crampton's
monograph and indicating the original distribution
of the *Partula* species. *Carnegie Institution of Washington.*

spent most of his adult life as a missionary in China and Japan, he
had, as a young man in his parents' parish, amassed an enormous
collection of Hawaiian achatinellids. Gulick came down strongly
for "internal factors" and against control by natural selection or
any other form of environmental influence. He could find no
correlation between the forms of shells and the local environments of their valleys. Places with apparently identical vegetation, moisture, and temperature might harbor shells of maximally different form.

Gulick, who so strongly opposed (and primarily for religious
reasons) the dominant determinism of the late nineteenth century, triumphantly concluded that the unpredictability of snail
shells could be fully generalized to an overall defense of contingency in history, including human free will:

> If my contention [that different forms arise in identical envi-
> ronments] is in accord with the facts, the assumption which
> we often meet that change in the organism is controlled in
> all its details by change in the environment, and that, there-

fore human progress is ruled by an external fate, is certainly contrary to fact. (From Gulick's famous 1905 treatise, *Evolution, Racial and Habitudinal*)

When I first quoted this line in my Ph.D. thesis of 1969, I did so with derision (and as a firm adaptationist). Twenty years later, I am not so sure that Gulick was wrong in his implication. I still feel that his personal religious motive has no place in science, but people often reach correct answers for wrong or illogical reasons. The contingency of history (both for life in general and for the cultures of *Homo sapiens*) and human free will (in the factual rather than theological sense) are conjoined concepts, and no better evidence can be provided than the "experimental" production of markedly different solutions in identical environments.

In any case, Gulick's conclusions drew a storm of protest from Darwinians. Alfred Russel Wallace, most committed of strict adaptationists, retorted (and not without justice) that Gulick's supposedly "identical" environments might only seem so to humans, but would appear markedly different to snails:

> It is an error to assume that what seem to us identical conditions are really identical to such small and delicate organisms as these land molluscs of whose needs . . . we are so profoundly ignorant. The exact proportions of the various species of plants, the numbers of each kind of insect or of bird, the peculiarities of more or less exposure to sunshine or to wind at certain critical epochs, and other slight differences which to us are absolutely immaterial and unrecognizable, may be of the highest significance to these humble creatures, and be quite sufficient to require some slight adjustments of size, form, or color, which natural selection will bring about.

In 1906, after reading Gulick's monograph, Henry Edward Crampton decided to enter the fray and to devote the remaining fifty years of his career to an immense study of *Partula* on Tahiti, Moorea, and surrounding islands. Crampton (1875–1956) had done excellent work in experimental embryology and studies of natural selection. This earlier effort led to a slight preference for adaptation, but Crampton maintained an open mind and was prepared to support Gulick's "internal factors" against Wallace's shaping by environment should the evidence warrant. Crampton made twelve expeditions to the Pacific and published three magnificent monographs—probably the finest work ever done on the evolution of land snails—collectively entitled *Studies on the Variation, Distribution, and Evolution of the Genus* Partula (Tahiti in 1917, other islands in 1925, and Moorea in 1932).

In short, and to summarize a half century of effort in a sentence, Crampton came down firmly on Gulick's side. He could find no evidence that the forms and colors of *Partula* could be predicted from surrounding environments. Identical climatic conditions seemed to evoke different solutions time after time.

Crampton interpreted the differences between snails in adjacent valleys as results of three major causes—isolation, mutation ("congenital factors" in his terminology), and adaptation by natural selection—with only a minor role for Darwin's favorite mechanism. He viewed the first factor, isolation, as a disposing precondition rather than an actual cause: Geographic separation produces nothing directly but establishes an independent population in which new features may spread. He saw the third factor, natural selection, as primarily negative. Once new features arise by some other mechanism, natural selection may eliminate them if they prove unworkable—but the source of creative change must lie elsewhere. Crampton, who was one of the first American biologists to recognize the importance of Mendel's work, located this source of creativity in his second factor of mutation, or "internally generated" change by congenital factors. In any environment, hundreds of possible anatomies might work—and the forms and colors of this particular population in that specific valley are fortuitous consequences of the largely nonadaptive mutations that happened to arise and spread in an isolated population.

The resulting pattern of differences among valleys is largely nonadaptive. Every local race must avoid elimination by natural selection (and is fit in this negative sense), but its particular features represent only one in a myriad of workable possibilities, and any particular solution arises by the happenstance of mutation in an isolated population, not by natural selection. Crampton contrasted the greater importance of mutation over selection in writing about *Partula* on Tahiti:

The role of the environment is to set the limits to the habitable areas or to bring about the elimination of individuals whose qualities are otherwise determined, that is, by congenital factors.

How can we assess the importance of Crampton's work sixty years after his last great monograph on *Partula* from Moorea (1932)? I am biased to be sure, for snail men (I am one) revere Crampton as a kind of patron saint, but I rank Crampton's *Partula* studies among the most important in the history of evolutionary biology for three major reasons. First, he was probably right in his central claim about the nonadaptive nature of most small-scale differences in form and color among snails of adjacent valleys. Evolutionary biology went through a phase of strong belief in strict adaptationism in the generation just following Crampton, and his works did suffer a temporary eclipse. But his three great monographs are winning new respect and attention in our current, more pluralistic climate of opinion.

Second, Crampton must gain our highest admiration, verging on appropriate awe, for the sheer dedication and effort of his immense labors. I spent only a day in a rented car on Moorea, and scarcely ventured out of the shade or off the paths—but I still nearly passed out from sunstroke. Crampton spent months on twelve separate expeditions, all in an age of ships and horses (not to mention shank's mare), rather than airplanes and rent-a-cars. In the charming understatement of conventional "objective" scientific prose, Crampton wrote but one small comment on working conditions:

> Field-work in such a region of Polynesia presents difficulties that are common to most tropical areas. . . . Steamship lines ply between only the principal ports, from which excursions to neighboring islands must be made by cutter, whaleboat, or canoe. . . . At times it is possible to procure horses. Almost without exception, however, the exploration of a valley can be accomplished only on foot, owing to the steep declivities to be traversed, the deep streams to be forded, and the absence of any trails whatsoever in the thick forest and undergrowth of the areas inhabited by Partulae.

But Crampton also recorded the countervailing pleasures that keep us all going:

> The experiences incidental to the active life necessitated by such work were many, varied, and interesting; but the present monograph is not the place for a description of the beautiful islands or of their delightful inhabitants. Suffice to say that the days and nights of arduous and sometimes dangerous effort included hours of keen enjoyment, for the island of Tahiti, especially, is of matchless beauty, while the chiefs and their families offered abundant hospitalities which it was a privilege to enjoy at the time as it is now a pleasure to acknowledge them.

Moreover, Crampton's labor only began with collecting. He then spent years measuring his snails (some 80,000 for the Tahiti monograph, and a whopping 116,000 for the Moorea work) and calculating statistics—all of which, incredibly (even for his day), he did personally and by hand! (No computers, no hand-held calculators; when Crampton speaks of "calculating machines," he means those old mechanical jobbies that performed division by successive subtraction and clanked away for minutes to perform simple operations.) Again, he wrote in understatement:

> The author is personally responsible for every direct measurement and for every detail of classification; hence the personal coefficient is uniform throughout the entire research. . . . In computing the standard deviations fractions were carried out to eight decimal places. . . . The length of time required for such quantitative analysis can be estimated only by those who themselves have engaged in such work. . . . These figures, together with a single line of text, may be all that represents two to eight weeks of mathematical drudgery. . . . Yet the employment of such methods is justified in the final results.

Third, and most important, ultimate judgment must reside in a criterion of utility. All good science is accumulative; no one can get everything right the first time. If Crampton's monographs were only monuments to past effort and ideas, they might still be

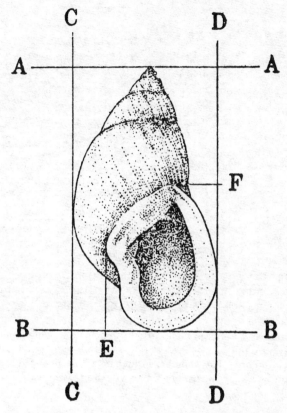

A figure from Crampton's monograph illustrating a *Partula* shell and some of the measurements that he made on each specimen. *Carnegie Institution of Washington.*

admired, but only as items of human paleontology. They are, in fact, precious mines for continuing revision and extension. I know this in the most personal way, for I have used Crampton's tables, the product of his years of "mathematical drudgery," in at least three of my technical papers.

To put this crucial point in another and stronger way, Crampton spent fifty years documenting the *current* geographic distribution and variation of *Partula* on Tahiti, Moorea, and nearby is-

Series	No.	Shell			Aperture			Length aperture + length shell, proportions
		Length	Width	Proportions	Length	Width	Proportions	
		mm.	mm.	p. ct.	mm.	mm.	p. ct.	p. ct.
Uhene, 1923, pallida	123	15.8061± .0443	10.0366±0.0252	63.4350±0.1446	8.6740±0.0245	6.7975±0.0205	78.2965±0.1153	54.7601±0.1233
fusca	130	15.7907± .0518	10.0015± .0253	63.3067± .1380	8.6677± .0255	6.7736± .0205	78.1016± .1439	54.8307± .1155
fulva	61	15.6402± .0658	9.9393± .0396	63.4181± .1904	8.5886± .0402	6.7557± .0308	78.3655± .1869	54.8274± .1718
sonata	18	16.3333± .1280	10.0222± .0688	61.4445± .4121	8.7000± .0653	6.8888± .0594	78.6111± .3817	53.3889± .2422
lyra	32	15.9781± .0887	10.0000± .0523	62.5000± .2615	8.7562± .0499	6.8625± .0397	78.4375± .2439	54.7187± .2284
All	364	15.8140± .0285	10.0038± .0152	63.2061± .0840	8.6660± .0151	6.7922± .0123	78.3215± .0834	54.2198± .0703
Valhere, 1909, pallida	11	15.8541± .1807	9.7000± .0901	61.3182± .3860	8.6818± .0648	6.6637± .0791	77.6818± .4050	55.2272± .3478
fusca	21	16.1071± .0940	9.8905± .0421	61.8809± .3443	8.6905± .0357	6.7664± .0580	77.5476± .3140	54.1191± .2634
fulva	12	15.8000± .0854	9.8500± .0695	62.7500± .2651	8.6500± .0618	6.6834± .0625	77.4167± .4458	54.8333± .2009
sonata	4	15.9000± .0876	9.6500± .0559	61.0000± .5058	8.6500± .0876	6.6000± .1012	79.7500± .8389	54.5000± .7154
lyra	5	16.2900± .1055	9.8200± .0482	60.3000± .2958	8.8200± .0615	6.7800± .0615	76.5000± .5047	54.5000± .4266
All	53	15.9877± .0595	9.8170± .0312	61.7453± .1878	8.6887± .0289	6.7151± .0287	77.6132± .2073	54.5755± .1586
Valhere, 1923, pallida	29	16.0121± .1027	9.9758± .0490	62.2931± .2191	8.7069± .0666	6.9000± .0558	79.0517± .3249	54.3276± .2612
fusca	49	15.8827± .0684	9.8143± .0423	61.7653± .1925	8.5980± .0393	6.7980± .0335	79.0102± .2353	54.0306± .1534
fulva	18	16.0000± .0944	9.9222± .0484	61.8888± .3899	8.7111± .0554	6.7333± .0322	77.1112± .3715	54.3333± .2962
sonata	6	16.1000± .1299	10.0666± .0587	62.3333± .2671	8.9000± .1090	7.0000± .0885	78.8333± .2596	54.5000± .3554
lyra	16	15.9437± .1151	9.8125± .0705	61.5625± .3119	8.6250± .0683	6.7375± .0599	78.3125± .3764	53.7500± .1520
All	118	15.9551± .0441	9.8831± .0267	61.9152± .1225	8.6560± .0271	6.8153± .0226	78.6271± .1531	54.1356± .1055
Valhalia north, 1924, pallida	29	15.2259± .0774	9.6241± .0540	63.1552± .2757	8.2862± .0526	6.5138± .0410	78.3621± .2563	54.3621± .2455
fusca	53	15.1575± .0556	9.6169± .0353	63.3868± .2316	8.2698± .0342	6.4925± .0283	78.5000± .2106	54.3679± .1955
fulva	25	15.0500± .0715	9.5720± .0509	63.5800± .2470	8.2840± .0444	6.4840± .0388	78.4200± .2016	54.9800± .2190
sonata	10	15.3800± .1443	9.5800± .0744	62.1000± .3719	8.4400± .0586	6.4800± .0554	76.5000± .5224	54.5000± .3984
lyra	14	14.7715± .1109	9.6000± .0664	63.6428± .3185	8.2000± .0678	6.4286± .0444	78.4286± .3570	55.3142± .2403
All	131	15.1278± .0364	9.5839± .0235	63.3015± .1338	8.2817± .0216	6.4878± .0176	78.2939± .1262	54.5916± .1136
Valhalia middle, 1909, pallida	52	15.1481± .0673	9.2693± .0387	61.0577± .2173	8.0539± .0329	6.2231± .0322	77.0962± .2537	52.9423± .1483
fusca	38	15.0895± .0765	9.3632± .0417	62.2368± .2110	8.1947± .0486	6.3158± .0380	77.2368± .3091	54.4210± .2373
fulva	24	15.3625± .0979	9.4416± .0556	61.4166± .3201	8.2500± .0472	6.3583± .0368	76.9584± .3204	53.7916± .2528
sonata	22	14.9000± .0953	9.1728± .0531	61.4546± .2987	8.0364± .0476	6.1909± .0456	76.9091± .3263	53.9545± .2663
lyra	12	14.9500± .1152	9.2333± .0768	61.7500± .4843	7.9834± .0604	6.1500± .0506	76.8333± .3009	53.4166± .3916
All	148	15.0946± .0394	9.3040± .0216	61.5338± .1250	8.1135± .0211	6.2581± .0183	77.0608± .1412	53.6486± .1063
Valhalia south, 1923, pallida	67	15.2783± .0539	9.5388± .0319	62.2762± .1683	8.3179± .0278	6.4015± .0249	76.7686± .2085	54.3060± .1432
fusca (Ⓞ)	85	15.1876± .0529	9.4459± .0267	62.1118± .1370	8.3329± .0297	6.5435± .0361	76.3706± .1582	54.7353± .1196
fulva	49	15.2093± .0615	9.0313± .0346	61.9062± .2004	8.3285± .0344	6.4234± .0261	77.0306± .1888	54.6428± .1697
sonata	20	15.3200± .0959	9.4500± .0485	61.7000± .2961	8.4100± .0560	6.5000± .0437	77.3000± .4065	54.8000± .2389
lyra	16	15.0875± .1129	9.2375± .0712	61.4375± .3495	8.2125± .0571	6.3125± .0453	77.0625± .3525	54.3125± .2746
All	237	15.2221± .0287	9.4595± .0170	62.0148± .0878	8.3261± .0163	6.3937± .0132	76.7447± .1016	54.5717± .0747
Matapoopoo middl., 1919, pallida	171	15.9289± .0394	9.2566± .0213	5*.0965± .1094	8.6860± .0237	6.4427± .0170	74.1625± .1120	54.3968± .1083
fusca	125	15.3180± .0502	9.2264± .0243	58.2760± .1317	8.6920± .0273	6.4608± .0196	73.9800± .1535	54.7880± .1158
fulva	69	16.0195± .0556	9.2392± .0312	57.6739± .0914	8.7348± .0308	6.4913± .0220	74.3651± .2293	54.4710± .1437
phase-purpurea	49	15.7296± .0678	9.1164± .0457	57.8266± .1901	8.6265± .0417	6.3531± .0327	73.4388± .3165	55.0102± .1785
fronata	5	15.8900± .1502	9.1800± .0615	57.3000± .4633	8.7800± .0818	6.4600± .0703	73.5000± .5047	55.1000± .3594
sonata	86	15.7616± .0617	9.1419± .0298	57.9419± .1573	8.6651± .0333	6.4093± .0234	73.4934± .2188	54.7558± .1400
lyra	47	15.7266± .0606	9.1085± .0349	57.9830± .2619	8.5341± .0428	6.3383± .0268	74.2447± .2327	54.0532± .2097
All	553	15.8540± .0226	9.2042± .0118	58.0118± .0626	8.6729± .0128	6.4266± .0091	74.0443± .0750	54.5814± .0566

One of nearly a hundred tables, most of comparable length and equally chock full of numbers, from Crampton's monograph on the *Partula* of Moorea. Each number in the chart is a calculated average based on many specimens, not simply a measurement. *Carnegie Institution of Washington.*

lands. This work has great and permanent value as a frozen snapshot, but Crampton's half century should be but a transient moment in the future history of *Partula*. Crampton devoted this lifetime of effort in order to *establish a baseline for future work. Partula* would continue to evolve rapidly, and Crampton's baseline would become a waystation of inestimable value. No scientist could view such dedication in any other light. Future changes have much more value than current impressions.

And Crampton's plan paid off—or so it seemed at first. Three of the world's finest biologists of land snails took up the study of *Partula* in the next generation, building explicitly on Crampton's work—Bryan Clarke of the University of Nottingham, Jim Murray of the University of Virginia, and Mike Johnson of the University of Western Australia. They have published numerous papers, in varying combinations of authorship, from the mid-1960s to the present day. Working primarily on everyone's favorite island of Moorea, they have made important revisions to Crampton's conclusions and have added great sophistication in mathematical procedures (now computerized) and genetical methods not available to Crampton. In 1980, Murray and Clarke ended an important paper, "The genus *Partula* on Moorea: Speciation in progress," with these words:

> Although we cannot yet reconstruct exactly the evolutionary history of the Moorean taxa, they have already revealed in exceptional detail the pattern of interactions between incipient species, and have presented some fascinating paradoxes. They offer both a museum and a laboratory of speciation.

Add snails to Burns' litany about the best laid plans of mice and men. Great expectations die quickly on the bonfires of human vanity. We are only a decade from these brave words of 1980, but Moorea is no longer a laboratory for studying active speciation in *Partula.* It has become a mausoleum.

Think of all the metaphors you know for little things made worse by attempted solutions that cascade to even greater problems, for you need this apparatus to grasp the extirpation of *Partula* on Moorea. Think of Pandora's box. Think of the old woman who swallowed a fly in the folk song. (She then swallowed a spider to catch the fly, a bird to catch the spider, a cat to catch the bird . . . and up the size range of the animal kingdom. Each successive verse gets longer as singers run through the full range of ingestions, but the last is stunningly brief: "There was an old lady who swallowed a horse. She died, of course.")

Partula eats fungi growing upon dead vegetation and poses no threat whatever to agriculture. Its only, and slight, impact upon the native economy is entirely positive, as women string the shells together to make leis for the tourist trade. But animals introduced onto isolated islands often play havoc both with native organisms and with agriculture, witness the rabbits of Australia and, to cite the most dangerous creature of all, the humans who wiped out so many species of moas on New Zealand. An introduced snail began the sad chain of destruction on Moorea.

In sharp contrast with the benign *Partula,* African tree snails of the genus *Achatina* are, in almost all cases, unmitigated disasters. First of all, they are gigantic (as snails go); second, they are voracious herbivores of living plants, including many agriculturally important species. With their clear record of destruction on island after island, I am amazed that people still introduce them purposefully. (They are brought in for food, for I'm told that they are succulent, and you do get a lot of meat per creature.) *Achatina* was first imported to the Indo-Pacific realm in 1803 by the governor of Réunion who brought them from Madagascar so that a lady friend could continue to enjoy snail soup. They escaped from his garden and devastated the island. By 1847, they had reached India. In the 1930s, they began to spread into South Pacific islands, usually by purposeful introduction for food.

Achatina fulica reached Tahiti in 1967 and soon spread to neighboring islands. By the mid-1970s, the infestation had become particularly serious on Moorea. The snails even invaded human dwellings; one report tells of a farmer who removed two wheelbarrow-loads of *Achatina* from the walls of his house. Admittedly, something had to be done. But *que faire,* as they say in this very French land?

The attempted solution, like the horse ingested to catch the fly, created greater havoc than the original problem. Biological control is a good idea in principle—better a natural predator than a chemical poison. But predators, particularly when introduced from alien places and ecosystems, may engender greater problems than the creature that inspired their introduction. How can you know that the new predator will eat only your problem animal? Suppose it prefers other creatures that are benign or useful? Suppose, in particular, that it attacks endemic species (often so vulnerable for lack of evolved defenses in the absence of native predators)?

Biological control should therefore be attempted only with the utmost caution. But, speaking of folk songs and citing a more recent composition than the old lady and the fly, "When will they ever learn?" In my personal pantheon of animals to hate and fear, no creature ranks higher than *Euglandina,* the "killer" or "cannibal" snail of Florida. *Euglandina* eats other snails—with utmost efficiency and voraciousness. It senses slime trails, locks onto them, and follows the path to a quarry then quickly devoured.

Euglandina has therefore developed a worldwide reputation as a potential agent of biological control for other snails. Yet, despite a few equivocal successes, most attempts have failed, often with disastrous and unintended side-effects, as *Euglandina* leaves the intended enemy alone and turns its attention to a harmless victim.

Forgive my prejudices, but I know what *Euglandina* can do in the most personal way (biologists can get quite emotional about the subjects of their own research). I spent the first big chunk of my career, including my Ph.D. dissertation, working on a remarkable Bermudian land snail named *Poecilozonites.* (This Darwin's finch among mollusks is the only large land snail that reached Bermuda. It radiated into a score of species in a great range of sizes and shapes. The fossil record is particularly rich, but at least three species survived and were thriving in Bermuda when I began my research in 1963.) *Euglandina* had been introduced in 1958 to control *Otala,* an imported edible snail that escaped from a garden and spread throughout the island as an agricultural pest (same story as *Achatina* and *Partula* on Moorea). I don't think that *Euglandina* has even dented *Otala,* but it devastated the native *Poecilozonites.* I used to find them by thousands throughout the island. When I returned in 1973 to locate some populations for a student who wanted to investigate their genetics, I could not find a single animal alive. (Last year, I relocated one species, the smallest and most cryptic, but the large *Poecilozonites bermudensis,* major subject of my research, is probably extinct.)

Thus, I feel the pain of Jim Murray, Bryan Clarke, and Mike Johnson. They had published papers on Moorean *Partula* since the mid-1960s. They never expected that their last pair of articles would be a wake.

Euglandina was introduced to Moorea on March 16, 1977, with the official advice and approval of the *Service de l'Economie Rurale* and the *Division de Recherche Agronomique*—despite easily available knowledge of its failures and devastations elsewhere.* *Euglandina* ignored *Achatina* and began a blitzkrieg, against *Partula*—more thorough, rapid, and efficient than anything that Hitler's armies ever accomplished. When my colleagues wrote their first article about this disaster in 1984 (see bibliography), *Euglandina* had already wiped out one of the seven *Partula* species on Moorea, and was spreading across the island at a rate of 1.2 km per year. Moorea is about 12 km across at the widest, and you quickly run out of island at that rate. My colleagues made the grim prediction that *Partula* would be completely gone by 1986.

One hates to be right about certain things. In 1988, Jim, Bryan, and Mike published another note with a brief and final title: "The extinction of *Partula* on Moorea." *Partula* is gone. My colleagues managed to collect six of the seven species before the end, and they have established captive breeding programs in zoos and biological research stations in several nations. Perhaps, one day, *Partula* can be reintroduced into Moorea. But *Euglandina* must be eliminated first, and no one knows how this can be done. Deep grafts, whether physical or emotional, are hard to extirpate—as Mary Martin discovered in her unsuccessful attempt to wash that man right out of her hair. Hope remains in Pandora's box, but how do you reenclose the bad guys?

---

*After this article first appeared in *Natural History,* Bryan Clarke wrote me a letter with the following interesting information on both frustration and hope:

> The story is more depressing than you know. I wrote to the French *Service de l'Economie Rurale* before *Euglandina* was introduced, asking them not to do so. They replied saying that they had not yet decided the matter, but would let me know. Of course they didn't. A similar correspondence took place between Jack Burch at Michigan and the USDA about introducing the Uglies into Hawaii. It got very heated.
>
> There are some little rays of hope. First, it seems that *Euglandina* has not yet reached Huahine, Raiatea or Tahaa, where *Partula* are still, it is said, surviving. I'm going this summer to check. There *may* still be *Partula* on Bora Bora. Second, the French are proposing to set up a snail conservation and study centre on Moorea (shutting the stable door . . .), and we are looking at ways of making 'cages' in the woods. There is also a faint hope of a refuge inside the crater of Mehetia (an extinct volcano about 60 miles SE of Tahiti).

Moorea may be the Bali Ha'i of our dreams, but life for *Partula* has become an unenchanted evening. Now night has fallen.

The story would be sad enough if only Moorea (and Bermuda) had fallen victim. But *Euglandina* is spreading just as rapidly on the larger, adjacent Tahiti, and *Partula* now survives in only two valleys. The even more diverse *Achatinella* is gone (or nearly so) on Oahu, largely for the same reason, although the spread of Honolulu hasn't helped either. More than half the species of bulimulids are extinct on the Galápagos.

It is so hard for an evolutionary biologist to write about extinctions caused by human stupidity. Emotions well up and extinguish rationality and writing. What can be said that hasn't been stated before—with great eloquence and little effect. Even the good arguments have become clichés—as corny as Kansas in August, as normal as blueberry pie.

Let me then float an unconventional plea, the inverse of the usual argument. Inverses often have a salutary effect in reopening pathways of thought. An undergraduate friend during my year in England, a brilliant debater, had to argue the affirmative in that tired old cliché of a subject: "This house believes that the monarchy should be eliminated." Instead of trotting out the usual points about the queen's expense account and the negative symbol of royalty in a democratic age, my friend claimed that the monarchy should be eliminated because it is unfair to monarchs and their families. All possibility of a normal private life evaporates. You can't have a date, drink a beer, or, God forbid, even belch in public without a headline in next day's scandal sheet.

The extinction of *Partula* is unfair to *Partula*. This is the conventional argument, and I do not challenge its primacy. But we need a humanistic ecology as well, both for the practical reason that people will always touch people more than snails do or can, and for the moral reason that humans are legitimately the measure of all ethical questions—for these are our issues, not nature's.

So I say, let us grieve for Henry Edward Crampton when we consider *Partula* on Moorea—for *Euglandina* and human stupidity have destroyed his lifetime's dedication. Crampton visited the Pacific a dozen times, when transportation was no aerial picnic. He tramped up and down the valleys, over the dangerous precipices, in intense tropical heat. He spent months and months measuring snails and toting up columns of figures *sans* computer—the worst sort of scientific drudgery. He published three great monographs on *Partula*.

The work is of great and permanent value in itself. But Crampton did not write for personal glory or to establish the frozen evolutionary moment of his own studies. He labored all his life to provide a baseline for future evolutionary work. *Partula* was a natural evolutionary laboratory, and he toiled to establish a starting point, with utmost care and precision, so that others could move the work forward and continue to learn about evolution by tracing the future history of *Partula*. What is more noble than a man's intellectual dedication—a lifetime of perseverance through the Scylla and Charybdis of all field biology: occasional danger and prolonged tedium? Crampton's work is now undone, even mocked. Grieve for his lost and lofty purposes.

Yet I also appreciate that we cannot win this battle to save species and environments without forging an emotional bond between ourselves and nature as well—for we will not fight to save what we do not love (but only appreciate in some abstract sense). So let them all continue—the films, the books, the television programs, the zoos, the little half acre of ecological preserve in any community, the primary school lessons, the museum demonstrations, even (though you will never find me there) the 6:00 A.M. bird walks.

Let them continue and expand because we must have visceral contact in order to love. We really must make room for nature in our hearts. Consider one last image of Ezio Pinza as Emil De Becque in *South Pacific*, and accept the traditional characterization of nature as female (if this convention offends you, then make nature male and fall "in love with a wonderful guy"). The words may be banal (and Pinza was only extolling Mary Martin, while I speak of all nature), but the emotional setting is incomparable and still can bring tears to any unjaded eye. Think of this greatest of bassos as he soars up to the tonic of his chord:

> Once you have found her, never let her go.
> Once you have found her, NEVER LET HER GO!

2 | The Golden Rule: A Proper Scale for Our Environmental Crisis

PATIENCE ENJOYS a long pedigree of favor. Chaucer pronounced it "an heigh vertu, certeyn" ("The Franklin's Tale"), while the New Testament had already made a motto of the Old Testament's most famous embodiment: "Ye have heard of the patience of Job" (James 5:11). Yet some cases seem so extended in diligence and time that another factor beyond sheer endurance must lie behind the wait. When Alberich, having lost the Ring of the Niebelungen fully three operas ago, shows up in Act 2 of *Götterdämmerung* to advise his son Hagen on strategies for recovery, we can hardly suppress a flicker of admiration for this otherwise unlovable character. (I happen to adore Wagner, but I do recognize that a wait through nearly all of the Ring cycle would be, to certain unenlightened folks, the very definition of eternity in Hades.)

Patience of this magnitude usually involves a deep understanding of a fundamental principle, central to my own profession of geology but all too rarely grasped in daily life—the effects of scale. Phenomena unfold on their own appropriate scales of space and time and may be invisible in our myopic world of dimensions assessed by comparison with human height and times metered by human lifespans. So much of accumulating importance at earthly scales—the results of geological erosion, evolutionary changes in lineages—is invisible by the measuring rod of a human life. So much that matters to particles in the microscopic world of molecules—the history of a dust grain subject to Brownian motion, the fate of shrunken people in *Fantastic Voyage* or *Inner*

41

Space—either averages out to stability at our scale or simply stands below our limits of perception.

It takes a particular kind of genius or deep understanding to transcend this most pervasive of all conceptual biases and to capture a phenomenon by grasping a proper scale beyond the measuring rods of our own world. Alberich and Wotan know that pursuit of the Ring is dynastic or generational, not personal. William of Baskerville (in Umberto Eco's *Name of the Rose*) solves his medieval mystery because he alone understands that, in the perspective of centuries, the convulsive events of his own day (the dispute between papacies of Rome and Avignon) will be forgotten, while the only surviving copy of a book by Aristotle may influence millennia. Architects of medieval cathedrals had to frame satisfaction on scales beyond their own existence, for they could not live to witness the completion of their designs.

May I indulge in a personal anecdote on the subject of scale? I loved to memorize facts as a child, but rebelled at those I deemed unimportant (baseball stats were in, popes of Rome and kings of England out). In sixth grade, I had to memorize the sequence of land acquisitions that built America. I could see the rationale behind learning the Louisiana Purchase and the Mexican Cession, for they added big chunks to our totality. But I remember balking, and publicly challenging the long-suffering Ms. Stack, at the Gadsden Purchase of 1853. Why did I have to know about a sliver of southern Arizona and New Mexico?

Now I am finally hoist on my own petard (blown up by my own noxious charge according to the etymologies). After a lifetime of complete nonimpact by the Gadsden Purchase, I have become unwittingly embroiled in a controversy about a tiny bit of territory within this smallest of American growing points. A little bit of a little bit—so much for effects of scale and the penalties of blithe ignorance.

The case is a classic example of a genre (environmentalists vs. developers) made familiar in recent struggles to save endangered populations—the snail darter of a few years back, the northern spotted owl vs. timber interests. The University of Arizona, with the backing of an international consortium of astronomers, wishes to build a complex of telescopes atop Mount Graham in southeastern Arizona (part of the Gadsden Purchase). But the old-growth spruce-fir habitat on the mountaintop provides the central range for *Tamiasciurus hudsonicus grahamensis*, the Mount Graham Red Squirrel—a distinctive subspecies that lives nowhere else, and that forms the southernmost population of the entire species. The population has already been reduced to some one hundred survivors, and destruction of 125 acres of spruce-fir growth (to build the telescopes) within the 700 or so remaining acres of best habitat might well administer a coup de grâce to this fragile population.

I cannot state an expert opinion on details of this controversy (I have already confessed my ignorance about everything involving the Gadsden Purchase and its legacy). Many questions need to be answered. Is the population already too small to survive in any case? If not, could the population, with proper management, coexist with the telescopes in the remaining habitat?

I do not think that, practically or morally, we can defend a policy of saving every distinctive local population of organisms. I can cite a good rationale for the preservation of species, for each species is a unique and separate natural object that, once lost, can never be reconstituted. But subspecies are distinctive local populations of species with broader geographical ranges. Subspecies are dynamic, interbreedable, and constantly changing; what then are we saving by declaring them all inviolate? Thus, I confess that I do not agree with all arguments advanced by defenders of the Mount Graham Red Squirrel. One leaflet, for example, argues: "The population has been recently shown to have a fixed, homozygous allele which is unique in Western North America." Sorry folks. I will stoutly defend species, but we cannot ask for the preservation of every distinctive gene, unless we find a way to abolish death itself (for many organisms carry unique mutations).

No, I think that for local populations of species with broader ranges, the brief for preservation must be made on a case by case basis, not on a general principle of preservation (lest the environmental movement ultimately lose popular support for trying to freeze a dynamic evolutionary world *in statu quo*). On this proper basis of individual merit, I am entirely persuaded that the Mount Graham Red Squirrel should be protected, for two reasons.

First, the squirrel itself: The Mount Graham Red is an unusually interesting local population within an important species. It is isolated from all other populations and forms the southernmost extreme of the species's range. Such peripheral populations, living in marginal habitats, are of special interest to students of evolution.

Second, the habitat: Environmentalists continually face the political reality that support and funding can be won for soft, cuddly, and "attractive" animals, but not for slimy, grubby, and ugly creatures (of potentially greater evolutionary interest and practical significance) or for habitats. This situation had led to the practical concept of "umbrella" or "indicator" species—surrogates for a larger ecological entity worthy of preservation. Thus, the giant panda (really quite a boring and ornery creature despite its good looks) raises money to save the remaining bamboo forests of China (and a plethora of other endangered creatures with no political clout); the northern spotted owl has just rescued some magnificent stands of old-growth giant cedars, Douglas fir, and redwoods (and I say Hosanna); and the Mount Graham Red Squirrel may save a rare and precious habitat of extraordinary evolutionary interest.

The Pinaleno Mountains, reaching 10,720 feet at Mount Graham, are an isolated fault block range separated from others by alluvial and desert valleys that dip to less than 3,000 feet in elevation. The high peaks of the Pinalenos contain an important and unusual fauna for two reasons. First, they harbor a junction of two biogeographic provinces: the Nearctic or northern by way of the Colorado Plateau and the Neotropical or southern via the Mexican Plateau. The Mount Graham Red Squirrel (a northern species) can live this far south because high elevations reproduce the climate and habitat found nearer sea level in the more congenial north. Second, and more important to evolutionists, the old-growth spruce-fir habitats on the high peaks of the Pinalenos are isolated "sky islands"—10,000-year-old remnants of a habitat more widely spread over the region of the Gadsden Purchase during the height of the last ice age. In evolutionary terms, these isolated pieces of habitat are true islands—patches of more northern microclimate surrounded by southern desert. They are functionally equivalent to bits of land in the ocean. Consider the role that islands (like the Galápagos) have played both in developing the concepts of evolutionary theory and in acting as cradles of origin (through isolation) or vestiges of preservation for biological novelties.

Thus, whether or not the telescopes will drive the Mount Graham Red Squirrel to extinction (an unsettled question well out-side my area of expertise), the sky islands of the Pinalenos are precious habitats that should not be compromised. Let the Mount Graham Red Squirrel, so worthy of preservation in its own right, also serve as an indicator species for the unique and fragile habitat that it occupies.

But why should I, a confirmed eastern urbanite who has al-ready disclaimed all concern for the Gadsden Purchase, choose to involve myself in the case of the Mount Graham Red Squirrel? The answer, unsurprisingly, is that I have been enlisted—in-voluntarily and on the wrong side to boot. I am fighting mad, and fighting back.

The June 7, 1990, *Wall Street Journal* ran a pro-development, anti-squirrel opinion piece by Michael D. Copeland (identified as "executive director of the Political Economy Research Center in Bozeman, Montana") under the patently absurd title: "No Red Squirrels? Mother Nature May Be Better Off." (I can at least grasp, while still rejecting, the claim that nature would be no worse off if the squirrels died, but I am utterly befuddled at how anyone could devise an argument that the squirrels inflict a posi-tive harm upon the mother of us all!) In any case, Mr. Copeland misunderstood my writings in formulating a supposedly scientific argument for his position.

Now scarcely a day goes by when I do not read a misrepresenta-tion of my views (usually by creationists, racists, or football fans in order of frequency). My response to nearly all misquotation is the effective retort of preference: utter silence. (Honorable intel-lectual disagreement should always be addressed; misquotation should be ignored, when possible and politically practical). I make an exception in this case because Copeland cited me in the service of a classic false argument—the standard, almost canoni-cal misuse of my profession of paleontology in debates about extinction. We have been enlisted again and again, in opposition to our actual opinions and in support of attitudes that most of us regard as anathema, to uphold arguments by developers about the irrelevance (or even, in this case, the benevolence) of modern anthropogenic extinction. This standard error is a classic exam-ple of failure to understand the importance of scale—and thus I return to the premise and structure of my introductory para-graphs (did you really think that I waffled on so long about scale only so I could talk about the Gadsden Purchase?).

Paleontologists do discuss the inevitability of extinction for all species—in the long run, and on the broad scale of geological time. We are fond of saying that 99 percent or more of all species that ever lived are now extinct. (My colleague Dave Raup often opens talks on extinction with a zinging one-liner: "To a first approximation, all species are extinct.") We do therefore identify extinction as the normal fate of species. We also talk a lot—more of late since new data have made the field so exciting—about mass extinctions that punctuate the history of life from time to time. We do discuss the issue of eventual "recovery" from the effects of these extinctions, in the sense that life does rebuild or surpass its former diversity several million years after a great dying. Finally, we do allow that mass extinctions break up stable faunas and, in this sense, permit or even foster evolutionary innovations well down the road (including the dominance of mammals and the eventual origin of humans, following the death of dinosaurs).

From these statements about extinction in the fullness of geological time (on scales of millions of years), some apologists for development have argued that extinction at any scale (even of local populations within years or decades) poses no biological worry but, on the contrary, must be viewed as a comfortable part of an inevitable natural order. Or so Copeland states:

Suppose we lost a species. How devastating would that be? "Mass extinctions have been recorded since the dawn of paleontology," writes Harvard paleontologist Stephen Gould. . . . The most severe of these occurred approximately 250 million years ago . . . with an estimated 96 percent extinction of species, says Mr. Gould. . . . There is general agreement among scientists that today's species represent a small proportion of all those that have ever existed—probably less than 1 percent. This means that more than 99 percent of all species ever living have become extinct.

From these facts, largely irrelevant to red squirrels on Mount Graham, Copeland makes inferences about the benevolence of extinction in general (though the argument only aplies to geological scales):

> Yet, in spite of these extinctions, both Mr. Gould and University of Chicago paleontologist Jack Sepkoski say that the actual number of living species has probably increased over time. [True, but not as a result of mass extinctions, despite Copeland's next sentence.] The "niches" created by extinctions provide an opportunity for a vigorous development of new species. . . . Thus, evolutionary history appears to have been characterized by millions of species extinctions and subsequent increases in species numbers. Indeed, by attempting to preserve species living on the brink of extinction, we may be wasting time, effort and money on animals that will disappear over time, regardless of our efforts.

But all will "disappear over time, regardless of our efforts"—millions of years from now for most species if we don't interfere. The mean lifespan of marine invertebrate species lies between 5 and 10 million years; terrestrial vertebrate species turn over more rapidly, but still average in the low millions. By contrast, *Homo sapiens* may be only 200,000 years old or so and may enjoy a considerable future if we don't self-destruct. Similarly, recovery from mass extinction takes its natural measure in millions of years—as much as 10 million or more for fully rekindled diversity after major catastrophic events.

These are the natural time scales of evolution and geology on our planet. But what can such vastness possibly mean for our legitimately parochial interest in ourselves, our ethnic groups, our nations, our cultural traditions, our blood lines? Of what conceivable significance to us is the prospect of recovery from mass extinction 10 million years down the road if our entire species, not to mention our personal lineage, has so little prospect of surviving that long?

Capacity for recovery at geological scales has no bearing whatever upon the meaning of extinction today. We are not protecting Mount Graham Red Squirrels because we fear for global stability in a distant future not likely to include us. We are trying to preserve populations and environments because the comfort and decency of our present lives, and those of fellow species that share our planet, depend upon such stability. Mass extinctions may not threaten distant futures, but they are decidedly unpleasant for species caught in the throes of their power. At the appropriate scale of our lives, we are just a species in the midst of such a moment. And to say that we should let the squirrels go (at our immediate scale) because all species eventually die (at geological scales) makes about as much sense as arguing that we shouldn't treat an easily curable childhood infection because all humans are ultimately and inevitably mortal. I love geological time—a wondrous and expansive notion that sets the foundation of my chosen profession—but such vastness is not the proper scale of my personal life.

The same issue of scale underlies the main contribution that my profession of paleontology might make to our larger search for an environmental ethic. This decade, a prelude to the millennium, is widely and correctly viewed as a turning point that will lead either to environmental perdition or stabilization. We have fouled local nests before and driven regional faunas to extinction, but we were never able to unleash planetary effects before this century's concern with nuclear fallout, ozone holes, and putative global warming. In this context, we are searching for proper themes and language to express our environmental worries.

I don't know that paleontology has a great deal to offer, but I would advance one geological insight to combat a well-meaning, but seriously flawed (and all too common), position and to focus attention on the right issue at the proper scale. Two linked arguments are often promoted as a basis for an environmental ethic:

1. We live on a fragile planet now subject to permanent derailment and disruption by human intervention;

2. Humans must learn to act as stewards for this threatened world.

Such views, however well intentioned, are rooted in the old sin of pride and exaggerated self-importance. We are one among millions of species, stewards of nothing. By what argument could we, arising just a geological microsecond ago, become responsible for the affairs of a world 4.5 billion years old, teeming with life that has been evolving and diversifying for at least three-quarters of this immense span. Nature does not exist for us, had no idea we were coming, and doesn't give a damn about us. Omar Khayyam was right in all but his crimped view of the earth as battered, when he made his brilliant comparison of our world to an eastern hotel:

> Think, in this battered caravanserai
> Whose portals are alternate night and day,
> How sultan after sultan with his pomp
> Abode his destined hour, and went his way.

This assertion of ultimate impotence could be countered if we, despite our late arrival, now held power over the planet's future. But we don't, despite popular misperception of our might. We are virtually powerless over the earth at our planet's own geological timescale. All the megatonnage in all our nuclear arsenals yields but one ten-thousandth the power of the 10 km asteroid that might have triggered the Cretaceous mass extinction. Yet the earth survived that larger shock and, in wiping out dinosaurs, paved a road for the evolution of large mammals, including humans. We fear global warming, yet even the most radical model yields an earth far cooler than many happy and prosperous times of a prehuman past. We can surely destroy ourselves, and take many other species with us, but we can barely dent bacterial diversity and will surely not remove many million species of insects and mites. On geological scales, our planet will take good care of itself and let time clear the impact of any human malfeasance.

People who do not appreciate the fundamental principle of appropriate scales often misread such an argument as a claim that we may therefore cease to worry about environmental deterioration, just as Copeland argued falsely that we need not fret about extinction. But I raise the same counterargument. We cannot threaten at geological scales, but such vastness has no impact upon us. We have a legitimately parochial interest in our own lives, the happiness and prosperity of our children, the suffering of our fellows. The planet will recover from nuclear holocaust, but we will be killed and maimed by billions, and our cultures will perish. The earth will prosper if polar icecaps melt under a global greenhouse, but most of our major cities, built at sea level as ports and harbors, will founder, and changing agricultural patterns will uproot our populations.

We must squarely face an unpleasant historical fact. The conservation movement was born, in large part, as an elitest attempt by wealthy social leaders to preserve wilderness as a domain for patrician leisure and contemplation (against the image, so to speak, of poor immigrants traipsing in hordes through the woods with their Sunday picnic baskets). We have never entirely shaken this legacy of environmentalism as something opposed to immediate human needs, particularly of the impoverished and unfortunate. But the Third World expands and contains most of the pristine habitat that we yearn to preserve. Environmental movements cannot prevail until they convince people that clean air and water, solar power, recycling, and reforestation are best solutions (as they are) for human needs at human scales—and not for impossibly distant planetary futures.

I have a decidedly unradical suggestion to make about an appropriate environmental ethic—one rooted, with this entire essay, in the issue of appropriate human scale vs. the majesty, but irrelevance, of geological time. I have never been much attracted to the Kantian categorical imperative in searching for an ethic— to moral laws that are absolute and unconditional, and do not involve any ulterior motive or end. The world is too complex and sloppy for such uncompromising attitudes (and God help us if we embrace the wrong principle and then fight wars, kill, and maim in our absolute certainty). I prefer the messier "hypothetical imperatives" that invoke desire, negotiation, and reciprocity. Of these "lesser," but altogether wiser and deeper principles, one has stood out for its independent derivation, with different words but to the same effect, in culture after culture. I imagine that our various societies grope towards this principle because structural stability (and basic decency necessary for any tolerable life) demand such a maxim. Christians call this principle the "golden rule"; Plato, Hillel, and Confucius knew the same maxim by other names. I cannot think of a better principle based on enlightened self-interest. If we all treated others as we wish to be treated ourselves, then decency and stability would have to prevail.

I suggest that we execute such a pact with our planet. She holds all the cards, and has immense power over us—so such a compact, which we desperately need but she does not at her own timescale, would be a blessing for us and an indulgence for her. We had better sign the papers while she is still willing to make a deal. If we treat her nicely, she will keep us going for a while. If we scratch her, she will bleed, kick us out, bandage up, and go about her business at her own scale. Poor Richard told us that "necessity never made a good bargain," but the earth is kinder than human agents in the "art of the deal." She will uphold her end; we must now go and do likewise.

3 | Losing a Limpet

DARWIN MARVELED at the abundance of giant Galápagos tortoises when he visited the islands in September, 1835, but he also noted a marked decline based on ease of human exploitation for food. (Ships would often take tortoises away by the hundreds, stacking them live in the hold to provide months of fresh meat "on the hoof." The tortoises were essentially defenseless. As a single barrier to capture, Darwin notes that ships usually sent out hunting parties in pairs, and two men could not lift the largest animals.) Darwin wrote in the *Voyage of the Beagle:*

> It is said that formerly single vessels have taken away as many as seven hundred of these animals and that the ship's company of a frigate some years since brought down two hundred to the beach in one day.

The species, though not threatened as a whole, is much depleted today, and several distinctive forms, once limited to single islands, have disappeared. I saw the saddest story of this legacy— Lonesome George, last survivor of the saddle-backed race from Pinta Island. No mate has been found for George, though the island has been scoured. He has been moved, for safe-keeping (and in apparently vain hope for salvation of his kind), to a research station on Santa Cruz Island, where I saw him several years ago. He is well fed and surely pampered, and he may live for another century or more; but his lineage, at least as a pure pedigree, is already extinct.

Every George must have his Martha. The last passenger pigeon, also a mateless vestige of a doomed race, died in the Cincinnati Zoo on September 1, 1914. Martha's body was taken to the Cincinnati Ice Company, suspended in a tank of water, frozen into a three-hundred-pound block of ice, and sent for extrication and stuffing to the Smithsonian Institution, where she resides today.

Galápagos tortoises were vulnerable and restricted in geography; their extreme reduction and partial extinction merits no special surprise. But how could the superabundant and widespread passenger pigeon crash and die within a century? By some estimates, they were once the most common bird in America. They migrated in huge flocks over most of eastern and central North America. Pioneer ornithologist Alexander Wilson estimated one such aggregation as containing more than 2 billion birds. One colony in Wisconsin spread out over 750 square miles. The famous testimony of Audubon himself, made in Ohio just one hundred years before Martha's death, not only identifies human rapacity as the cause of eventual decline, but also depicts the fabulous abundance:

> As the time of the arrival of the passenger pigeons approached, their foes anxiously prepared to receive them. Some persons were ready with iron pots containing sulphur, others with torches of pine knots; many had poles, and the rest, guns. . . . Everything was ready and all eyes were fixed on the clear sky that could be glimpsed amid the tall tree-tops. . . . Suddenly a general cry burst forth, "Here they come!" The noise they made, even though still distant, reminded me of a hard gale at sea, passing through the rigging of a close-reefed vessel. The birds arrived and passed over me. I felt a current of air that surprised me. Thousands of the pigeons were soon knocked down by the polemen, whilst more continued to pour in. The fires were lighted, then a magnificent, wonderful, almost terrifying sight presented itself. The pigeons, arriving by the thousands, alighted everywhere, one above another, until solid masses were formed on the branches all around. Here and there the perches gave way with a crack under the weight, and fell to the ground, destroying hundreds of birds beneath. . . . The scene was one of uproar and confusion. . . .

Even the gun reports were seldom heard, and I was made aware of the firing only by seeing the shooters reloading. . . .

The picking up of the dead and wounded birds was put off till morning. The pigeons were constantly coming and it was past midnight before I noticed any decrease in the number of those arriving. The uproar continued the whole night. . . .

Towards the approach of day, the noise somewhat subsided. Long before I could distinguish them plainly, the pigeons began to move off. . . . By sunrise all that were able to fly had disappeared. . . . Eagles and hawks, accompanied by a crowd of vultures, took their place and enjoyed their share of the spoils. Then the author of all this devastation began to move among the dead, the dying and the mangled, picking up the pigeons and piling them in heaps. When each man had as many as he could possibly dispose of, hogs were let loose to feed on the remainder.

In 1805, passenger pigeons sold for a penny a piece in markets of New York City. By 1870, birds were reproducing only in the Great Lakes region. Hunters used the newly invented telegraph to inform others about the location of dwindling populations. Perhaps the last large wild flock, some 250,000 birds, was sighted in 1896. A gaggle of hunters, alerted by telegraph, converged upon them; fewer than 10,000 birds flew away. The last wild passenger pigeon was killed in Ohio in 1900. The few zoo colonies dwindled, as keepers could never induce the birds to breed regularly. By 1914, only Martha remained.

These sad, oft-told tales are canonical stories of the extinction saga: defenseless populations composed of individuals that are easy to find and profitable to kill. Restricted compass on an island is the surest path to destruction—the dodo or tortoise model. But even a large, continental spread will not save a vulnerable population—the passenger pigeon model.

One environment, however, has been seen as a refuge for at least most kinds of organisms—the open ocean. Here, or so the argument goes, geographic ranges are usually large enough, and ecological tolerances sufficiently broad, to prevent rapacious humanity (or any other agent of extinction) from getting every last one. Populations may be beaten way back, but a few survivors will always find a refugium.

This claim is as old as modern biology itself. In the first great document of evolutionary theory, published in 1809, Lamarck tried to deny extinction altogether. (In his theory of creative response to perceived needs, and inheritance of characters thus acquired, organisms should evolve fast enough to overcome any environmental danger.) But Lamarck did allow an exception for conspicuous species on land. Even "Lamarckian" response cannot be quick or extensive enough to overcome the most powerful and efficient agency of environmental disturbance—human depredation. Lamarck wrote: "If there really are lost species, it can doubtless only be among the large animals which live on the dry parts of the earth; where man exercises absolute sway, and has compassed the destruction of all the individuals of some species which he has not wished to preserve or domesticate." But small inconspicuous oceanic species should be immune from our influence: "Animals living in the waters, especially the sea waters, . . . are protected from the destruction of their species by man. Their multiplication is so rapid and their means of evading pursuit or traps are so great, that there is no likelihood of his being able to destroy the entire species of any of these animals."

We would downplay Lamarck's optimism about the oceans today. Conspicuous species of large organisms with small populations are vulnerable—and several fish and marine mammals, including Steller's sea cow, have succumbed. But Lamarck's distinction and prognostication has held. The ledgers of death in historic times do not include marine invertebrates.

Extinction has certainly received its fair share of attention in our newspapers and TV specials. We are so used to tales of destruction, so inured or even numbed, that we expect almost any species, anywhere in the world, to be the next victim. We have engraved the notion of fragility upon our souls.

But step back from all these accounts of death and think for a moment: Have you ever heard about the extinction of a marine invertebrate species, even among widely exploited lobsters, scallops, or conchs. We may drive a local population of marine invertebrates to death, but never an entire species. In *The Panda's Thumb,* a previous volume in this series published in 1980, I told the sad story of *Cenobita diogenes,* a large Bermudian hermit crab, now apparently doomed because the only shell large enough to hold its body, the whelk *Cittarium pica,* was eaten to destruction on Bermuda. (The crabs now eke out a tenuous existence within fossil *Cittarium* shells eroded from Bermudian hillsides.) I was deluged with suggestions for salvation. One man offered to design a plastic *Cittarium* and to ship them by the thousands for distribution on Bermudian beaches. I was touched by his ingenuity and generosity, but a much simpler and more effective solution exists. *Cittarium pica* is extinct on Bermuda, but not elsewhere. This species is eaten with gusto on most West Indian islands, and piles of empty shells are stacked on beaches and roadsides. Any enterprising savior of *Cenobita* could easily fill a boat and bring the real McCoy back to Bermuda.

In short, Lamarck was right, and his distinction of sea and land has much to teach us about the general phenomenon of species death. By our records and reckoning, no marine invertebrate species has become extinct during historic times. (Geological extinctions, of course, occur at characteristic rates over millions of years—thus illustrating the immensity of difference between our time and earth time). Geerat Vermeij, a leading expert on oceanic life and its vulnerability, wrote in 1986 that "marine invertebrates are relatively immune from extinction." So Lamarck was right—that is, until 1991.

The first issue of the *Biological Bulletin* for 1991, the technical journal published by the Marine Biological Laboratory at Woods Hole, contains the following article by James T. Carlton and four other authors (including my good friend Gary Vermeij who must now eat his words or be happy that he wrote the disclaimer "*relatively* immune from extinction"): "The first historical extinction of a marine invertebrate in an ocean basin: The demise of the eelgrass limpet *Lottia alveus*" (see the bibliography).

Limpets are snails with an unusual mode of growth. Snail shells are cones that expand slowly and wind around an axis during growth, producing the conventional corkscrew of increasing width. But the limpet cone expands so rapidly that the shell never winds around its axis for more than a fraction of a whorl. Thus, a limpet shell looks like a Chinese hat of the old caricatures. The large open end clamps tightly down upon a rock, or a food source, and this power of adhesion has made the limpet a symbol of tenacity and stubbornness in many languages and cultures. In

England, for example, limpets are (according to the *OED*), "officials alleged to be superfluous but clinging to their offices."

Lottia alveus, the eelgrass limpet of the western Atlantic, once lived in fair abundance from Labrador to Long Island Sound. Although this geographic range might have been broad enough to win the usual marine immunity from extinction, two peculiar features placed the eelgrass limpet into a rare category of vulnerability. First, as its name implies, the eelgrass limpet lived and fed only on a single species of plant, *Zostera marina.* (*Zostera,* a fascinating biological oddity in its own right, is one of the few marine genera of angiosperms, or ordinary flowering plants. Most people assume that all marine plants are algae, but a few "advanced" flowering land plants have managed to invade the oceans, usually forming beds of "sea grass" in shallow waters.)

Lottia alveus had a long and narrow shell, just wide enough to fit snugly over a *Zostera* blade. The limpet fed exclusively on epithelial cells of the sea grass. The fate of *Lottia* therefore depended upon the health of *Zostera.* Moreover, *Lottia* had an un-

A specimen of *Lottia alveus,* first species of marine invertebrate organisms to become extinct in historical times. Left: side view. Right: top view.

usually narrow range of physiological tolerance, particularly for changes of salinity. This limpet could not survive any marked departure from normal oceanic salinity of some thirty-three parts per thousand, whereas *Zostera* spans a much broader range and can live in brackish waters of much reduced salinity.

Both *Zostera* and *Lottia* originated in the Pacific Ocean and invaded the North Atlantic in late Tertiary times (just before the ice ages), as many other species did, through the Bering Strait and along the Arctic Ocean. Populations of *Lottia* remain in both the eastern and western Pacific, so the entire genus is not lost. (Taxonomists are still debating the status of *Lottia*. Some regard the three populations—eastern Pacific, western Pacific, and extinct Atlantic—as fully separate species, others as subspecies of a single, coherent group. In either case, the Atlantic populations were distinct in form and color, and clearly represent some degree of genetic differentiation.)

Atlantic specimens of *Lottia* were reported as abundant by all collectors from first extensive descriptions of the mid-nineteenth century through the 1920s. The last living specimens were reported in 1933. The five authors have searched diligently for *Lottia alveus* throughout its former range from the early 1970s through 1990. They also looked in fourteen major museum collections for specimens gathered during the past fifty years—and found none that could be verified. They even searched through several herbarium collections of *Zostera*, hoping that dried limpet shells might be found on the pressed sheets. (Museums often preserve reference collections of plants by flattening them onto sheets of paper then bound into books.) Again, no *Lottia* could be found. This diligence certainly wins our assent to their preliminary conclusion that *Lottia alveus* is truly extinct in the Atlantic.

But why did *Lottia* gain the dubious honor of first marine invertebrate to disappear during historic time? Carlton and coauthors have also provided a coherent and satisfying explanation that neatly combines a specific historical event with the general biology of *Lottia*. Between 1930 and 1933, *Zostera* virtually disappeared from both the eastern and western Atlantic Ocean. (This species of sea grass has suffered numerous declines throughout its recorded history, but none nearly so severe as this accidental correlate with economic depression on adjacent lands.) The cause of this near wipe-out has been debated for years, with disease and environmental fluctuation as leading contenders. A series of articles published during the 1980s has decisively implicated a marine protist, the slime mold *Labyrinthula* (a unicelled creature that aggregates to form temporary colonies), as chief culprit. Infestation by the pathogenic species of *Labyrinthula* leads to formation of small black patches on *Zostera* leaves. The patches spread, eventually causing death and detachment of the entire blade.

This massive die-off of *Zostera* led to marked changes in associated ecosystems, including great reduction in migratory waterfowl populations and loss of commercial scallop fisheries. But neither *Zostera,* nor any of these ecological dependents, became entirely extinct. *Zostera* itself tolerates a much wider range of salinities than does the pathogen *Labyrinthula.* All populations of normal marine salinity were completely destroyed, but *Zostera* can also live in brackish water, while *Labyrinthula* cannot. Thus, relict *Zostera* populations hunkered down in low salinity refugia, and the species survived. Other species associated with *Zostera* also pulled through, either because they could also tolerate the low salinities of *Zostera* refuges, or because they could live on other resources, though often with much smaller populations, while *Zostera* was absent. When *Zostera* returned after the *Labyrinthula* epidemic subsided, these other species reflowered as well.

But consider the cruel fate of poor *Lottia alveus.* This limpet lacked the flexibility of all other species associated with *Zostera.* *Lottia* could not hunker down with *Zostera* in the low-salinity refugia because this limpet could only live in normal marine waters. And *Lottia* could not switch to another host species because it ate only the epithelial cells of *Zostera* blades. For *Lottia,* the total disappearance of *Zostera* in waters of normal salinity spelled complete destruction. *Zostera* returned, but no *Lottia* remained to greet the renewed bounty.

Does the story of *Zostera* and *Lottia* bear a message for our chief parochial concern with the subject of extinction—anthropogenic assault on the biosphere and consequent loss of biodiversity? In a literal sense, the answer must be "rather little." *Lottia* was no Galápagos tortoise or passenger pigeon—creatures hounded to death by human hunters. *Lottia* didn't even fall victim to some unintended consequence of human disruption in natural habitats. In fact, *Homo sapiens* probably played no role whatsoever in the death of *Lottia* (I doubt that one person in a million ever laid eyes on the creature). *Lottia*'s extinction was an ordinary natural event, the kind that, summed through geological ages, produces the basic pattern of life's history. Epidemics are as intrinsic as water and sunshine in the history of life. They don't usually wipe out a species entirely (as the *Labyrinthula* epidemic spared *Zostera*). Yet species clearly have differential susceptibility to extinction, and some factors of weakness enhance the possibility of death in epidemics. Natural selection can only work for immediate reproductive advantages and cannot overtly protect a species against unexpected vicissitudes of time. Many strengths of the moment engender a potential for later extinction as an unintended and detrimental side consequence. So long as *Zostera* bloomed and seas stood at normal salinity, intense specialization may have aided individual *Lottia*. But such narrowly committed forms are usually the first to go when unusual circumstances wipe out a highly specific habitat, even temporarily. Marine species are "relatively immune" from extinction because few have such narrow commitments, but *Lottia,* as an exception for its intense specialization, paid the ultimate price.

Lottia does bear a symbolic message for the anthropogenic theme, however. As the first species to die (during historic times) in the one habitat that, from Lamarck to 1991, seemed free of such danger, *Lottia* must stand as a warning and an emblem—as the Crispus Attucks of a potential wave in the most protected arena, if our environmental assaults worsen. Didn't British power laugh at the ragtag rebellion when Attucks and four others died in the Boston Massacre of 1770? Most crises start with something small, something virtually beneath notice. But whispers soon grow to whirlwinds. Limpets, with their low profiles and large apertures (often serving as suction cups for attachment), are metaphors for tenaciousness, for hunkering down in times of trouble. How appropriate, then, as a warning against complacency, that a real version of this symbol should be the first species to die in a realm of supposed invulnerability.

2 | Odd Bits of Vertebrate Anatomy

4 | Eight Little Piggies

RICHARD OWEN, England's greatest vertebrate anatomist during Darwin's generation (see Essay 5), developed the concept of an archetype to explicate the evident similarities that join us with frogs, flamingoes, and fishes. (An archetype is an abstract model constructed to generate an entire range of anatomical design by simple transformation of the all-inclusive prototype.) Owen was so pleased with his conception that he even drew a picture of his archetype, engraved it upon a seal for his personal emblem, and, in 1852, wrote a letter to his sister Maria, trying to explain this arcane concept in layperson's terms:

> It represents the archetype, or primal pattern—what Plato would have called the "divine idea" on which the osseous frame of all vertebrate animals—i.e. all animals that have bones—has been constructed. The motto is "the one in the manifold," expressive of the unity of plan which may be traced through all the modifications of the pattern, by which it is adapted to the very habits and modes of life of fishes, reptiles, birds, beasts, and human kind.

Darwin took a much more worldly view of the concept, substituting a flesh and blood ancestor for a Platonic abstraction from the realm of ideas. Vertebrates had a unified architecture, Darwin argued, because they all evolved from a common ancestor. The similar shapes and positions of bones record the historical happenstance of ancestral form, retained by inheritance in all later species of the lineage, not the abstract perfection of an ideal shape in God's realm of ideas. Darwin burst Owen's bubble with a marginal note in his personal copy of Owen's major work, *On the Nature of Limbs*. Darwin wrote: "I look at Owen's archetype as more than idea, as a real representation as far as the most consummate skill and loftiest generalization can represent the parent form of the Vertebrata."

However we construe the concept of an organizing principle of design for major branches of the evolutionary tree—and Darwin's version gets the modern nod over Owen's—the idea remains central to biology. Consider the subset of terrestrial vertebrates, a group technically called Tetrapoda, or "four-legged" (and including amphibians, reptiles, birds, and mammals in conventional classifications). Some fly, some swim, and others slither. In external appearance and functional role, a whale and a hummingbird seem sufficiently disparate to warrant ultimate separation. Yet we unite them by skeletal characters common to all tetrapods, features that set our modern concept of an archetype. Above all, the archetypal tetrapod has four limbs, each with five digits—the so-called pentadactyl (or "five-fingered") limb.

The concept of an archetype does not require that each actual vertebrate display all canonical features, but only that uniqueness be recognized as extreme transformations of the primal form. Thus, a whale may retain but the tiniest vestige of a femur, only a few millimeters in length and entirely invisible on its streamlined exterior, to remind us of the ancestral hind limbs. And although a hummingbird grows only three toes on its feet, a study of embryological development marks them as digits two, three, and four of the full ancestral complement. The canonical elements are starting points and generating patterns, not universal presences.

In the tetrapod archetype, no feature has been more generally accepted than the pentadactyl limb, putative source of so many deep and transient human activities, from piano playing to touch typing, duck shooting, celebratory "high fives," and decimal counting (twice through the sequence of "this little piggy . . ."). Yet this essay will challenge the usual view of such a canonical number, while not denying its sway in our lives.

The great Swedish paleontologist, Erik Jarvik, closed his two-volume magnum opus on vertebrate structure and evolution with a telling point about pentadactyl limbs and human possibilities.

He noted how many "advanced" mammals modify the original pattern by loss and specialization of digits—horses retain but one as a hoof; whales practically lose the whole hind limb. Jarvik noted that an essential coupling of a multidigited hand, fit for using tools, with an enlarging brain, well suited to devising new and better uses for such technology, established the basis and possibility of human evolution. If the ancestor of our lineage had lost the original flexibility of the "primitive" pentadactyl limb and evolved some modern and specialized reduction, human intelligence would never have developed. In this important sense, we are here because our ancestors retained the full archetypal complement of five and had not substituted some new-fangled, but ultimately more limiting, configuration. Jarvik writes:

> The most prominent feature of man is no doubt his large and elaborate brain. However, this big brain would certainly never have arisen—and what purpose would it have served—if our arm and hand had become specialized as strongly as has, for instance, the foreleg of a horse or the wing of a bird. It is the remarkable fact that it is the primitive condition, inherited from our osteolepiform ancestors [fishes immediately ancestral to tetrapods] and retained with relatively small changes in our arm and hand, that has paved the way for the emergence of man. We can say, with some justification, that it was when the basic pattern of our five-fingered hand for some unaccountable reason was laid down in the ancestors of the osteolepiforms that the prerequisite for the origin of man and the human culture arose.

I don't dispute Jarvik's general point: The retention of "primitive" flexibility is often a key to evolutionary novelty and radiation. But is the five-fingered limb a constant and universal tetrapod archetype, interpreted in Darwin's evolutionary way as an ancestral pattern retained in all descendant lineages?

Erik Jarvik is maximally qualified to address this question (his rationale, of course, for raising it in the first place), for he has done by far the most extensive and important research on the earliest fossil tetrapods—the bearers and perpetrators of the five-fingered archetype in any evolutionary interpretation. (Fish fins are constructed on different principles, although the lobe-finned ancestors of tetrapods built a bony architecture easily transformable to the fore and hind limbs of terrestrial verte-brates. In any case, fish do not display the pentadactyl pattern, and this central feature of canonical design arose only with the evolution of the Tetrapoda.)

The oldest tetrapods were discovered in eastern Greenland by a Danish expedition in 1929. They date from the very last phase of the Devonian period, a geological interval (some 390 to 340 million years ago) often dubbed the "age of fishes" in books and museum exhibits that follow the silly chauvinism of naming time for whatever vertebrate happened to be most prominent. The Swedish paleontologist Gunnar Säve-Söderbergh collected more extensive material in 1931 and directed the project until his un-timely death in 1948. Erik Jarvik then took over the project and, during the 1950s, published his extensive anatomical studies of two genera that share the spotlight of greatest age for tetra-pods—*Ichthyostega* and *Acanthostega*. Although no specimens pre-served enough of the fingers or toes for an unambiguous count, Jarvik (see figure) reconstructed the earliest tetrapods with the canonical number of five digits per limb.

Our confidence in this evidence-free assumption of an initial five began to crumble in 1984, when the Soviet paleontologist O. A. Lebedev reported that the newly discovered early tetrapod *Tulerpeton,* also of latest Devonian age, bore six digits on its limbs. This find led anatomist and embryologist J. R. Hinchliffe to sug-gest in 1989, prophetically as we have just learned, that five digits represents a secondary stabilization, not an original state. Hinch-liffe entitled his article "Reconstructing the archetype: Evolution

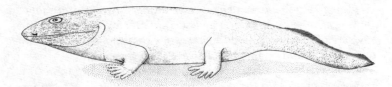

The standard reconstruction of *Ichthyostega* from Jarvik's 1980 book. Note the five digits on each limb. *From* Basic Structure and Evolution of Vertebrates, *vol. 1, p. 235.*

of the pentadactyl limb," and ended with these words: "Restriction to the pentadactyl form may have followed an evolutionary experimental phase."

Hinchliffe's suspicion has now been confirmed—in spades. In October 1990, M. I. Coates and J. A. Clack reported on new material of *Ichthyostega* and *Acanthostega*, collected by a joint Cambridge-Copenhagen expedition to East Greenland in 1987 (see bibliography). Some remarkable new specimens—a complete hindlimb of *Ichthyostega* and a forelimb of *Acanthostega*—permit direct counting of digits for the first time.

In an admirable convention of scientific writing that maximizes praise for past work done well and minimizes the disturbing impact of novelty, Coates and Clack write: "The proximal region [closest to the body] of the hindlimb of *Ichthyostega* corresponds closely with the published description, but the tarsus [foot] and digits differ." In fact, the back legs of *Ichthyostega* bear, count 'em, seven toes!—with three smallish and closely bound digits corresponding to the hallux ("big toe" in human terms) of ordinary five-toed tetrapods (see figure). *Acanthostega* departs even more strongly from a model supposedly common to all; its forelimb bears eight digits in a broad arch of increasing and then decreasing size (see figure).

The conclusion seems inescapable, and an old "certainty" must be starkly reversed. Only three Devonian tetrapods are known. None has five toes. They bear, respectively, six, seven, and eight digits on their preserved limbs. Five is not a canonical, or archetypal, number of digits for tetrapods—at least not in the primary sense of "present from the beginning." At best (for fans of pentadactyly) five is a later stabilization, not an initial condition.

Moreover, in the light of this new information, an old fact may cast further doubt on the primacy of five. The naive "ladder of life" view depicts vertebrate evolution as a linearly ascending series of amphibian–reptile–mammal–human (with birds as the only acknowledged branch). But ladders are culturally comforting fictions, and copious branching is the true stuff of evolution. Tetrapods had a common ancestor to be sure, but modern amphibians (frogs and salamanders) represent the termini of a large branch, not the inception of a series. Moreover, no fossil amphibian seems clearly ancestral to the lineage of fully terrestrial verte-

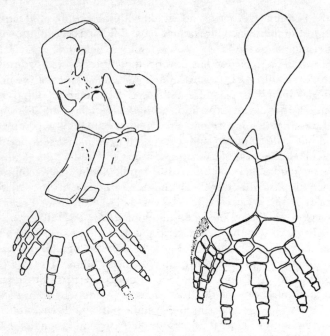

Left: forelimb of *Acanthostega* with 8 digits. Right: hindlimb
of *Ichthyostega* with seven digits. *Reprinted by permission from* Nature *vol.
347, p. 67; Copyright © 1990 Macmillan Magazines Limited.*

brates (reptiles, birds, and mammals), called Amniota to honor
the "amniote" egg (with hard covering and "internal pond"), the
evolutionary invention that allowed, in our usual metaphors,
"complete conquest of the land" or "true liberation from water."
(The point is tangential to this essay, but do pause for a moment
and consider the biases inherent in such common "descriptions."
Why is the ability to lay eggs on land a "liberation"; is water
tantamount to slavery? Why is exclusive dwelling a "conquest"?
Who is fighting for what? Such language only makes sense if life is
struggling upward towards a human pinnacle—the silliest and
most self-centered view of evolution that I can imagine.)

The first fossil reptiles are just about as old as the first amphibians clearly in the group that eventually yielded our modern frogs
and salamanders. Thus, rather than a ladder from amphibian to

reptile, both the fossil record and the study of modern vertebrate anatomy suggest an early branching of the tetrapod trunk into two primary limbs—the Amphibia and the Amniota (reptile, bird, and mammal).

And now, the point about pentadactyly and its limits: The Amniota do, indeed, show the canonical pattern of five toes upon each limb (or some modification from this initial state). But Amphibia, both living and fossil, have five toes on the hindlimbs and *only four* on the front limbs. Anatomists have known this for years of course, but have always assumed that this reduction to four proceeded from an initial and canonical five. This conclusion must now be challenged. If all the earliest tetrapods had more than five digits, and if amphibians have been separate from amniotes since the beginning of terrestrial life, why assume that the four toes of the amphibian forelimb descended from a primary five? All modern stabilizations probably proceeded from more than five. Perhaps the amphibian forelimb went from this higher number directly to four, without any pentadactyl stage between. If so, then pentadactyly crumbles on two grounds: (1) It does not represent the original state of tetrapods (as six-, seven-, and eight-toed earliest forms show); and (2) it may not mark the canonical state in one of the two great living lineages of tetrapods.

A key to understanding these new views may be found in a brilliant paper (see bibliography) on the embryological development of limbs, based on work done just down the hall from my office and published in 1986 by Neil H. Shubin (now at the University of Pennsylvania) and Pere Alberch (now director of the Natural History Museum in Madrid). Shubin and Alberch try to depict the complexity of the tetrapod limb as the outcome of interactions among three basic processes: branching (making two series from one), segmentation (making more elements in a single series), and condensation (union between elements). The limb builds from the body out—shoulder to fingers, thigh to toes. The process begins with a single element extending from the trunk—humerus for the arm, femur for the leg. A branching event produces the next elements in sequence—radius and ulna for the arm, tibia and fibula for the leg. The branching (to wrist bones) sets the distinctive pattern that eventually makes fingers. This key bifurcation is markedly asymmetrical, as one bone ceases to branch (and yields but a single row of segments as the limb continues to develop), while the other serves as a focus for all subsequent multiplication of elements, including the production of digits. Oddly enough, the bone that does not branch is the larger of the two elements—the radius of the arm and the tibia of the leg. The hand and foot are made by branching from the smaller element—the ulna of the arm and the fibula of the leg. (A glance at the accompanying figure should make these anatomical arcana clear.)

These basic facts have long been appreciated. Shubin and Al-

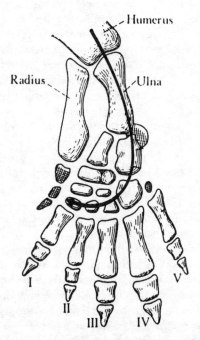

Standard anatomy of a tetrapod forelimb showing the axis of embryological development according to Shubin and Alberch. *From* Basic Structure and Evolution of Vertebrates, *vol. 1, p. 235.*

berch make their outstanding contribution in providing a new account of subsequent branching. The classical view holds that a central axis continues from the ulna (or fibula), and that the subsequent branches project from this axis (much like the persistent midvein and diverging lateral veins of a leaf). In this view, the roots of the digits represent different branches. Under this model, largely unchallenged for more than one hundred years, debate focused on the identity of the main axis and its position relative to the digits. T. H. Huxley, for example, argued that the main axis passed through digit three; the British vertebrate paleontologist D. M. S. Watson favored digit four, while the American W. K. Gregory advocated a position between digits one and two.

Shubin and Alberch do not deny the idea of a central axis, but they radically reorient its position. Instead of passing through a particular digit (with remaining digits branching to one side or the other), Shubin and Alberch's axis passes through the basal bones of *all* the digits in sequence, from back to front (again, a glance at the figure will make this argument clear).

The elegant novelty of this switch may not be evident in the simple change of position for the axis. Consider, instead, the question of timing. Under the old view, one might talk about a dominant digit (focus of the central axis) and subordinate elements (products of increasingly distant branching), but no implications of timing could be drawn. Under Shubin and Alberch's revision, the array of digits becomes a sequence of timing: Spatial position is a mark of temporal order. Back equals old; front is young. The piggy that "cried wee, wee, wee all the way home" comes first, the one that went to market is last. The thumb and big toe may be functionally most important in humans, but they are the last to form.

As always in natural history, nothing is quite so simple, or free from exceptions, as its cleanest and most elegant expression. Actually, the penultimate digit always forms first—ironically, the piggy that had none—and the sequence then proceeds from back to front with one exception in a reverse branch to digit five. Moreover, this generality meets a fascinating exception in the urodeles (the amphibian group of newts and salamanders, although the other major amphibian lineage of Anura, the frogs, forms digits in the usual back-to-front sequence). Uniquely among tetrapods, urodeles work from front to back (although they also follow the rule of penultimate first, beginning with digit two and then proceeding on towards five). Some zoologists have used this basic difference to argue that urodeles form an entirely separate evolutionary line of tetrapods, perhaps even arising from a different group of fish ancestors. But most (including me) would respond that embryonic patterns are as subject to evolutionary change as adult form, and that an ancestor to the urodele lineage—for some utterly unknown and undoubtedly fascinating reason—shucked an otherwise universal system in tetrapods and developed this "backwards" route to the formation of digits.

But why bring up this innovative model for embryological formation of digits in the context of new data on the multiplicity of fingers and toes in the earliest tetrapods? I do so (as did Coates and Clack in their original article) because the Shubin and Alberch model suggests a simple and obvious mechanism for a later stabilization of five from an initial lability that yielded varying numbers of supernumerary digits. If digits form from back to front in temporal order, then reduction can be readily achieved by an earlier shut-down. The principle is obvious and pervasive: Stop sooner. We can reduce population growth if families halt at two children. You can cut down on smoking or drinking by setting a limit and stopping each day at the reduced number (easier said than done, but the principle is simple enough to articulate). Evolution can reduce the number of fingers by stopping the back-to-front generating machine at five. What we now call digit one (and view as the necessary limit of an invariant archetype) may only be the stabilized stopping point of a potentially extendable sequence.

This perspective makes immediate sense of some old and otherwise unexplained data of natural history. Many lineages in all tetrapod groups reduce the original complement of five to some smaller number—sometimes right down to one, as in horses. As a general principle of reduction, known since Richard Owen's time, digit one is the first to go. Owen wrote in 1849:

> To sum up, then, the modifications of the digits: they never exceed five in number on each foot in any existing vertebrate animal above the rank of Fishes. . . . The first or innermost digit, as a general rule, is the first to disappear.

Under Shubin and Alberch's model, the reason behind this rule is obvious: last formed, first gone (the natural analog of the economic maxim: last hired, first fired).

The opposite phenomenon of polydactylous mutations (producing more than five digits) also supports the Shubin and Alberch model. In humans, most polydactylous mutations produce a sixth finger as a simple duplication (subsequent to initial branching) of one member in the usual sequence of five—a phenomenon outside the scope of Shubin and Alberch's concerns. But, in several other species, the supernumerary elements of multifingered mutants arise by extension as digits continue to form after the branching of digit number one, the usual terminus of the series. J. R. Hinchliffe writes in 1989: "Many polydactylous mutants . . . have an array of five normal digits, with the supernumerary digits added preaxially [that is, after formation of digit one]." Moreover, Hinchliffe cites some experimental data on inhibition of DNA synthesis during embryology of the lizard *Lacerta viridis*. With less material available for building body parts, digits may be lost. The last-formed digit is always the first to go. Data from both sides therefore support the idea that digits form in temporal series, back to front, and that spatial position is a mark of order in embryological timing: Extra digits are added to, and old digits are lost from, the temporal end-point of the canonical sequence—digit number one.

The pleasure of discovery in science derives not only from the satisfaction of new explanations, but also, if not more so, in fresh (and often more difficult) puzzles that the novel solutions generate. So for the Shubin and Alberch model and our new discoveries on multiplicity of digits in the earliest tetrapods. We used to think of five digits as invariant and canonical, and our chief question was always: Why five? But if five is a secondary stabilization, a stopping point in a temporal sequence with other potential (but unrealized) terminations, we must ask a very different, and in many ways more interesting, question: Why stop at this point? What, if anything, is special about five?

Since five seems to possess a certain arbitrariness under the new views, the tenacity of its stabilization in tetrapods seems all the more enigmatic. The embryological apparatus remains capable of producing more than five (at least in many species), as mutational and experimental data show. But these polydactylous mutations remain as anomalies of individuals or of small and evanescent family lines. They never stabilize within a larger group, and no vertebrate species has more than five digits generated from the back-to-front axis of the Shubin and Alberch model.

The best proof of this assertion lies in apparent (but not actual) exceptions of several tetrapod species with six functional digits. Yes, Virginia, several species do grow six fingers as a rule, not an exceptional state of mutant individuals. Yet this sixth finger is always generated in a different manner and not by the obvious (and apparently easy) mechanics of simple extension past digit one on the Shubin-Alberch series. Frogs, for example, often have six digits on their hind feet (or five on their normally four-fingered front feet). But this extra digit forms in a unique manner by extension of the unbranched sequence of bones leading out from the radius or tibia—the limb bones that never serve as foci for branches and therefore do not (in any other tetrapod species) participate in the production of digits. Anatomists have long recognized the anomalous character of these unique digits by naming them *prepollux* (for the forelimb) or *prehallux* (for the hindlimb). (*Pollux* and *hallux* are technical names for digit number one—our thumb and big toe. Prepollux and prehallux therefore designate an anomalous digit, located in front of the usual front and formed in a different manner.)

A few mammals also possess a functional sixth digit—the panda, whose false "thumb" has been a staple of these essays, and several species of moles. But these false thumbs are formed from extended wrist bones, and are not true digits at all. These facts seem to heighten the oddity (and rigidity) of stabilization at five in a sequence that was once extendable, remains so now for mutations and experimental manipulations, but seems recalcitrant in setting a maximum of five as a normal state in all tetrapod species. When six functional digits form, the extra item must be built in another way.

So why five? Of two major approaches to this question, the conventional Darwinian, or adaptationist, strategy tries to discern a marked advantage, or even an inevitability, for five in terms of utility for an organism's habitat (an advantage that might promote this configuration by natural selection). A plausible case can be made in terms of benefits for terrestrial life. Creatures that evolve from water to land face many novel challenges, none more severe than the new force of gravity and the consequent need for support in the absence of buoyancy previously supplied by water. The transition from fins to limbs provides the basis for this support, and an old argument holds that five might be an optimal configuration for weight-bearing—a central axis running through on digit three, with adequate and symmetrical buttressing on each side (one or three toes might not provide enough lateral support against wobbling, while seven toes might be superfluous and interfere with locomotion). On this argument, tetrapods have five toes because support and locomotion demand (or at least strongly encourage) this configuration as optimal.

The argument is not implausible, and surely gains credence from the probability that five digits evolved twice on hind limbs— separately, that is, in the two great divisions of tetrapods. The most obvious counterargument may also be support in disguise: Why, if five be best on land, do the earliest tetrapods bear six, seven, and eight toes respectively? A paradoxical retort holds that these first tetrapods evolved their limbs for locomotion in water and remained predominantly, if not entirely, aquatic. *Ichthyostega,* as long recognized, maintained a small tail fin and lateral-line canals on the skull. (Lateral line organs "hear" sound by sensing vibrations propagated through water, a method that does not work in thin air—see Essay 6.) Coates and Clack's restoration of *Ichthyostega* and *Acanthostega* limbs add support to this interpretation in a streamlined shape and a limit to rotation that might keep the limb horizontal, in fin position, rather than rotated downward to support a body on land (at least for *Acanthostega,* though the *Ichthyostega* forelimb seems fully load-bearing).

But strong elements of doubt also plague this adaptationist view. First, as stated above, members of one tetrapod lineage, the amphibians, grow but four toes on their front legs, and we have no evidence for an initial five—so pentadactyly may not be a universal stage in terrestrial vertebrates. Second, if five (with symmetry about a strong central toe) be the source of advantage, then why does our favorite species, the traditional measure of all things—*Homo sapiens*—retain five, require great strength in using only two limbs against gravity, but construct the end-member first toe as the main weight bearer?

The second major approach—historical contingency in my favored terminology (see my recent book, *Wonderful Life*)—argues that five was not meant to be, but just happens to be. Other configurations would have worked and might have evolved, but they didn't—and five works well enough. The obvious supports for this alternative view lie scattered throughout this essay. If five is so good, why do so many species devise such curious and devious means to produce six (prepollux or converted wrist bone)? If five is so predictable, why does one of two lineages grow but four? (I should say right up front that neither of these two positions—adaptation or contingency—really addresses the greatest puzzle of all: the recalcitrant stability of five once it evolves. I suspect that this is a question for embryologists and geneticists; phylogenetic history may offer little in the way of clues. Why should five, once attained by whatever route and for whatever reason, be so stubbornly intractable as an upper limit thereafter, so that any lineage, again evolving six or more, must do so by a different path? The inquiry could not be more important, for this issue of digits is a microcosm for the grandest question of all about the history of animal life: Why, following a burst of anatomical exploration in the Cambrian explosion some 550 million years ago, have anatomies so stabilized that not a single new phylum [major body plan] has ever evolved since?)

But the greatest boost to contingency lies in the discovery that prompted this essay in the first place—seven digits in *Ichthyostega* and eight in *Acanthostega*. If tetrapods had five at the beginning, and always retained five thereafter, then some predictability or inevitability could legitimately be maintained. (At the very least, no fuel would exist for an alternative proposal.) But if the first members of the lineage had six, seven, or eight toes, then alternative possibilities are legion, and an eventual five may be a happenstance, not a necessity.

Embryologist Jonathan Cooke, in a commentary written to accompany Coates and Clack's paper, agrees with me that possible contingency of pentadactyly is the most interesting implication of the new discovery. But he makes a curious statement in his advocacy. Cooke writes:

> But for most of us, philistine enough to accept the historically contingent nature of evolution, there is nothing specially deep about the number five. Pianists should ponder

the challenge that our motor cortexes would have been set had Bach or Scarlatti sported eight deeply and ineffably named fingers per hand.

I love the idea, but I decry the apology and abnegation implied by the designation of "philistine" for contingency. This unnecessary humility follows an unfortunate tradition of self-hate among scientists who deal with the complex, unrepeatable, and unpredictable events of history. We are trained to think that the "hard science" models of quantification, experimentation, and replication are inherently superior and exclusively canonical, so that any other set of techniques can only pale by comparison. But historical science proceeds by reconstructing a set of contingent events, explaining in retrospect what could not have been predicted beforehand. If the evidence be sufficient, the explanation can be as rigorous and confident as anything done in the realm of experimental science. In any case, this is the way the world works: No apologies needed.

Contingency is rich and fascinating; it embodies an exquisite tension between the power of individuals to modify history and the intelligible limits set by laws of nature. The details of individual and species's lives are not mere frills, without power to shape the large-scale course of events, but particulars that can alter entire futures, profoundly and forever.

Consider the primary example from American history. Northern victory was not inevitable in the Civil War, for the South was not fighting a war of conquest (unwinnable given their inferiority in manpower and economic wealth), but a struggle to induce war weariness and to compel the North to recognize their boundaries. The Confederacy had almost succeeded in 1863. Their armies were deep into Pennsylvania; draft riots were about to break out in New York City; Massachusetts was arming the first regiment of free black volunteers—not from an abstract sense of racial justice but from an urgent need for more bodies. In this context, the crucial Battle of Gettysburg occurred in early July. Robert E. Lee made a fateful error in thinking that his guns had knocked out the Union battery, and he sent his men into the nightmare of Pickett's Charge. Suppose we could rerun history and give Lee another chance. This time, armed with better intelligence perhaps, he does not blunder and prevails. On this replay, the South might win the war, and all subsequent American history becomes radically different. The actual outcome at Gettysburg is no minor frill in an inevitable unrolling of events, but a potential setting point of all later patterns.

Never apologize for an explanation that is "only" contingent and not ordained by invariant laws of nature, for contingent events have made our world and our lives. If you ever feel the slightest pull in that dubious direction, think of poor Heathcliff who would have been spared so much agony if only he had stayed a few more minutes to eavesdrop upon the conversation of Catherine and Nelly (yes, the book wouldn't have been as good, but consider the poor man's soul). Think of Bill Buckner who would never again let Mookie Wilson's easy grounder go through his legs—if only he could have another chance. Think of the alternative descendants of *Ichthyostega,* with only four fingers on each hand. Think of arithmetic with base eight, the difficulty of playing triple fugues on the piano, and the conversion of this essay into an illegible Roman tombstone, for how could I separate words withoutathumbtopressthespacebaronthistypewriter.

5 | Bent Out of Shape

WE ALL DREAM about retirement projects that might recapture the lost pleasures of youth, or perfect what we had, perforce, abandoned when the practicalities of making a living and supporting a family intervened. Some day, in a rosy future after the millennium, I will take out my old stamp album or sit down at the piano and finally progress beyond the first of Bach's two-part inventions and the Prelude in C Major from Book 1 of the Well-Tempered Clavier.

Charles Darwin, my hero and role model, achieved this exquisite pleasure, so I may yet have hope for emulation. His last book, published a year before his death, treated the apparently arcane, but vitally important subject of earthworms and their role in forming the topsoil of England. This wonderful and disarming book unites Darwin's end, in the calmness of old age, secure in the knowledge of accomplishment, with his more tumultuous youth, sparked by the fires of unrealized ambition. For Darwin wrote the précis of his worm book in 1838, just two years after the *Beagle* docked—a brilliant five-page article, presenting the entire argument that would fill a book more than forty years later. Darwin concluded:

> The explanation of these facts, . . . although it may appear trivial at first, I have not the least doubt is the correct one, namely, that the whole operation is due to the digestive process of the common earth-worm.

Odd juxtapositions always intrigue me. I do not grant them deep meaning, and firmly believe that they represent nothing more than coincidence. Nonetheless, we do take notice, if only because we must find patterns to tell stories. Darwin published his paper in the fifth volume of the second series of the *Transactions of the Geological Society of London* in 1838. I was reading this paper a few months ago, and couldn't help turning the last page to note the subsequent article, a four-page "Note on the dislocation of the tail at a certain point observed in the skeleton of many Ichthyosauri," written by Richard Owen.

Richard Owen, then a young man, became England's greatest comparative anatomist and first director for the Natural History division of the British Museum when the collections finally escaped the shadows of the Elgin Marbles at Bloomsbury and won their own magnificent home in South Kensington (one of the world's great Victorian buildings and an essential stop on any visit to London).

Owen and Darwin had a long and problematical relationship. Darwin originally courted Owen's friendship and support. (Owen, at Darwin's request, formally described for publication the fossil mammals that Darwin had collected on the *Beagle*. Darwin's famous *Toxodon*, for example, was named, described, and illustrated by Owen.) But the relationship inevitably soured, in part because Owen's vanity could not bear Darwin's successes. Legend holds that Owen's rejection of evolution prompted their final break, but such a falsehood only records our propensity for simplifying stories told in the heroic mode, thus making "bad guys" both nasty and stupid. Owen did reject natural selection, and with vigor, as an excessively materialistic theory depending too much on external environments and too little on laws of organic structure, but he embraced evolution as a guiding principle in natural history.

In any case, the juxtaposition of worm and ichthyosaur dates from 1838, an early period of their friendship. I couldn't help noticing another link more interesting than mere spatial proximity. Darwin wrote, as quoted above, that his subject seemed trivial but really unleashed a cascade of implications leading to substantial importance. Owen then made the very same point, arguing that an apparently broken tail in an ichthyosaur might seem entirely devoid of interest, but that close study yielded generalities of more than passing concern. Since the conversion of detail to wide message, through links of tangential connection, forms the stock-in-trade of these essays, I could hardly avoid such a double invitation to discourse at greater length on the tail bend of ichthyosaurs.

Ichthyosaurs are a group of marine reptiles with bodies so fishlike in external form that they have become the standard textbook example of "convergence"—evolved similarity from two very different starting points as independent adaptive responses to a common environment and mode of life (wings of birds and bats, eyes of squids and fishes). Ichthyosaurs are not closely related to dinosaurs, though they arose at about the same time and became extinct before the great wipeout that ended the dinosaurs' reign some 65 million years ago. (The god-awful spelling of their names, with its unpronounceable sequence *chth,* only records an orthographic convention in converting Greek letters to Roman. This four-consonant sequence represents two Greek letters, *chi* and *theta,* one transliterated *ch,* the other as *th.* Both belong to a five-letter Greek word for fish, and ichthyosaur means "fish lizard." We meet the same orthographic problem in such words as ophthalmology. But never despair and remember that things could always be worse. What would you do if that four-letter sequence came right at the beginning of a word—as it does in a common barnacle with the most forbidding name of *Chthamalus.*)

In considering the convergence of ichthyosaur upon fish, we marvel most at the form and location of fins and paddles—the machinery of swimming and balancing. The fore and hind paddles are, perhaps, least remarkable, for ancestral structures are clearly present as front and back limbs of terrestrial forebears— and these can be modified, as whales and dolphins have done, to forms better suited for sculling than for walking. But the dorsal (back) and caudal (tail) fins are boggling in their precision of convergence with analogous structures in fishes. For the terrestrial ancestors of ichthyosaurs obviously possessed neither back nor tail fin, and ichthyosaurs therefore evolved these structures from scratch—yet they occupy the position, and maintain the form, that hydrodynamic engineers deem optimal for propulsion and balance.

Yet just as ichthyosaurs themselves developed these fishlike

The classic painting of an ichthyosaur by Charles R. Knight. Note the fish-like position of the fins. *Courtesy of Department of Library Services, American Museum of Natural History.*

features in a graduated transition from terrestrial ancestors, so too did our understanding of their extensive convergence grow piece by piece. To be sure, the basic similarity with fishes had never been doubted. In fact, the first two published references, both in 1708, mistook ichthyosaur vertebrae for the backbone of a fish. Both the celebrated Swiss naturalist J. J. Scheuchzer, in his *Querelae Piscium* (Complaints of the Fishes) and the German J. J. Baier, in his work on fossils from the area of Nuremberg, presented figures of ichthyosaur vertebrae for a most interesting purpose: to maintain that fossils are true remains of creatures that once lived, and not some manifestation of a plastic force inherent in rocks and ordained to establish global order by eliciting parallel forms in the organic and inorganic realms (an idea that strikes us as absurd today, but that made lingering sense within a neo-Platonic ideology not yet fully dispersed by the causal theories of Newton and Descartes).

Both Scheuchzer and Baier argued that these "fish" fossils recorded the devastation of Noah's flood. Scheuchzer's work is written as a humorous conversation among fossil fishes annoyed at humans who do not recognize their organic nature and affinity with living relatives. As for Baier, I recently had the pleasure of purchasing a copy of his rare work, without the slightest expectation that I would soon, or ever, have any practical or immediate use for this beautiful book. What a pleasure, then, to read his two page discussion of "ichthyospondyli" (fish vertebrae), with its conclusion that we must view them *"pro piscibus vere petrificatis . . . pro universalis Diluvii reliquiis"*—as truly petrified fish, remains of the universal flood.

Better evidence, primarily from bones of the skull and paddles, revealed the reptilian nature of ichthyosaurs by the early nineteenth century, but strong convergence upon fishes remained the prevailing theme of most writing. Nonetheless, though skeletons revealed the streamlined body and fishlike paddles, two missing pieces conspired to prevent any full appreciation for the true (and awesome) extent of convergence—for the back and tail fins, as soft structures, had not been discovered. All the early reconstructions—by Buckland, Conybeare, de la Beche, Hawkins, and other worthies of early English geology—showed a slithering serpent without back or tail fins, not the reptilian embodiment of a swordfish. How, then, did the two key pieces fall into the piscine puzzle?

Richard Owen's note of 1838 stands as the chief document in this resolution. Thanks largely to keen insight and uncanny field work from Mary Anning, and to support from the demented and

Caricature of ichthyosaurs by Henry de la Beche, made in the early nineteenth century before the back and tail fins had been discovered.
Courtesy of Department of Library Services, American Museum of Natural History.

eccentric Thomas Hawkins (whose monographs of 1834 and 1840 must rank as the craziest documents ever written in paleontology), many good skeletons of ichthyosaurs were collected in England during the early nineteenth century. Owen had noticed an apparent peculiarity in one fine specimen—a sharp downward bend in the sequence of rear vertebrae at about two-thirds the distance from the back flippers to the end of the tail. Owen gave little thought to this tailbend, reasoning that it only represented an anomaly (probably a postmortem artifact) of a single specimen. But when skeleton after skeleton showed a tailbend in the same position, Owen realized that he had stumbled upon a phenomenon worthy of explanation. Owen wrote:

> Having recently examined many saurian skeletons now in London, the greater part of which have been disencumbered of their earthy shroud by the chisel of Mr. Hawkins, a

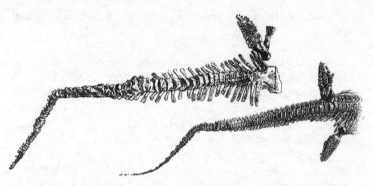

Illustration of ichthyosaur tail bends taken from Richard Owen's 1838 article. *Courtesy of Department of Library Services, American Museum of Natural History.*

condition of the tail which, on a former occasion, in a single instance had arrested my attention, but without calling up any theory to account for it, now more forcibly engaged my thought, from observing that it was repeated, with scarcely any variation, in five instances [boy, did they love to write back then, as in Owen's "disencumbered of their earthy shroud" for our modern "dug out of the rock"]. The condition to which I allude is an abrupt bend or dislocation of the tail . . . the terminal portion continuing, after the bend, almost as straight as the portion of the tail preceding it. In short, the appearance presented is precisely that of a stick which has been broken, and with the broken end still left attached, and depending [that is, hanging] at an open angle.

Owen then drew the right conclusion for the wrong reason and correctly inferred the existence of a tail fin. He argued that the constant position of the tailbend must record an attachment of some structure at this point. He rightly conjectured that this organ must be a tail fin, and he even predicted its vertical position (as in fishes) rather than a horizontal orientation (as in whales). But he wrongly assumed that the bend must represent a dislocation (probably after death) of an originally straight vertebral column—perhaps because the tail fin bloats with gas as the animal begins to decay, thereby fracturing the vertebral column at the front border of the fin. Owen then added other conjectures, and wrote:

> The appearance in the tail of the *Ichthyosaurus* . . . is too uniform and common to be due entirely to an accidental and extrinsic cause. I am therefore disposed to attribute it to an influence connected with some structure of the recent animal; and most probably to the presence of a terminal . . . caudal fin, which, either by its weight, or by the force of the waves beating upon its extended surface, or by the action of predatory animals of strength sufficient to tug at without tearing it off, might . . . give rise to a dislocation of the caudal vertebrae immediately proximal to its attachment.

The puzzle finally achieved its solution in the 1890s when the perennial, but rarely granted, prayer of all paleontologists was answered by the powers that be. Ichthyosaurs with preserved soft parts were discovered in the Holzmaden deposits near Stuttgart. These sediments are so rich in organic oils and bitumen that they actually burn. (One fire raged beyond control from 1668 to 1674 and another from 1937 to 1939). Details of internal organs are not retained in these bitumen beds, but body outlines remain intact as black films upon the light gray rock. (Most of the fine specimens displayed at museums throughout the world come from the Holzmaden beds, and many readers are no doubt familiar with ichthyosaur body outlines preserved as blackened films on the rock under and behind the bones.)

The Holzmaden ichthyosaurs finally proved the extent of external convergence upon the stereotypical form of a free-swimming fish. The dorsal fin, with no bony support at all, was revealed for the first time. And the caudal fin, correctly inferred by Owen from the tailbend, now stood out for all to see. The fin was vertical, as Owen had surmised, and composed of two nearly equal and symmetrical lobes. The vertebral bend did mark, again as Owen had conjectured, the anterior border of the fin—but as an item of normal anatomy, not a postmortem artifact or dislocation. The vertebral column bent naturally down to follow the lower border of the lower lobe of the tail right to the animal's rear end. No other vertebrate displays this orientation. In fishes, the

A fossil ichthyosaur with characteristic and excellent preservation from the Holzmaden deposits. Note the outlines of back and tail fins, and also the bending of the vertebral column into the lower lobe of the tail. *Courtesy of* Natural History.

vertebral column either stops at the inception of the tail or extends, as in sharks, into the upper border of the upper lobe. No wonder that the ichthyosaur tailbend had provoked such confusion for more than fifty years.

Nearly a century has passed since the Holzmaden discoveries revealed the true nature of the ichthyosaur tailbend by exposing its enclosure within the caudal fin. Yet the tailbend continues to provoke commentary and controversy for two main reasons as outlined by Chris McGowan of the Royal Ontario Museum, Toronto, and the world's leading expert on ichthyosaurs (my thanks to Dr. McGowan not only for his many illuminating articles, but especially for enduring a long phone call of inquiry during my research for this essay). First, and positively, the location, angle of downturn, and length of the vertebral column after the bend specify both the size and form of the caudal fin (only the Holzmaden ichthyosaurs preserve the fin itself as a carbonized film; all other specimens are bones alone, and the tailfin must be inferred from the vertebral column).

Second, and representing yet another dubious triumph of expectation over observation (perhaps the most common of human foibles), many classic specimens have been reconstructed on the assumption that tailbends must be present. I raise no issue of fraud or delusion. In many specimens, the vertebrae (particularly the small items at the rear end) lie scattered over the rock surface. The wonderfully expert and professional Holzmaden preparators adopted the custom of removing these bones entirely from the matrix and then resetting them in the inferred position of the living animal—that is, with a tailbend. We have no doubt that several ichthyosaur species developed a pronounced tailbend, since perfect specimens with preserved body outlines clearly show the tail vertebrae extending into the lower lobe of the caudal fin. But perhaps other species (particularly the earliest forms) lacked a tailbend, and perhaps preparators have tended to exaggerate the amount of inclination in reconstructing their specimens.

If the actual tailbends of most specimens on display are thus infected with doubt, how can we be confident about the existence and form of the caudal fin in most species? And, since this information is crucial to our understanding of swimming and maneuvering in ichthyosaurs, how can we hope to reconstruct the ecology of these fascinating animals? Obviously, we need a criterion of confirmation separate from the bend itself. Fortunately, McGowan has been able to establish such a criterion and to devise an ingenious way of putting it into practice.

How can an angular bend be produced in a basically linear structure (like the vertebral column), built from a sequence of disks that must follow, one behind the other, without large spaces between? As the accompanying sketch shows, tailbends imply a change in the shape of the crucial vertebral disks at the bend itself—from their usual form (with upper and lower borders of equal width) to a wedge with a wider border on top and a narrower edge below. A succession of wedge-shaped disks will inevitably cause the tail to bend, and the greater the difference in width between upper and lower borders, the more pronounced the bend. In fact, by a simple construction akin to the problems we all worked in high-school plane geometry, the angle of the bend can be inferred from the number of wedge-shaped disks and their intensity of wedging.

But how can this wedging be assessed? The vertebrae of most skeletons are at least partially embedded in rock, and both ends are rarely exposed to reveal the extent of wedging (while mu-

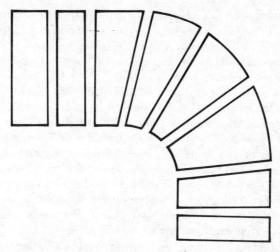

Shape of vertebral disks in ichthyosaurs with tail
bends. Note the necessary wedge shape at the bend
itself. *Ben Gamit.*

seums rarely look kindly upon requests for sufficient mayhem
upon their specimens to dig the vertebrae out of the enclosing
matrix). McGowan solved this problem with a boost from modern medical technology—computed tomography as provided by
a CT-scanner. These marvelous, donut-shaped x-ray devices can
take a photographic slice right through a human body in any
orientation (so long as the body fits into the donut-hole of the
machine). Well, an ichthyosaur in its matrix is often about the
same size as a human body. Why not take a CT-scan of vertebrae
at the tailbend, thus producing a photographic image of the vertebral disks while still embedded in their matrix? (McGowan
didn't initiate the application of CT-scanning to paleontological
material. Several successful attempts have been made in the past
few years, including the resolution of cranial capacities and form
of unerupted teeth in some important skulls of the human fossil
record.)

McGowan used a CT-scanner to affirm that *Leptopterygius
tenuirostris,* an early ichthyosaur with an uncertain tailbend cur-

rently subject to hot dispute, did grow a series of six wedge-shaped vertebral disks in the crucial region—not strongly wedged to be sure (none producing more than a five degree bend), but yielding in their ensemble a modest tailbend of some 25 degrees (see McGowan's article, "The ichthyosaurian tail-bend: A verification problem facilitated by computed tomography," in the bibliography). Somehow, I feel a great sense of satisfaction in the affirmation of this continuity in human striving for knowledge through time—to think that a discussion beginning in two Latin treatises written in 1708, proceeding through the keen observations of England's greatest anatomist in the 1830s, and on to the discovery of preserved body outlines in a famous German locality during the 1890s should be resolved, as we begin our last decade's countdown towards the millennium, by the latest device of medical machinery!

Yet, however satisfying the particular resolution, this tale (and tail) would convey no message or meaning (to those outside the tiny coterie of ichthyosaurian aficionados) if the problem of the ichthyosaur tailbend did not illuminate something central in evolutionary theory. Ichthyosaurs are most celebrated for their convergence upon the external form of superior swimmers among fishes. Since English traditions in natural history place primary emphasis on the concept of adaptation, these similarities of fish and marine reptile have won the lion's share of written attention—for we know how the threefold combination of flippers, backfin, and tailfin work in efficient hydrodynamic coordination, and we are awed that two independent lineages evolved such uncanny resemblance for apparently similar function. This awe even predates evolutionary theory, for an earlier attribution to God's benevolent care inspired as much admiration as our current respect for the power of natural selection. William Buckland, Owen's close colleague, had a special affection for ichthyosaurs. He also wrote the greatest paean of the 1830s to adaptation as proof of God's benevolence. In *Geology and Mineralogy Considered with Reference to Natural Theology*, written in 1836, Buckland invoked the precise convergence of ichthyosaur and fish as a proof of God's goodness. Buckland acknowledged that an ordinary reptile would be in severe trouble at sea, but ichthyosaurs have been granted by divine fiat (read "endowed by natural selection" for a modern version of the same argument):

. . . a union of compensative contrivances, so similar in their relations, so identical in their objects, and so perfect in the adaptation of each subordinate part, to the harmony and perfection of the whole; that we cannot but recognize throughout them all, the workings of one and the same eternal principle of Wisdom and Intelligence, presiding from first to last over the total fabric of the Creation.

Yet, in our complex world of natural history, almost any profuse enthusiasm also elicits its mitigating opposite. (Such a cautionary splash of cold water may then emerge as a primary theme with more enlightening implications in itself.) Yes, ichthyosaur convergences are remarkable; only a soulless curmudgeon could fail to be impressed by the fishlike form of this descendant from ordinary terrestrial reptiles. Only the most militant denigrator of Darwin and the entire English tradition could fail to utter the word *adaptation* with both confidence and admiration.

But another perspective demands equal attention—and Owen, the much misunderstood proponent of a continental tradition that viewed adaptation as superficial, and sought regularities of form underneath a garb of immediate design, discussed ichthyosaurs primarily in the light of this alternative. What are the limits to adaptation imposed by the disparate anatomical designs underlying a convergence (fishes and reptiles in this case)? To what extent must the ichthyosaur remain in the thrall of its past, quite unable to mimic the form of a fish exactly because the historical legacy of a reptilian body plan precludes a large set of favorable options? To what degree, in short, must an ichthyosaur remain an easily identified reptile in marine drag?

For a primary statement of this alternate theme (limits imposed by inherited design), we must look to the largely forgotten work of the great Belgian paleontologist Louis Dollo (1857–1931). Dollo gave his name to an evolutionary principle known as *irreversibility* (often called Dollo's Law). In one of the cruel ironies often imposed by history, many fine thinkers win their posthumous recognition only by eponymous linkage with a principle so widely misunderstood that true views turn into their opposite. Many evolutionists interpret Dollo's Law as an antiquated statement about inherent, directional drives in evolution—a last gasp of a mystical vitalism that the Darwinian juggernaut finally defeated. In fact, Dollo was a convinced mechanist, and a Darwinian in basic orientation (with some interesting wrinkles of disagreement).

To Dollo, irreversibility epitomized the nature of history under simple conditions of mathematical probability (Dollo had obtained an extensive education in mathematics and attributed his formulation of irreversibility to this training). Evolutionary transformations are so complex—involving hundreds of independent changes—that any complete reversal to a former state becomes impossible for the same reason that you will never flip 1,000 heads in a row with an honest coin. No mysticism, no vitalism, only the ordinary operation of probability in a complex world. A simple change (increase in size, mutation in a single gene) may be reversed, but the standard transformations that form the bread and butter of paleontology (origin of flight in birds, evolution of humans from apelike ancestors) cannot run backwards to recover an ancestral state exactly.

History is irrevocable. Once you adopt the ordinary body plan of a reptile, hundreds of options are forever closed, and future possibilities must unfold within the limits of inherited designs. Adaptive latitude is impressive, and natural selection (metaphorically speaking) is nothing if not ingenious. A terrestrial reptile may return to the sea and converge upon fishes in all important aspects of external form. But the similarity can only be, quite literally, skin deep and truly superficial. The convergence must be built with reptilian parts, and this historical signature of an evolutionary past cannot be erased. Dollo explicitly linked his principle of irreversibility with a concept that he called "indestructibility of the past."

When we look again at the three great convergences of ichthyosaurs—the flippers, the dorsal fin, and the caudal fin—but this time from the alternate perspective of limits imposed by irrevocable starting points, we find that these features beautifully illustrate the three most important principles of irreversibility as a signature of history.

1. *The flippers, or you must use parts available from ancestral contexts little suited to present environments.* The flippers, by external form, are well adapted for swimming and balancing. But their internal bony structure reveals a terrestrial reptile under the marine adaptation. The front flipper begins with a stout humerus, followed by

a shortened and flattened radius and ulna, side by side. The carpals and metacarpals (hand bones) and phalanges (finger bones) follow in a similar flattened modification. In an interesting change (still related to an irrevocable ancestral state), the phalanges are multiplied into long rows that mimic the rays of fish fins. Humans have three phalanges per finger (two for the thumb); ichthyosaurs can grow more than twenty per finger.

2. *The dorsal fin, or you can't get there from here.* The dorsal fin of fishes generally contains a strengthening set of bony rays. Similar structures might well have benefited ichthyosaurs, but their terrestrial ancestors built no recruitable body parts along the back. Ichthyosaurs therefore evolved a boneless dorsal fin (that would have eluded us altogether if we had never discovered the Holzmaden specimens).

3. *The caudal fin and its tailbend, or you must always build a converging structure with some distinctive difference, due to irrevocable ancestry, from the original model.* The vertebral column of fishes, as noted above, either stops at the inception of the tail or extends into the upper lobe. Only in ichthyosaurs do the vertebrae bend down into the lower lobe of the tail. We do not know why ichthyosaurs developed this strikingly different and unique internal structure (I would need another essay to discuss the interesting structural and functional explanations that have been proposed), but all convergences evolve with distinctive differences based on a thousand quirks of disparate ancestries.

Louis Dollo has long been one of my private heros. I meant to cite his views on irreversibility as a centerpiece of this essay, but I didn't know, until a chance discovery in the midst of my research, that he had written an entire paper on the caudal fin of ichthyosaurs—and at a most interesting time, in 1892 just after the discovery of fin outlines in the Holzmaden specimens. Dollo rejoiced that these beautifully preserved specimens had resolved *"la curieuse dislocation de la colonne vertébrale, signalée depuis longtemps"* (the curious dislocation of the vertebral column, recognized for so long). And he proposed an explanation rooted in uniqueness imposed by irrevocable history. I doubt that he was right in detail, but his conjecture is ingenious, and entirely in the spirit of an important and insufficiently appreciated principle of historical reconstruction. He argued that the tailbend arose because the two-lobed caudal fin of ichthyosaurs evolved from a skin-fold along the back (source of the dorsal fin as well), which extended itself in a posterior direction to form the upper lobe of the tailfin and then pushed the vertebral column down to form the lower lobe. Since several modern reptiles maintain such a skin fold along the back, but never along the belly, new fins could only evolve along the dorsal edge of the body, and the vertebral column could only be pushed down to form a two-lobed tailfin. But ancestral fishes maintained a fin-fold along both back and belly, and a two-lobed tailfin could evolve as a lower lobe pushed the vertebral column up.

Richard Owen, in contrast with his adaptationist colleague Buckland, appreciated the primacy of maintained reptilian design as the main lesson of ichthyosaur convergence. He wrote in his great monograph on British fossil reptiles (published between 1865 and 1881, and anticipating Dollo's concerns):

> The adaptive modification of the Ichthyopterygian skeleton, like those of the Cetacean [whale] relate to their medium of existence; [but] they are superinduced, in the one case upon a Reptilian, in the other upon a Mammalian type.

At about the same time, and in a more pointed commentary on the same theme of irrevocability in history, W. S. Gilbert (in *Princess Ida*) then penned a crisp epitome to remind his audiences of evolution's major lesson:

> Darwinian man though well behaved
> At best is only a monkey shaved.

6 | An Earful of Jaw

THE MOST SUBLIME of all beauties often proceed from the softest or the smallest—the quadruple pianissimos of Schubert's "Schöne Müllerin," as sung by Fischer-Dieskau (and penetrating with brilliant clarity to the last row of the second balcony, where I once sat for the greatest performance I ever witnessed) or the tiny birds of brilliant plumage depicted in the marginalia of medieval manuscripts. But even the most refined and intellectual character may succumb without shame to the sheer din employed now and then by great composers to overwhelm the emotions by brute force rather than ethereal loveliness—Ravel's orchestration of the "Great Gate of Kiev" at the end of Moussorgsky's *Pictures at an Exhibition,* or the last scene of Wagner's *Die Meistersinger.*

I once had the privilege of singing with the Boston Symphony at Tanglewood in the midst of *numero uno* among musical dins— the *Tuba mirum* of Berlioz's *Requiem.* I had listened to the piece all my adult life; we had rehearsed (without orchestra) for weeks. I knew exactly what was coming as the dress rehearsal began. The four supplementary brass choirs enter one after the other, building and building to a climax finally joined by the timpani—eight pair, I think, although they seemed to extend forever in an endless row before the choral risers. And against this ultimate crescendo, the basses alone (including me) must sing the great invocation of the last judgment:

> Tuba mirum spargens sonum
> Per sepulchra regionum
> Coget omnes ante thronum

(The wondrous sound of the trumpet goes forth to the tombs of all regions, calling all before the throne.)

So it should go, and so it went—but not for me. I had devolved into tears and spinal shivers—not in ecstasy at the beauty, but in awe at the volume. (Forewarned is forearmed; I was fine at the performance itself.) Great composers have every right to exploit the physiology of emotional response in this way, but only sparingly, for timing is the essence (and most of Berlioz's *Requiem* is soft).

My memory of this extraordinary incident in my emotional ontogeny focuses upon a curious highlight of mixed modalities. The sound of the brass assaulted my ears, but the thunder of the timpani followed another, unexpected route. It entered the wooden risers under my feet and rose from there to suffuse my body; sound became feeling.

I am no disciple of Jung, and I do not believe in distant phyletic memory. Yet, in an odd and purely analogical sense, I had become a fish for a moment. We (and nearly all terrestrial vertebrates) hear airborne sound through our ears; fish feel the vibrations of waterborne sound through their lateral line organs. Fish, in other words, "hear" by feeling—as I had done through a set of wooden risers with a density closer to water than to air.

For an optimal combination of fascination with excellent documentation, no saga in the history of terrestrial vertebrates can match the evolution of hearing. Two major transitions, seemingly impossible but then elegantly explained, stand out at opposite ends. First, at the inception of terrestrial life: How can creatures switch from feeling vibrations through lateral lines running all over their bodies to hearing sounds through ears? How, in other words, can new organs arise without apparent antecedents? Second, at the last major innovation in vertebrate design: How can bones that articulate the upper and lower jaws of reptiles move into the mammalian ear to become the malleus and incus (hammer and anvil) in the chain of three bones that conduct sound from the eardrum (the tympanum in anatomical parlance, recalling my Berlioz story in the singular) to the inner ear? How, in other words, can organs switch place and function without destroying an animal's integrity as a working creature? How can we even imagine an intermediary form in such a series? You can't eat with an unhinged jaw. Creationists have used this difference between reptiles and mammals to proclaim evolution impossible a priori—I mean, really, how can jawbones become ear bones? Get serious! Yet, we shall see, once again, that the domain of conventional thought can be much narrower than the capabilities of nature—although ideas should be able to extend and soar beyond reality.

The key to the riddle of both these transitions lies in the major theme of my Berlioz story—multiple modalities and dual uses. You can pat your head and rub your stomach, walk and chew gum at the same time (most of us, at least), feel and hear sound, chew and sense with the same bones.

Nature writing in the lyrical mode often exalts the apparent perfection and optimality of organic design. Yet, as I frequently argue in these essays, such a position plunges nature into a disabling paradox, historically speaking. If such perfection existed as a norm, you might revel and exult all the more, but for the tiny problem that nature wouldn't be here (at least in the form of complex organisms) if such optimality usually graced the products of evolution.

I recently made my first trip to Japan to deliver a lecture at the opening of an annual series that will bring one American scholar to Japan and a Japanese counterpart over here to speak on a common topic. I was both pleased and intrigued by our assigned theme for this initial year (largely at Japanese request)—creativity. (Some Japanese apparently fear—although my superficial impressions included nothing to sustain such anxiety—that their scholars and industrialists excel at efficiency and alteration, but not at innovation.)

I had no words of wisdom on Japanese life (I would not dare, not even by the old criterion that experts are folks who have been in a country either more than twenty years or less than two days); nor do I understand the sources of creativity in the human psyche of any culture. So, following the fine maxim that a shoemaker must stick to his last (a wooden model of a foot, not a final goal), I spoke on the evolutionary meaning of creativity—specifically, on the principles that permit major transitions and innovations in the history of life. I don't know that my message was well received in this land of supreme artistry in the efficient use of limited space, for I held that the watchwords of creativity are sloppiness, poor fit, quirky design, and above all else, redundancy.

Bacteria are marvels of efficiency, simple cells of consummate workmanship, with internal programs, purged of junk and slop, containing single copies of essential genes. But bacteria have been bacteria since life first left a fossil record 3.5 billion years ago—and so shall they probably be until the sun explodes. Such optimality provokes wonder but provides no seeds for substantial change. If each gene does one, and only one, essential thing superbly, how can a new or added function ever arise? Creativity in this sense demands slop and redundancy—a little fat not for trimming but for conversion; a little overemployment so that one supernumerary on the featherbed can be recruited for an added role; the capacity to do several things imperfectly with each part. (Don't get me wrong. Bacteria represent the world's greatest success story. They are today and have always been the modal organisms on earth; they cannot be nuked to oblivion and will outlive us all. This time is their time, not the "age of mammals" as our textbooks chauvinistically proclaim. But their price for such success is permanent relegation to a microworld, and they cannot know the joy and pain of consciousness. We live in a universe of trade-offs; complexity and persistence do not work well as partners.)

To build a vertebrate along the tortuous paths of history, evolution must convert the poet's great metaphor into flesh and bones. "I hear it," writes Yeats, "in the deep heart's core." I don't mean to be excessively literal, but if creatures couldn't occasionally hear with their lungs (as some snakes do) or with their jaws (as our immediate reptilian ancestors probably did), we would not now have ears so cleverly wrought that they fool us into the attractive but untenable vision of organisms as objects of optimal design. Consider the first and last major steps in anatomical construction of the mammalian middle ear—for we know no better or more intriguing story in the evolution of vertebrates.

1. *The origin of hearing bones in the first terrestrial vertebrates.* The hearing of sound in thin air poses a major physical problem: How can low-pressure airborne waves be converted into high-pressure waves suitable for transmission by fluids in the cochlea of the inner ear? Terrestrial vertebrates use two major devices to make the necessary conversion. First, on the "stiletto heel" principle (quoting a metaphor from my colleague T. S. Kemp), they collect sound on the relatively large area of the eardrum but eventually

transmit the waves into the inner ear through a much smaller opening called the *fenestra ovalis* (oval window). Second, they pass the vibrations from eardrum to oval window along a bone or series of bones, called in mammals, the malleus, incus, and stapes, or hammer, anvil, and stirrup to honor a truly uncanny resemblance. These bones act as levers to increase the pressure as sound waves travel toward the brain.

Fish have an inner ear, but no eardrum or middle-ear bones; they "hear" primarily through their lateral line organs by detecting the movement of water produced by sound waves in this dense medium. How then could middle-ear bones arise in terrestrial vertebrates, apparently from nothing?

The first vertebrates had no jaws. Modern lampreys and hagfishes survive as remnants of this first vertebrate radiation; their formal name, Agnatha (or jawless), embodies their anatomy (or partial lack thereof). In agnathans, a series of gill openings lies behind the boneless mouth—and this arrangement foreshadows the evolution of jaws. In the first jawed fishes, gills are supported by a series of bones, one set for each gill slit. Each set includes an upper and lower bar, pointing forward and hinged in the middle. Obviously, this arrangement, although evolved for supporting gills, looks uncannily like the upper and lower jaws of a typical vertebrate. We do not know for certain whether jaws arose from a functioning gill arch that moved forward to surround the mouth or whether jaws and gill arches just represent two specializations, always separate, but generated from the same system of embryological development. In either case, we do not doubt that gill supports and jaws are homologous structures (that is, evolved from the same source and representing the "same" organ in different forms—like arms and legs or fingers and toes). The evidence for homology is multifarious and overwhelming: (1) the embryo builds both jaw precursors and gill arches not from mesoderm, the source of most bones, but from migrating neural crest cells of the developing head; (2) both structures are made of upper and lower bars, bending forward and hinged in the middle; (3) the muscles that close the jaw are homologues of those that constrict the gill slits.

If vertebrate jaws represent an anterior gill arch, then another crucial element of the skull also derives from the gill supports just behind. The upper bar of the next gill arch in line becomes the

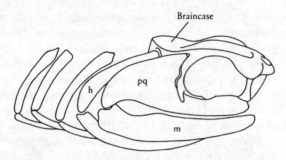

A classic figure of homologies between gill arch and
jaw bones, taken from R. L. Carroll's *Vertebrate
Paleontology and Evolution*. The upper and lower jaws
(pq and m) have the same position and form as all
the gill arches behind them. Note also that the
upper element of the gill arch just behind the jaws
articulates with the braincase. This bone becomes
the hyomandibular (h) and later the stapes in
terrestrial vertebrates. *Courtesy of Department of Library Services,
American Museum of Natural History.*

hyomandibular of jawed fishes, a bone that functions in support
and coordination by linking the jaws to the braincase.

All this detail may seem distant from the origin of hearing
bones, but we are closing in quickly (and shall arrive before the
end of this paragraph). Mammals have three middle-ear bones—
hammer, anvil, and stirrup, or stapes. And the stapes is the homo-
logue of the hyomandibular in fishes. In other words—but how
can it happen?—a bone originating as a gill support must have
evolved to brace the jaws against the braincase, and then changed
again to function for transmission of sound when water ceded to
air, a medium too thin to permit "hearing" by the lateral line.

As usual in a world of encumbrances, we must flush away an old
and conventional concept before we can understand how such an
"inconceivable" transition might actually occur without impediment in theory or practice. We must forget the old models of
horses and humans mounting a chain of improvement in functional continuity—from small, simple, and not-so-good to larger,
more elaborate, and beautifully wrought. In these models, brains

are always brains and teeth always teeth, but they get better and better at whatever they do. Such schemes may work for the improvement of something already present, for a kind of stately continuity in evolution. But how can something original ever be made? How can organisms move to a truly novel environment, with needs imposed for functions simply absent before? We require a different model for major transitions and innovations, for King Lear was correct in stating that "nothing will come of nothing."

We need, in other words, a mechanism of recruitment and functional shift. Evolution does not always work by enlarging a rudiment. It must often take a structure functioning perfectly well in one capacity and shift it to another use. The original middle-ear bone, the stapes, evolved by such a route, changing from a stout buttressing bone to a slender hearing bone.

If each organ had only one function (performed with exquisite perfection), then evolution would generate no elaborate structures, and bacteria would rule the world. Complex creatures exist by virtue of slop, multiple use, and redundancy. The hyomandibular, once a gill support, then evolved to brace jaw and braincase. But this bone happens to lie right next to the otic capsule of the inner ear—and bone, for reasons incidental to its evolution, can transmit sound with reasonable efficiency. Thus, while functioning primarily as a brace, the hyomandibular also acquired other uses. Skates and rays take in water through a round opening, called the spiracle, located in front of the other gill slits. The hyomandibular then helps to pump this water into the mouth cavity, and thence out and over the gill slits. Closer to our phyletic home, the hyomandibular may help to ventilate the lungs of modern lungfishes.

I have wanted to write about the origin of middle-ear bones ever since I began this series, for we have no finer story in vertebrate evolution. But I like to wait for a handle in new information, and one recently came my way (see J. A. Clack, in bibliography, on finding the oldest stapes). The first known tetrapods (four-legged terrestrial vertebrates) hail from eastern Greenland in rocks 360 million years old (see Essay 4). They have been known for some time under the names *Ichthyostega* and *Acanthostega,* but their stapes had not been well resolved before. Clack found six stapes of *Acanthostega,* four preserved in their life positions.

Clack suggests not only a dual but a triple function for the stapes of these first land vertebrates. The bone is stout and dense, not slender and delicate as in stapes adapted largely for hearing. This original stapes must still have functioned in its earlier role as a brace (other early tetrapods, including mammalian ancestors, also had stout stapes). Clack also advocates a supplementary role in respiration. Finally, she makes a key observation based on the stapes's location: "The stapes is likely to have had some auditory function because of the close association between the footplate [a part of the stapes] and the otic [ear] capsule."

Such a multifarious bone nearly bursts with evolutionary potential. The stapes may have braced for a hundred million years, but it also worked for respiration and hearing if only in an incipient or supplementary way. When the cranium later lost its earlier mobility, and the braincase became firmly sutured to the skull—as occurred independently in several lineages of terrestrial vertebrates—the stapes, no longer needed for support, used its leverage and amplified a previously minor role in hearing to a full-time occupation.

2. *The origin of mammalian middle-ear bones.* The odyssey of the stapes (stirrup) is extraordinary enough, a tale worthy of Scylla, Charybdis, and all the wiles of Circe—from gill support to a brace between jaws and braincase to a hearing bone for airborne sound. Yet the other two bones of the mammalian middle ear, named long ago by an age that knew the blacksmith's forge, have an even more curious history. The hammer and anvil (malleus and incus), as elements of the gill arch in front of the hyomandibular, became parts of the jaw in early vertebrates. In fact, they took up the central role of connecting and articulating the upper and lower jaws—as they still do in modern amphibians, reptiles, and birds. The quadrate bone of the reptilian upper jaw became the incus of mammals, while the articular bone of the lower jaw became the malleus. The transition, so improbable in bold words, is beautifully documented in the fossil record and in the embryology of all modern mammals.

The homology of reptilian jawbones to mammalian ear bones was discovered by German anatomists and embryologists well before the advent of evolutionary theory. In 1837, C. B. Reichert made the key observations and expressed the surprise that this tale has elicited ever since. With these words, Reichert introduced his section on the *Entwicklungsgeschichte der Gehörknöchelchen* (developmental history of the little hearing bones). (German looks so god-awful for its massive words. But these tongue twisters are usually made of little words compounded, and the system becomes beautifully transparent, even charming, once you break the big items into their elements. The Germans have preferred to construct technical terms as compounds of their ordinary words, rather than from fancy and foreign Latin or Greek. A rhinoceros is a *Nashorn,* or "nose horn" as rhinoceros actually says in Greek; a square is a *Viereck,* or four-corner. Our technical literature refers to the hammer, anvil, and stirrup as "auditory ossicles"; don't you prefer the German *Gehörknöchelchen,* or little hearing bones?) In any case, Reichert wrote: "Seldom have we met a case, in any part of animal organization, in which the original form of an early [embryological] condition undergoes such extensive change as in the ear bones of mammals. We would scarcely believe it. . . . Nevertheless, it happens in fact."

Reichert recognized all key outlines of the story: that all the ear bones derive from the first two sets of gill-arch bones, the hammer and anvil from the first arch (forming the jaw of vertebrates), and the stirrup from the second arch (forming the hyomandibular of fishes). He noted that the lower jaw first forms with a precursor called Meckel's cartilage (in honor of another great German anatomist of the generation just before, J. F. Meckel). The mandible or jawbone then ossifies on the side of Meckel's cartilage. Meanwhile, the posterior end of Meckel's cartilage, forming the back end of the jaw in the early pig embryo, ossifies and then detaches to become the malleus of the middle ear. One could hardly ask for more direct evidence, and Reichert's observations have been affirmed thousands of times since.

(As a tangential comment in my continuing campaign against textbook copying, the accompanying illustration shows Reichert's original figure of a developing lower jaw in the embryonic pig; h and i represent parts of the future malleus forming at the back end of Meckel's cartilage (g); the ossifying mandible (a) begins to surround and supplant the cartilage. Meanwhile, the incus (k) and the stapes (n) form as bones separate from the lower jaw. This figure has been copied and degraded, like xeroxes of xeroxes, ever since this 1837 original. I last saw its clone in a vertebrate anatomy textbook published in 1971. Two

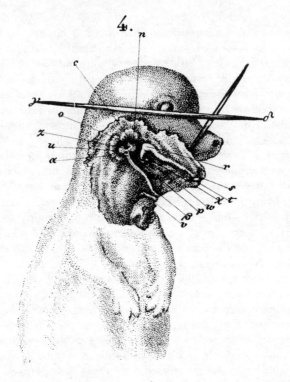

Two illustrations from Reichert's classic article of 1837, containing his discovery of the homologies of mammalian middle-ear bones. See text for explanation. *Courtesy of Department of Library Services, American Museum of Natural History.*

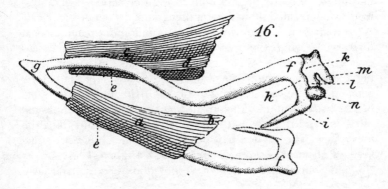

bits says that the author of this text [who undoubtedly copied the figure from a book just slightly older than his] would be shocked to learn that his picture dates from 1837. This time, everyone lucks out because Reichert was a great anatomist and his figure is basically correct; but think of the capacity for compounded error inherent in this procedure of mindless copying. I also include, to give an interesting [if gory] flavor of common styles of illustration during the early nineteenth century, one of Reichert's graphic preparations of a pig embryo, dissecting pins and all.)

Thus, every mammal records in its own embryonic growth the developmental pathway that led from jawbones to ear bones in its evolutionary history. In placental mammals, the process is complete at birth, but marsupials play history postnatally, for a tiny kangaroo or opossum enters its mother's pouch with future ear bones still attached to, and articulating, the jaws. The bones detach, move into the ear, and the new jaw joint forms—all during early life within the maternal pouch.

Paleontological and functional evidence join the embryological data to construct a firm tripod of support, giving this narrative pride of place among all transitions in the evolution of vertebrates by combining strength of documentation with fascination of content. One theme stands as the coordinating feature of this narrative (and of my entire essay): redundancy and multiple use as the handmaidens of creativity.

We might employ this theme to make an abstract prediction about the character of intermediary forms in the fossil record. Contrary to creationist claims that such a transition cannot occur in principle because hapless in-betweens would be left without a jaw hinge, the principle of redundancy suggests an obvious solution. Modern mammals hinge their jaws between squamosal (upper jaw) and dentary (lower jaw) bones; other vertebrates between quadrate (upper jaw) and articular (lower jaw) bones destined to become the incus and malleus of the mammalian ear. Suppose that mammalian ancestors developed a dentary-squamosal joint while the old quadrate-articular connection still functioned—producing an intermediary form with a double jaw joint. The old quadrate-articular joint could then be abandoned, as its elements moved to the ear, while the jaw continued to function perfectly well with the new linkage already in place.

Our woefully inadequate fossil record is not brimming with

intermediary forms, for reasons often discussed in these essays. But the origin of mammals represents a happy case of abundant evidence. The abstract predictions of the last paragraph (actually advanced by paleontologists before the discoveries, so I am not just making a rhetorical point here) have been brilliantly verified in abundant fossil bone. The cynodont therapsids, our ancestral group among the so-called mammallike reptiles, show numerous trends to reduction and loosening of both quadrate and articular bones in the old reptilian jaw joint. Meanwhile, the dentary of the lower jaw enlarges and extends back to contact the upper jaw. (In mammals, the dentary forms the entire lower jaw; reptilian jaws contain several postdentary elements, all reduced and then suppressed or dispersed in mammals.) Many cynodonts develop a second articulation between the squamosal and a postdentary element of the lower jaw called the surangular. (This joint is not the later mammalian dentary-squamosal link, but its formation illustrates a multiple evolution of the intermediacy proclaimed impossible by creationists.) Finally, two or three genera of advanced cynodonts develop a second articulation of truly mammalian character between the dentary and squamosal. One such genus (although the evidence has been disputed) bears the lovely and distinctive name *Diarthrognathus,* or two-jointed jaw.

Moreover, the earliest true mammals do not yet have a fully independent malleus and incus. These bones remain affixed to the jaws and continue to participate in articulation, in both *Morganucodon* and *Kuehneotherium,* the two best known early mammals. "In this sense," wrote Edgar F. Allin in 1975, "the earliest mammals did not yet possess a 'mammalian middle ear.'" By Upper Jurassic times, still well within the early days of mammalian life in a world dominated by dinosaurs, these bones had entered the ear, and an exclusively dentary-squamosal joint had formed.

Embryology and paleontology provide adequate documentation of the "how," but we would also like more insight into the "why." In particular, why should such a transition occur—especially since the single-boned stapedial ear seems to function quite adequately (and, at least in some birds, every bit as well as the three-boned mammalian ear)? We are nowhere near the full answer to this complex question, but one hint conveys special interest and also illustrates the principle of redundancy one more time.

Pelycosaurs, those sail-backed creatures included in every set of plastic dinosaurs and every box of chocosaurus cookies, are not dinosaurs at all, but our distant ancestors—forebears of the therapsid reptiles that eventually evolved into mammals. The stapes of pelycosaurs lies in close contact with the quadrate bone of the upper jaw (forerunner of the incus that now articulates with the stapes in the mammalian middle ear). This linkage continues and sometimes intensifies in descendant therapsids—the more immediate ancestors of mammals. This anatomical connection strongly suggests that the quadrate of mammalian ancestors, while functioning primarily in jaw articulation, already played a subsidiary role in the transmission of sound. Allin argues: "From the nature of its junction with the stapes, the cynodont quadrate obviously took part in sound conduction."

Unfortunately, we cannot experiment on extinct animals and have no direct evidence for quadrate hearing in the actual ancestors of mammals. However, we do know that reptilian quadrates can transmit sound while still acting as part of a jaw joint, for several modern reptiles use an important quadrate path to their inner ear. (These creatures are not mammalian ancestors to be sure, but they do demonstrate the possibility, indeed the actuality, of this crucial multiple modality in the evolution of mammalian hearing.) Snakes, for example, have no external ear or eardrum, and many scientists had considered them entirely deaf, until recent studies illustrated sensitivity to sound over most of the body, especially around the large lung that can transmit vibrations to the inner ear. But another route offers special advantages to a creature so close to the substrate by God's direct decree: "... upon thy belly shalt thou go, and dust shalt thou eat all the days of thy life." Snakes hear primarily by placing their heads on the ground and passing vibrations from the lower jaw to the quadrate and finally to the stapes—thus closely following the eventual mammalian pathway. In addition, direct experiments on several lizards and on the tuatara of New Zealand show that vibrations directed at the quadrate are passed to the stapes and recorded in the brain.

May I confess an ulterior motive in closing—for complex and

abstract excursions can be mere glosses upon simpler aims. In-jokes have delicious qualities because they are inaccessible to all but the initiated. But sometimes, in-jokes are so good that we long to share them, yet despair for the volume of background required. Well, this essay can be read as nothing but an extended pony for understanding one of my favorite humorous poems. My colleague John Burns, a lepidopterist now in Washington but formerly at Harvard, used to introduce our weekly seminars with his punny doggerel. We loved the poems and came more to hear his introductions than to suffer through the subsequent speeches. John finally published his verses in a volume called *Biograffiti* (Demeter Press, 1975), with an introduction by yours truly. My favorite is a pithy epitome of mammalian ear evolution, entirely incomprehensible to 99 percent of the population, but now vouchsafed to you, my dear readers, as a small reward for your persistence and as a dessert after this ponderous dissertation:

Evolution of Auditory Ossicles

With malleus
Aforethought
Mammals
Got an earful
of their ancestors'
Jaw.

7 | Full of Hot Air

FIORELLO LA GUARDIA may be destined to go down in history primarily as godfather to an airport. But he was a great mayor for New York in tough years of depression and war. (My birth certificate even bears his signature—well, at least a stamped version.) He also possessed in abundance the trait that we find most welcome, but encounter all too rarely, in people of accomplishment—a willingness to acknowledge occasional and inevitable error. In his most famous quip, La Guardia once remarked, "When I make a mistake it's a beaut!"

Scientific "misconduct" is now a hot topic, both for journalists and members of Congress. In this somewhat frantic climate, we should pause to consider the essential distinction between fraud and error—for the two concepts are diametrically opposed, although self-appointed watchdogs sometimes make the tragic mistake of uniting them as graded forms of malfeasance. Fraud is a social and psychological pathology, although science must learn to police itself. Error is the inevitable byproduct of daring—or of any concentrated effort for that matter. You might as well legislate against urination after beer drinking.

No great work of science has ever been free of error, and any extensive or revolutionary work must contain a few of La Guardia's beauts. Intellectual progress is a complex network of false starts and excursions into trial and error. Darwin's *Origin of Species,* for example, sprinkles numerous errors into its ocean of reforming validity. The errors are so frequent, and so varied, that we might even try to establish categories.

Darwin, first of all, commits several errors of fact. Here I would skip the dull and quotidian misreporting of information and concentrate on the far more interesting errors based on predictions from theoretical premises that turn out to be false or exaggerated. Darwin's commitment to gradualism, for example, led him to make at least two prominent, and outstandingly wrong, conjectures: (1) He gave a time of more than 300 million years for the "denudation of the Weald" (the erosion of the region, forty miles wide, between the north and south Chalk Downs in southern England), based on his belief in the steady, grain-by-grain character of geological erosion. But alteration need not proceed so slowly, or so continuously, and the actual time is one-third to one-fifth of Darwin's generous allotment. (2) Multicellular animal life begins with geological abruptness at the "Cambrian explosion" some 550 million years ago. Darwin, who rejected biological rapidity even more zealously than the geological variety, predicted that the "explosion" must be illusory and that the pre-Cambrian history of multicellular animal life must be as long as, or longer than, the 570 million years of success ever since. We now have an excellent record of pre-Cambrian life—and no multicellular animals arise until just before the Cambrian explosion.

A second category might be labeled errors of judgment: political miscalculations really. The savvy Darwin made few mistakes in this mode, but he slipped occasionally by giving free rein to fatuous speculations in a treatise that gained its power by sinking a weighty anchor in sober fact and avoiding the fanciful conjectures of previous writing about evolution. In a passage that he would later rue, and that gave aid, comfort, rhetorical advantage, and belly laughter to the enemy, Darwin wrote:

> In North America the black bear was seen by Hearne swimming for hours with widely open mouth, thus catching, like a whale, insects in the water. . . . If the supply of insects were constant, and if better adapted competitors did not already exist in the country, I can see no difficulty in a race of bears being rendered, by natural selection, more and more aquatic in their structure and habits, with larger and larger mouths, till a creature was produced as monstrous as a whale.

(Later editions of the *Origin* kept the first factual sentence and expunged all the rest.) A statement like this need not be false (indeed, as a speculation, we cannot tell); the important thing, as Machiavelli would have said, is to avoid the appearance of silliness.

A third category, perhaps the most revealing, includes mistakes that most of us don't recognize because we make them ourselves. Call them errors of thoughtless convention. I include here those passive repetitions of standard cultural assumptions stated so automatically, or so deeply (and silently) embedded within the structure of an argument, that we scarcely detect their presence. Darwin may have been the greatest intellectual revolutionary of the nineteenth century, but he made a few outstanding errors in this category, most related to his ambiguity on the great bugbear of progress—a concept that had no place in the basic mechanics of natural selection, but that Darwin, as an eminent Victorian, could not abandon entirely.

Consider Darwin's treatment of the evolution of vertebrate lungs and their relationship with the swim bladders of bony fishes—an example that Darwin obviously viewed as important to his general argument because he repeats the story half a dozen times in the *Origin.* Darwin begins by noting, correctly, that the lung and swim bladder are homologous organs—different versions of the same basic structure, just as a bat's wing and a horse's foreleg share a common origin indicated by the similar arrangement of bones in body parts that now work in such different ways. But Darwin then draws a false inference from the fact of homology. He claims, with increasing confidence ending in certainty, that lungs evolved from swim bladders:

> All physiologists admit that the swim bladder is homologous . . . in position and structure with the lungs of the higher vertebrate animals; hence there seems to me to be no great difficulty in believing that natural selection has actually converted a swim bladder into a lung, or organ used exclusively for respiration. I can, indeed, hardly doubt that all vertebrate animals having true lungs have descended by ordinary generation from an ancient prototype, of which we

know nothing, furnished with a floating apparatus or swim bladder.

Many readers will be puzzled at this point, as I have perplexed several generations of students by presenting the argument in this form. What can be wrong with Darwin's claim? The two organs are homologous, right? Right. Terrestrial vertebrates evolved from fishes, right? Yes again. So lungs must have evolved from swim bladders, right? Wrong, dead wrong. Swim bladders evolved from lungs.

I love this example, especially as a pedagogical tool, because an outstandingly counterintuitive assertion—the evolution of swim bladders from lungs—becomes the favored hypothesis with sudden and stunning clarity as soon as we shed a common, disabling assumption and start considering the question in a different light. The problem lies with a chronic confusion—abetted by cultural prejudice in this case—between structural sequence and branching order.

The literature of experimental psychology often reports comparative data of performance on various tests for learning in, say, a planarian worm, a crab, a carp, a turtle, and a dog. These are often reported as an "evolutionary sequence" of mental advance. Such statements make evolutionary biologists howl in rage or, if our mood be better, merely with laughter. This motley crew of animals represents no evolutionary sequence at all: vertebrates did not arise from arthropods; mammals did not evolve from turtles; and carp are further from the fishes that did give rise to terrestrial descendants than aardvarks are from humans. However, although the psychologists are dead wrong in their terminology of "evolutionary order," their sequence may have some validity as a structural series—worm, bug, fish, turtle, and dog might express some increasing property of neurological functioning.

When we turn to another common sequence—fish, amphibian, reptile, mammal, monkey, human—the problems intensify, for now we cannot even speak of a legitimate structural sequence. Frogs live in different places, but are they "higher" than swordfishes or sea horses? What odds would you put on a ground sloth going up against a *Triceratops*? Fine, you say; no necessary progress here, but surely this venerable lineage records *the* path of vertebrate evolution. And now we come to the crux of the error about lungs and swim bladders. If this sequence is *the* path of vertebrate change, then swim bladders must evolve to lungs, as Darwin said—for the canonical fish, the first member of the series, has a swim bladder, while all of us at the top have lungs.

But we wallow in a double confusion when we make this "intuitively obvious" assertion—first, the false assumption of progress, which makes the lung a "higher" organ than a swim bladder and thus unfit for creatures on the bottom; second, and more seriously, the confusion of ladders and bushes, or sequences and branching orders. Fish-amphibian-reptile-mammal is not *the* road of change among vertebrates; it represents only one path-

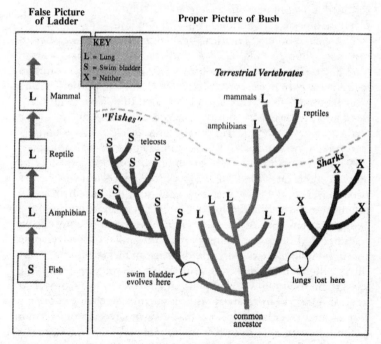

The correct sequence of lung evolving to swim bladder is almost inconceivable on the false evolutionary model of a ladder in vertebrate evolution. But, with the proper iconography of a bush, the sequence becomes clear. *Joe LeMonnier. Courtesy of Natural History.*

way among thousands in the complexly branching bush of vertebrate evolution (the accompanying figure should make my argument clear). All the other pathways lead to creatures that we continue to call "fish" in the vernacular. In terms of variety in anatomical design, we find far more diversity among the creatures called fish than among all the terrestrial vertebrates put together. The terrestrial line is a single branch, with astounding success to be sure, but with limited diversity in underlying anatomical structure (whatever the outward variety of flying birds, slithering snakes, and thinking people). By contrast, fishes are astoundingly disparate in basic design and include lineages that separated a hundred million years before any terrestrial vertebrate arose. Consider the jawless lampreys, the boneless sharks (also lacking either lung or swim bladder), and the odd coelacanth; don't confine your image to the canonical creature impaled on a hook at the end of your rod and line.

Yes, that canonical creature—called a teleost, or member of the vast group of "higher" bony fishes—generally has a swim bladder. But teleosts, although they include almost all common fishes today, are evolutionary latecomers, arising in the sea long after mammals first evolved on land. Yes, they have swim bladders, and they are fish—but they are not ancestors to any terrestrial vertebrate. Their status, as late and derived, leaves entirely unresolved the issue of what came first: swim bladders or lungs.

A reconstruction of vertebrate branching order gives a clear answer to this question: Darwin was wrong; ancestral vertebrates had lungs. (For details of this argument, see the article by Karel F. Liem, cited in the bibliography). The first vertebrates maintained a dual system for respiration: gills for extracting gases from seawater and lungs for gulping air at the surface. A few modern fishes, including the coelacanth, the African bichir *Polypterus,* and three genera of lungfishes, retain lungs. One major group, the sharks and their allies, lost the organ entirely. In two major lineages of derived bony fishes—the chondrosteans and the teleosteans—lungs evolved to swim bladders by atrophy of vascular tissue to create a more or less empty sac and, in some cases, by loss of the connecting tube to the esophagus (called the trachea in humans and other creatures with lungs). Some fishes retain the connection of swim bladder with esophagus; they can inflate their swim bladders by gulping air at the surface. Fishes with separate swim bladders usually extract gases from blood flowing through an extremely fine and rich system of vessels surrounding the bladder and possessing one of the loveliest technical names in all biology—the *rete mirabile,* or "wondrous network."

I would not wish to issue overt praise for mistakes, but Darwin's error on the swim bladder falls into the category that we welcome as particularly instructive, for correction involves a sudden shift from the "can't be" to the completely obvious—that almost thrilling property of scales falling from eyes. The agent of correction, moreover, is not a new and pristine fact, but a change in an underlying conceptual structure.

Let us then praise Darwin's fruitful error on this basis, but also for another, and even more important, reason. Darwin may have gotten his sequence backward, but he was using the story to illustrate a vital and widely misunderstood principle of evolutionary theory—and the illustration works just as well whether swim bladders evolve to lungs or vice versa. Why, then, was Darwin so interested in this issue in the first place?

One common argument against evolution held (and still holds among the lingering opposition) that small changes within a "basic kind" might occur, producing the range from Chihuahua to Great Dane, or Shetland pony to old dobbin hauling the Budweiser truck. But transitions between types are forbidden because fundamental novelties cannot arise by evolution. The classic form of this argument holds that since "novel" structures often arise (or so evolutionists claim) from ancestral organs with strikingly different functions, transitional forms would be inviable because they would exist in the never-never land of utter unworkability, with one key function degenerated and another not quite established. To cite a classic case (with an elegant resolution as we shall see), how could reptiles evolve into mammals if bones that articulate the reptilian jaw must evolve to the malleus and incus (hammer and anvil) of the mammalian middle ear (see Essay 6)? No intermediary form could live without a jaw articulation, as the leisurely earward transition occurred. In other words, both the "before" and the "after" make sense as functional organisms, but the "in between" doesn't work.

Lung and swim bladder represent a classic example of this dilemma, whichever way the sequence proceeded. The organs are homologues, and one presumably evolved to the other. But how could the transitional form have survived, either stuck like a lead weight on the bottom as buoyancy failed, while breathing required access to the surface; or raring to float but gasping for breath?

Darwin begins by warning us against a priori claims of impossibility in principle, for multifarious nature so often gets the last laugh over this particular form of human vanity: "We should be extremely cautious in concluding that an organ could not have been formed by transitional gradations of some kind." Darwin's ingenious solution involves a double linkage of one-for-two, with two-for-one—mysterious when stated so abstractly but beautifully simple by illustration, with lungs and swim bladders as a primary example. First, Darwin tells us, single organs often perform more than one function—one-for-two:

> Numerous cases could be given . . . of the same organ performing at the same time wholly distinct functions. . . . In such cases natural selection might easily specialize, if any advantage were thus gained, a part or organ, which had performed two functions, for one function alone, and thus wholly change its nature by insensible steps.

The primitive swim bladder, Darwin argues (and we may reverse the argument for lungs), may also have worked in a subsidiary way in gas exchange—and this latter role may have been intensified as the original use dropped out, in the evolution of the lung. But the one-for-two principle cannot resolve the problem of intermediary stages—for how could a fish breathe as the original lungs lost their primary function?

Darwin therefore calls upon his second, coupled principle of two-for-one. Many vital functions are performed by two or more organs, and one can change so long as the other continues to play the needed role. We can breathe through both our nose and mouth—and thank goodness, or we would all be dead of colds:

> Two distinct organs sometimes perform simultaneously the same function in the same individual. . . . In these cases, one of the two organs might with ease be modified and perfected so as to perform all the work by itself; . . . and then

this other organ might be modified for some other and quite distinct purpose.

We can now understand why Darwin liked the example of lungs and swim bladders so much. He had made a reasonable conjecture about one-for-two in arguing for supplementary respiration in swim bladders, and he had definite evidence about two-for-one in the presence of numerous living fishes with dual systems of breathing—gills and lungs. (The official taxonomic name of the lungfishes, Dipnoi, means "two breathing.") Thus, using lungs and swim bladders as his key example in a central defense of large-scale evolution, Darwin concluded:

> For instance, a swim bladder has apparently been converted into an air-breathing lung. The same organ having performed simultaneously very different functions, and then having been specialized for one function; and two very distinct organs having performed at the same time the same function, the one having been perfected whilst aided by the other, must often have largely facilitated transitions.

Readers might fairly balk at this point. The argument coupling one-for-two and two-for-one is logically sound, but doesn't it smack of special pleading and gross improbability? How often can you expect to find such a combination? Perhaps both situations are uncommon; their conjunction would then be nearly incomprehensible. Rare times rare equals rare squared, or effectively impossible.

But we now come to the true beauty of Darwin's argument. Neither situation is rare, and the two phenomena—one-for-two and two-for-one—are not really separate at all. Both are expressions of a deeper, and profoundly important, principle—*redundancy* as the ground of creativity in any form. They are two sides of the same coin—and the coin, although priceless in intellectual value, is as common as a penny.

The notion that organs are "for" particular things, ideally suited for one and only one job, is a vestige of old-style creationism—the idea that God made each creature, fully formed and perfect in function. If each organ existed explicitly for a single role, then I suppose that one organ doing more than one thing would be rare, and that two organs doing the same thing might be even rarer. But organs were not designed for anything; they evolved—and evolution is a messy process brimming with redundancy. An organ might be molded by natural selection for advantages in one role, but anything complex has a range of other potential uses by virtue of inherited structure—as we all discover when we use a dime for a screwdriver, a credit card to force open a door, or a coat hanger to break into our locked car (not someone else's, let us hope, and surely not, let us pray, for ending unwanted pregnancies in our newly dawning era of restrictions). Any vital function narrowly restricted to one organ gives a lineage little prospect for long-term evolutionary persistence; redundancy itself should possess an enormous advantage. (Redundancy in this form solves the otherwise intractable problem of evolution in mammalian jaws, as outlined above. Intermediary forms, as shown by direct evidence of fossils, not abstract conjecture, developed a second articulation between dentary and squamosal bones [the current mammalian jaw joint], and elements of the old articulation could then lose their former function and pass into the ear.)

In fact, the swim bladder itself provides an excellent example of multiple possibilities as a norm. The swim bladder is primarily an organ of buoyancy in teleost fishes. By filling the bladder with gas, an animal that would otherwise sink becomes neutrally buoyant and can rest without expending energy in the midst of the water column. (In a related function, fishes at neutral buoyancy gain more power in forward motion because they need not divert energy into supplying lift to counteract sinking—see R. McNeill Alexander in the bibliography. Interestingly, some sharks are pelagic (floating) in habit; how can they stay up, since their entire lineage lost the organ that becomes a lung or swim bladder in other fishes? These sharks have enormous livers constructed largely of a hydrocarbon with a density considerably less than seawater—another good example of multiple use as a norm.

But the swim bladder performs at least three other important but secondary functions in many species of teleost fishes:

(1) Most curiously, perhaps, the swim bladder has reacquired a supplementary respiratory function in several lineages of fishes, all living in swampy or stagnant waters, where gulping air at the

surface might be an important alternative to breathing with gills.

(2) Many teleosts use their swim bladder as an organ of sensa-tion. Since gas is so responsive to changes in pressure, some fishes can judge their depth in the water column with receptors embedded in the wall of their swim bladder. Many other fishes use their swim bladder as an accessory organ of hearing. Gases are more compressible than water, and sound vibrations may be recorded more sensitively in their impact upon swim bladder gases than upon any other part of the body. Supplementary hear-ing has evolved in at least two strikingly different ways. Some fishes have developed thin forward extensions of the swim blad-der; these pass through openings of the skull and make direct contact with the ear. In another major group, the cypriniforms (including most of the world's freshwater fishes), vibrations from the swim bladder are transmitted to the ear via a chain of three separate bones located on either side of the vertebral column and called Weberian ossicles to honor the German scientist who rec-ognized their mode of operation in 1820. (Darwin used this ex-ample of multiple function in the *Origin of Species.*)

(3) Sound production: Again, several lineages use the swim bladder either to enhance sounds made by other parts of the body or as a direct agent of production. (Some fishes are essen-tially silent, but many make sounds, especially in courtship or in aggressive displays.) The triggerfish *Balistes* (another lovely name) stridulates by rubbing its postclavicle bone against its cleithrum—but this otherwise minor sound is greatly amplified by resonance from the adjacent swim bladder. Another group of fishes grates its pharyngeal teeth and also turns a little rumble into a modest roar by resonance of the swim bladder. In other fishes, the swim bladder produces sounds directly by expulsion of gas bubbles. T. H. Huxley once wrote a special note to *Nature* (in 1881) to describe what can only be called herring farts. These fish expel gas in pulses from the swim bladder out an orifice adjacent to the anus. In the oh-so-proper style of scientific reporting, a British review article of 1953 described Huxley's suggestion "that the mouse-like squeaks made by captured herring might be caused by the escape of gas through the posterior opening."

If I may move, in conclusion, from minor end rumblings to a renewed assault upon the high ground, I don't know if the *Origin of Species* contains an argument more general or more important than Darwin's recognition that pervasive redundancy makes evolution possible. If animals were ideally honed, with each part doing one thing perfectly, then evolution would not occur, for nothing could change (without losing vital function in the transition), and life would quickly end as environments altered and organisms did not respond.

But rules of structure, deeper than natural selection itself, guarantee that complex features must bristle with multiple possibilities—and evolution wins its required flexibility thanks to messiness, redundancy, and lack of perfect fit. Human creativity is no different, for I think we are dealing with a statement about the very nature of organization—something so general that it must apply to any particular instance.

How sad then that we live in a culture almost dedicated to wiping out the leisure of ambiguity and the creative joy of redundancy. These days, even the most complex concepts must be reduced to photo opportunities and sound bites, and elections are decided by fifteen-second images of men surrounded by flags and alleged criminals walking through symbolic revolving doors. We may be creating a generation of sheep—and although these pleasant mammals outnumber New Zealanders by almost twenty-five to one, I rather suspect that *Homo sapiens,* properly nurtured by redundancy and ambiguity, will continue to prevail.

Redundancy, and its counterpart of ambiguity in multiple meaning, are our way, our most precious, most human way. We rail at computers because for all their awesome power, they do not grasp our essential ambiguities. They cannot adequately translate one of our languages into another, and we must speak to them in a way utterly unnatural for us—that is, without ambiguity (hence an entire industry devoted to debugging). Faced with La Guardia's or Darwin's errors, they grind to a halt. We adjust, we parry, we prevail, we transcend. It could be one hell of a partnership, so long as we keep the upper hand. I shall place my bets on the shepherds of New Zealand and hope that the analogy holds.

3 | *Vox Populi*

Evolving Visions

8 | Men of the
Thirty-Third Division:
An Essay on Integrity

MY FATHER, like so many men of his generation, lost the opportunity for a college education through a conspiracy of circumstances—first the depression, then a stint in the war, and finally too much time gone and too many kids to feed when the external hurdles finally dropped. But Leonard Gould was a man of great intellect, keen perception, and broad interests. Late in his life, and largely in response to my concerns, he became enthralled with human evolution and spent much of his retirement reading—with his characteristic, consummate care—the literature, both popular and technical, of paleoanthropology.

One day, during a visit, he approached me in great frustration. Spread before him were the latest books of America's two greatest anthropologists. "Look," he said, "Professor Uno ridicules Professor Due for believing such a silly idea, but Due really says something quite different; look right here on the page. Meanwhile Due excoriates Uno for speaking nonsense, but look here, Uno doesn't say any such thing. Now what am *I* misunderstanding?" My heart would sink whenever my father attributed the carelessness of scholars to his own ignorance based on lack of professional training. I could never get him to understand that advanced degrees and letters after a name guarantee no new level of wisdom and that, in the end, there is no substitute for old-fashioned careful reading. I could never convince him that he had a far better chance than Uno or Due to grasp the integrity of another man's argument. After all, he had the prerequisites of basic intelligence and adequate knowledge of jargon; and he possessed, in addition and in abundance, two cardinal traits rarely encountered in active scholars: time to read carefully, and lack of distorting preconceptions. I read the two books. He was right again. Uno and Due were ripping apart the nonexistent caricatures of each other's ideas.

I've been in this business for nearly a quarter century now, and nothing depresses me more than the rampant, seemingly inveterate mischaracterization that lies at the core of nearly every academic debate. We are not incapable of arguing about intellectual substance and empirical reality, but we sure seem to prefer misunderstanding as a subject for invective. The root of this lamentable behavior can only lie in careless habits of reading and thinking (or, worse, in our willingness to argue without reading at all).

The foundation of my distress is a moral position traditionally precious to the world of intellectuals. What do we have for judgment, worth, and honor but the integrity of our ideas—using integrity both in the moral meaning of honesty in argument and, even more importantly, in the literal sense of uniting our various notions into coherent intellectual structures? We may define brilliance in scholarship as the surprise, power, and beauty of such integrity at its finest. The truly awesome intellectuals in our history have not merely made discoveries; they have woven variegated, but firm, tapestries of comprehensive coverage. The tapestries have various fates: Most burn or unravel in the footsteps of time and the fires of later discovery. But their glory lies in their integrity as unified structures of great complexity and broad implication.

Yet, in our harried world of sound bites and photo ops, we focus on anecdote rather than structure, and scholars are identified by items rather than by their precious tapestries. Lavoisier discovered oxygen, Darwin evolution; an apple fell in Newton's sight, and Mozart had a dirty mind. Bad enough in the hagiographical mode, when good intentions accompany items meant as emblems of valor. But what of the dark side, so pervasive in academic life? What can be more destructive of our fragile community than the mode of criticism that slices a jagged hunk out of the tapestry, misreads and simplifies the item as a strawman in a campaign of destruction, and then tries to define the scholar by the misappropriated patch? He who steals my purse does steal trash, but he who slices and dices my tapestry cuts out my heart.

May I honor my father's respect for integrity by correcting a legend rooted in the slice-a-patch method of commentary, and involving the founder of the field represented by squabbling Professors Uno and Due—the study of human fossils.

Only two substantial discoveries of human fossils predate Darwin's great book on the *Descent of Man* (1871), and neither offered any evidence for human ancestry. The Neanderthals, widely misinterpreted at first as modern skeletons with deformation or disease, represent a very late side-branch in the tale of human evolution, while the Cro-Magnons, living even closer to the present time, are already us in all crucial aspects of anatomy and brain size.

Eugène Dubois, a medical officer in the Dutch army, found the first fossils of an old and truly ancestral human in 1891 and 1892, ten years after Darwin's death. The tale of his discovery ranks among the best in the annals of scientific perseverance and perspicacity. Dubois (1858–1940), a physician and lecturer in anatomy at Amsterdam University, became enthralled with the new science of evolution and longed to find the greatest desideratum of all, the closest paleontological equivalent to the holy grail—the "missing link" (to use the antiquated terminology of Dubois's day). Accepting the usual view that tropical Asia offered greatest promise, and taking advantage of his nation's colonial presence in Indonesia, Dubois resigned his post in Amsterdam and signed up for an eight-year hitch as medical officer second class in the Royal Dutch East Indies Army, with the clear ulterior motive of using every spare moment in the search for human ancestors. He was disappointed on Sumatra from 1887 to 1890, but then struck pay dirt in Java two years later along the Solo River near the village of Trinil. There, in October 1891, Dubois's workmen found a skull cap and later, in August 1892, a femur of a primate with marked human affinities. The thigh bone could scarcely be distinguished from our own, but the skull cap belonged to a creature with a brain of some 900 cm^3, or roughly two-thirds modern size. (The romantic image of field work depicts our hero, sleeves rolled up, pickaxe in hand, and sweating bullets in the scorching sun as he unearths his precious find. The realities of colonial life and complex expeditions dictate less inspiring but more realistic scenarios. Dubois entrusted the day-to-day digging to convict laborers commanded by two army sergeants. The fossils, packed in teak leaves, then accumulated on the veranda of Dubois' headquarters, where he presumably first saw his quarry.)

Dubois reconstructed this greatest of nineteenth-century discoveries as a human of about our bulk and build, fully erect in posture (judging from the femur), but with a brain two-thirds our size—in short, an excellent "missing link" on the theory, already popular in Dubois's time and now well established, that upright posture preceded, and may have triggered, the enlargement of the human brain. Dubois named his creature *Pithecanthropus erectus,* or the ape-man who walked upright. We continue to regard Dubois's species as our direct ancestor, though a revised notion of even closer affinity led to a redesignation as *Homo erectus.* This species has now been found throughout the Old World, both elsewhere in Asia (initially as the famous Peking Man, discovered during the 1920s) and, more recently, in its original African home. Dubois's beginning, a skullcap and femur from Java, has blossomed into a well-documented ancestor, widely spread over three continents.

Such an auspicious start might have brought only honor and further triumph, but a witches' brew of scientific contentiousness, temper of the times, and quirks in Dubois's own psyche soon derailed any pleasant development and turned Dubois's bounty into bitterness. He returned from Java in 1895 and began to display his specimens at scientific meetings. He received much warm support and several overt testimonials in medals and honorary doctorates. But he also generated a firestorm of doubt and protest in this perennially contentious field. Some labeled his find "merely" an ape, others a diseased modern skeleton, still others a false jumble of a modern human femur with an ape's skullcap.

At this point, we encounter the canonical legend of Dubois— the story that I shall attempt to refute by replacing the slashed patch of its content back into Dubois's intellectual tapestry. The basic facts are not in dispute; the traditional interpretation, however, is not only wrong, but perversely backwards.

At the turn of the new century, Dubois withdrew the Trinil bones, locked them away, and refused access to all scientists for nearly a quarter century. Finally, in 1923, spurred by entreaties and pressured from high places (inquiries in Parliament and representations from the Royal Dutch Academy of Sciences), Dubois relented and brought his specimens back to scientific scrutiny. But now, he pulled the ultimate volte-face and declared in his final publications that the Trinil skullcap and femur belonged to a giant gibbon!

From these bare facts, the obvious interpretation (and morality play) flows in three statements: First, that Dubois locked up the bones in anger as a desperate rearguard defense against a growing and withering attack; second, that Dubois festered in bitterness during a quarter century of unalloyed funk; third, that his ultimate reversal and redesignation of Java Man as a giant gibbon stands as the quintessential sick joke of science—a final discharge of acrid disgust from a bitter and dyspeptic old man. Dubois's descent into the psychic maelstrom therefore becomes a canonical tale of penalties for dogmatism and, especially, for failure to follow the norms of collegial cooperation.

All popular books on human evolution present this interpretation. None other than Professor Uno (of my initial tale) wrote in 1974, supporting the first statement: "The uproar of dissent was so great that Dubois withdrew his Java finds from scientific exhibition and locked them away in a museum strong box." On the second statement, a 1982 source tells us: "For a quarter of a century he remained almost silent on the subject, the fossils buried beneath the floor of his dining room" (note the disagreement about the venue of Dubois's hiding place).

But no part of the tale inspires such unanimity of interpretation, and such cluck-clucking at Dubois's personal, self-inflicted misfortune, as the final and perverse dénouement—the degrading of Java Man to a giant gibbon. Let us begin by giving Professor Due his due (from a 1949 book): "One voice alone now cried that the Java Man was not a man, but a giant, tree-walking gibbon. . . . And here it was that *Pithecanthropus* felt the unkindest cut of all. For the voice was the voice of Dr. Dubois himself." Or this from 1981: "He attempted to show that *Pithecanthropus* was in fact a very large ape of gibbon-like appearance. . . . But by then events had overtaken Dubois and his interpretation. . . . These papers . . . reflect the acrimony of a weary old man." And one more from 1982: "While his visitors concluded that the bones really were those of an early form of man, Dubois responded that they had in fact belonged to a giant gibbon. . . . Dubois maintained this absurd belief right up to his death in 1940."

In these distortions of his true view, Dubois has fallen victim to what I like to call the Wambsganss effect: October 10, 1920, men on first and second, no one out; fifth inning of the fifth game of the World Series. With runners going, second baseman Bill Wambsganss spears a line drive, steps on second, and tags the runner coming in from first to complete, in but a second or two, the only unassisted triple play in the history of the World Series. It wasn't even a great play, just a highly unusual circumstance with an almost automatic result. Yet Wambsganss is forever identified with his moment of fortuitous glory. Mention his name, and a fan's only conceivable reaction, aside from a lament about spelling and pronunciation, will be: "unassisted triple play, 1920 World Series." No matter that Wamby (as they called him for ease in spelling and fit into box scores) had a solid twelve-year career at second base. He has become an eternal prisoner to his moment of triumph.

So too for Eugène Dubois. He found the first genuinely ancestral human fossil, and while no one can take it away from him, he has also become so affixed to his episode of transcendent success that no one ever asks what he did for the rest of his life. The Wambsganss effect has slashed out the patch of *Pithecanthropus*, buried the rest of Dubois's intellectual tapestry, and interpreted his entire life in the light of his great moment. Yet by so doing, we have grossly misserved him. Dubois had a vision—wrong as could be in many respects, but an integrated vision nonetheless. The Java bones fit into his vision and his life's work; we cannot understand Dubois interpretations if we look no further than his skullcap and femur. Paleontologists laugh at the common legend that we can reconstruct an entire animal from a toe bone. Of course, we can't. Why then do we dare to commit the same fallacy and steal a man's intellectual birthright by reading his worldview from one item of a rich lifetime?

What, then, was Dubois doing during those twenty-five years of his supposed grand funk? Did he just lock up the bones, mutter the Dutch equivalent of "sod off," and go home to tend his dikes and tulips? We must begin by correcting the first statement of the legend (as admirably done by Bert Theunissen in an excellent recent biography, the first to treat Dubois's tapestry). Dubois did withdraw the bones, but not, as Theunissen proves, in retreat before a growing storm of denial. In fact, although the Trinil specimens had generated warm debate, Dubois was clearly gaining support when he made his peculiar move. (Theunissen's discovery makes Dubois's action all the more difficult to understand. We must search, he argues, in Dubois's psyche. Theunissen writes: "He had an infinite capacity for annoyance with any scientist whose viewpoint differed from his own. . . . His suspicious nature [verged] on the paranoid." Needless to say, I am not defending, but only trying to understand, Dubois's markedly uncollegial action. He could not bear rebuke and paid the price for his intemperance in ultimate dismissal by his colleagues.)

If Dubois, therefore, did not lock up the bones in scared retreat, but rather in proud defiance, we may question the second statement of the legend as well. Did he just spend those twenty-five years dithering in the agony of defeat, or did he occupy his time with something more positive and productive? And now we come to the crux of the story, and to the ultimate basis for refuting the giant gibbon legend.

Dubois continued to publish at high volume during this entire time. He retained posts as professor at Amsterdam and curator of the Teylers Museum in Haarlem. His numerous papers are matters of public record, readily available in most good technical libraries of natural history. He devoted his work during these years to a series of studies on the distribution and evolution of brain size in mammals. He wrote far more on this subject than he ever published on *Pithecanthropus*. Moreover, he did not pursue these studies as a sidelight or a fresh start after a Trinil disaster. He began his work on mammalian brains as part of a deliberate program to understand the evolution of human intelligence. In particular, he sought to develop a quantitative method for interpreting the brain size of *Pithecanthropus* and its role in human evolution. The brain work was tapestry, not escape.

(I must confess to a particular and personal stake in this issue. At the beginning of my career, I published several papers on the evolution of brain size. I pursued the same basic mathematical strategy that Dubois had pioneered, beginning in 1897—plotting by power functions. I stand with a tiny handful of scientists who first encountered Dubois as an initiator in the study of brain size by power functions, not as the discoverer of *Pithecanthropus*. My strong interest in the history of science also led me to track down the relationship of this work to Java Man and particularly to the giant gibbon legend.)

Dubois began by confronting the problem that had faced scientists since Cuvier made the key observation just after the French revolution: In some meaningful sense, and not from the mere vanity of our cosmic arrogance, humans are the brainiest creatures on earth. But how shall we measure our superiority? Absolute brain size will not do, for whales and elephants beat us by virtue of their larger bodies, and the consequent need for larger brains to coordinate such bulk. But relative brain size (ratio of brain to body weight) will not work either because brain size increases more slowly than body size along the mammalian spectrum from mouse to elephant—and a shrew, by this false criterion, would be the paragon of mammalian intellectual achievement.

Dubois understood, as others had before him, that the solution to this problem must lie in quantifying the normal or expected relationship between brain weight and body weight in mammals—the so-called mouse-to-elephant curve. We know that brain weight increases more slowly than body weight, but by what amount and in what relationship? Dubois's predecessors had tried ordinary arithmetic plotting, with little success. In 1891, Otto Snell found the basic solution that we still accept and use today: Relationships between brain and body weight are best described by power functions, yielding a straight line when the logarithms of brain and body are plotted against each other. Snell proposed an exponent of two-thirds for the power function— meaning that brains increase roughly two-thirds as fast as bodies along the mouse-to-elephant curve.

Once a standard mouse-to-elephant curve has been established, the right criterion for assessing human superiority (or the status of any particular species among mammals as a whole) becomes clear: Compare the actual brain weight of a species in question with the expected brain weight for an average mammal of the same body weight on the mouse-to-elephant curve. By this proper criterion, no mammal exceeds the positive deviation of *Homo sapiens.*

Snell had established the basis for a solution in 1891, but Dubois's classic paper of 1897 provided the first extensive data. By comparing seven pairs of closely related mammalian species differing substantially in body size, Dubois recalculated Snell's exponent at five-ninths—that is, Dubois argued that brains increase five-ninths as fast as bodies along the mouse-to-elephant curve. (Dubois was quite wrong in this conclusion, and his error set the basis for his later problems as we shall see. No single number can capture the range of variation among mammalian rates; the evolution of the brain is not a problem in point masses and frictionless surfaces from a Physics 1 laboratory exercise. Moreover, Snell's two-thirds generally works better and more often than Dubois's five-ninths). Using five-ninths as a standard value, Dubois then compared the brains of "intelligent" species with expected brain sizes at their own body weights, thereby establishing a scale of cranial capacity that rightly placed *Homo sapiens* at the pinnacle.

Now Dubois did not begin these studies for idle curiosity or for solace or novelty after anger at criticism of *Pithecanthropus*. He wanted to know the standard relationship of brain and body in order to calculate the intermediary status of Java Man between modern apes and humans. In 1935, he reminisced:

> It was to obtain a better insight into this new organism that, soon after the discovery, I undertook the search for laws which regulate cerebral quantity in Mammals, a study which indeed furnished evidence as to the place of *Pithecanthropus* in the zoological system, and with which I am still intensively occupied, on account of its great biological significance.

At the Fourth International Anthropological Congress, meeting in Cambridge, England, in 1898, Dubois used the formula from his studies on brain weight to judge the status of *Pithecanthropus*. At a cranial capacity of 855 cm³ inferred from the skullcap (Dubois later revised this figure upward to about 900 cm³), an average modern ape would weight 230 kg and an average modern human but 19 kg. Since *Pithecanthropus*, as reconstructed from the femur, clearly stood between these figures in weight, the Trinil hominid must have carried a brain intermediate in size between modern apes and human. Dubois concluded in 1898: "From these considerations it follows that *Pithecanthropus erectus* is an intermediate form between man and the apes . . . a most venerable ape-man, representing a stage in our phylogeny."

Two years later, Dubois locked up the bones, but he never

stopped his work on brain size. He amassed a large collection of brain casts, most made by himself, at the Teylers Museum, and he continued to publish substantial papers with important data. At the same time, he developed an evolutionary perspective based on his brain work and hostile to Darwinism.

Dubois began with two assumptions, long common in continental traditions of evolutionary thought, but foreign to English procedures. Dubois held, first of all, that sequences of species with increasingly positive deviations from the mouse-to-elephant curve indicated an intrinsic push towards higher intelligence in the evolution of mammals—an inherent drive that could not be explained by something so base as Darwinian natural selection and adaptation to changing environments. Dubois wrote in 1928:

> Here is a law of evolution come forth out of the nature of the living being itself, not imposed by the surroundings. . . . It is self-evident that this perfecting, this steady progression cannot have been caused by factors outside the animal, to which Darwin ascribed phylogenesis.

As a second anti-Darwinian conviction, Dubois followed his countryman and former mentor Hugo DeVries in arguing that evolution occurs by sudden leaps, not gradual transformation. "Again and again," Dubois writes in striking metaphor, "we find pillars of the expected bridges, never arches." Applying this assumption to the human brain, Dubois held in 1935: "There was a leap from the anthropomorphous [ape] level to the pithecanthropus level, not the gradual slow ascent presumed by the darwinistic hypothesis."

Dubois made his fateful error in applying these assumptions to his data on brain sizes. He scanned his tables of deviations from the mouse-to-elephant curve, and managed to delude himself into believing that he could arrange these values into a few discrete groups differing by factors of two. In other words, Dubois thought that he could identify a group of species with a common level of minimum brain size. From this foundation, brains increased by sudden doubling, with the next group at twice the basal brain weight, the next at four times, the next at eight times, et cetera. Starting from the top, and setting the human maximum at one, Dubois claimed standard values of one-quarter for great apes, one-eighth for most carnivores and hoofed herbivores, one-sixteenth for Leporidae (rabbits), one-thirty-second for Muricidae (mice), and, at the bottom, one-sixty-fourth for Soricidae (shrews).

Dubois then took his final step and proposed a neurological and embryological basis for enlargement of the brain by leaps. Neurons are formed in the embryo and do not divide after birth; most mammals enter the world with their full complement of brain cells. Each division of neurons should double the weight of the brain (to grasp Dubois's theory, we must bypass the obvious, and valid, objection that all neurons do not divide at once on the beat of a metronome). Thus, evolution progresses by inserting one more division into the program of embryological development—one extra doubling towards the human pinnacle. Dubois wrote in 1928: "One segmentation more or less determines the degree of development in the central organ of the animal life, the extent of the outer and inner world of the animal."

By estimating the adult complement of neurons, Dubois even felt that he could specify the number of divisions from an initial cell. Shrews had undergone twenty-seven divisions. Counting up the chart of doublings, modern humans stand at the apex of intrinsically driven brilliance with thirty-three divisions.

Alert readers will have detected a gap in Dubois' series and my arguments. Humans rank as one, apes as one-quarter; humans have thirty-three divisions, apes thirty-one. What happened to one-half (and to thirty-two divisions) in this supposedly unbroken sequence of perfection by doubling? Clearly, the creature with half a human brain and thirty-two divisions must be the holy grail of human paleontology—our direct ancestor, the true "missing link" in a world of evolution by sudden leaps.

And now, having presented Dubois's tapestry, I may finally correct the last and most insidious claim of the standard legend— the dénouement of the morality play as Dubois, in aged despondency and defeat, redesignates his once-proud ancestor as nothing but a giant gibbon. With the tapestry in place, Dubois's argument inverts to sensibility. Dubois used the proportions of a gibbon to give *Pithecanthropus* a brain at exactly half our level, thereby rendering his man of Java, the pride of his career, as the direct ancestor of all modern humans. He argued about gibbons to exalt *Pithecanthropus,* not to demote the greatest discovery of his life.

The argument is beautifully clear, adamantine even (however wrong in retrospect), once you understand its basis in Dubois's work on brain size—the five-ninths scaling law of the mouse-to-elephant curve, evolution towards perfection by doubling, and the missing value of one-half human size and thirty-two doublings. Dubois desperately wanted *Pithecanthropus* as a direct ancestor under his evolutionary view. But the brain of Java Man ranked with embarrassing bulk at some 900 cm³, or two-thirds human volume. The Trinil femur does not differ in size from our own and suggests a modern human body weight for *Pithecanthropus* in a reconstruction based on human proportions.

Dubois's hopes seemed stymied. A human creature of our weight, but with two-thirds our brain size, could not be an ancestor in a world of necessary evolution by sudden doubling. *Pithecanthropus* would have to settle for reduced status as a sterile side-branch of the human trunk. But then Dubois thought of a brilliant way out. His initial reconstruction of the femur as essentially human had always been challenged by many scientists. Rudolf Virchow, the great German pathologist, had argued in 1895 that the femur looked more like a gibbon's than a human's.

Now suppose that Virchow had been right after all? Then *Pithecanthropus* would have to be reconstructed with the proportions of a gibbon—particularly, with longer arms and a greatly expanded chest and upper body. Such a giant gibbon would weigh, by Dubois's calculation, just over 100 kg. Backing down the five-ninths curve to an average human weight of some 65 kg—the proper point of comparison with the brain of modern humans—a 900 cm³ brain on a 100 kg creature becomes, you guessed it, a brain of exactly half our size at appropriate human weight. *Pithecanthropus* had been deftly rescued from the ultimate limbo of an evolutionary dead end, and raised again to the most exalted status of our direct ancestor. Dubois wrote triumphantly in 1932:

> *Pithecanthropus* was not a man, but a gigantic genus allied to the gibbons, however superior to the gibbons on account of its exceedingly large brain volume and distinguished at the same time by its faculty of assuming an erect attitude and gait. It had the double cephalization of the anthropoid apes

in general and half that of man. . . . I still believe, now more firmly than ever, that the *Pithecanthropus* of Trinil is the real "missing link."

In other words, Dubois never said that *Pithecanthropus* was a gibbon (and therefore the lumbering, almost comical dead end of the legend); rather, he reconstructed Java Man with the proportions of a gibbon in order to inflate the body weight and transform his beloved creature into a direct human ancestor—its highest possible status—under his curious theory of evolution.

Eugène Dubois is no hero in my book, if only because I share the spirit of his unorthodoxies, but disagree so strongly with his version, and regard his supporting arguments as so weakly construed and so willfully blind to opposing evidence (the dogmatist within is always worse than the enemy without). Nonetheless, I step forward in his cause because he has been so badly served by careless reading and neglect of his primary work (theories of brain size) in favor of his moment of fame (the discovery of *Pithecanthropus*). As a bitter consequence, Dubois's ingenious attempt to retain *Pithecanthropus* as a direct human ancestor has been widely misread in a precisely opposite manner as an ultimate surrender, almost comical in its transmogrification of a human forebear into a giant gibbon. With the patch of *Pithecanthropus* alone, the legend of the giant gibbon easily takes root; you need the tapestry of Dubois's evolutionary theories about brain size, the integrating vision of his intellectual life, to grasp his true intent.

Good scholars struggle to understand the world in an integral way (pedants bite off tiny bits and worry them to death). These visions of reality—I have called them tapestries in this essay—demand our respect, for they are an intellectual's only birthright. They are often entirely wrong and always flawed in serious ways, but they must be understood honorably and not subjected to mayhem by the excision of patches.

I think that Dubois was wrong in his basic approach, and surely incorrect in his ingenious argument for *Pithecanthropus* remade as a human ancestor with a gibbon's build. He was a Platonic numerologist in a messy world. He longed for absolutes, for invariable laws based on clean and simple numbers. He never understood that nothing generates the shadows on Plato's cave.

The shadows *are* reality, and they are diffuse, with penumbras spreading out and intersecting in complex patterns. Arthur Keith, the great British anthropologist, made the necessary observation with adequate gentleness in an obituary notice for Dubois: "He was an idealist, his ideas being so firmly held that his mind tended to bend facts rather than alter his ideas to fit them." Still, just as the old cliché tells us about love, it is surely better to have a dream and to be wrong, than never to dream at all.

9 | Darwin and Paley Meet the Invisible Hand

THE FRENCH REVOLUTIONARY government defined their new basis of measurement—the meter—as one ten-millionth of the quadrant of the earth's circumference from pole to equator. While I do appreciate both the democratic intent and objectivity of such a choice, I must confess my continuing fondness for older units of explicitly human scale. Monarchs may not merit a role as standard-bearer in this sense, but at least we can empathize with a yard defined (in one common legend at least) as the distance from King Edgar's nosetip to outstretched middle finger; or the foot as King John's regal proclamation (after stamping his print on wet ground at a time of peace, rather than Magna Carta rebellion, with his nobles): "Let it be the measure from this day forward"; or an inch as the length of the knuckle on King Edgar's thumb.

But when inches required subdivision, our large frames failed to supply obvious reference points, and our forebears sought agricultural aids: Three (or sometimes four) barleycorns made an inch, and five poppy seeds a barleycorn.

I mention these arcana to explicate a quotation from William Paley's *Natural Theology* (1802). When the good reverend cites the value of a barleycorn, he means "damned little." Paley, out to prove the existence and benevolence of God from the good design of organisms, faced a puzzle in analyzing behavior. How, in God's well-designed world, can organisms spend so much time and energy engaged in behavior for purposes they cannot understand? Birds must copulate to reproduce and must reproduce to perpetuate their kind, but birdbrains cannot grasp this chain of logic:

> When a male and female sparrow come together, they do not meet to confer upon the expediency of perpetuating their species. As an abstract proposition, they care not the value of a barley-corn whether the species be perpetuated, or not. They follow their sensations; and all those consequences ensue, which the wisest counsels could have dictated, which the most solicitous care of futurity, which the most anxious concern for the sparrow world, could have produced. But how do these consequences arise?

The problem, Paley tells us, has a clear solution in such cases. Sex, after all, feels good; birds indulge for pleasures of the moment, while their benevolent creator implants the bonus of his own intent in perpetuating one of his created species:

> Those actions of animals which we refer to instinct, are not gone about with any view to their consequences . . . but are pursued for the sake of gratification alone; what does all this prove, but that the prospection [that is, knowledge of ultimate benefit], which must be somewhere, is not in the animal, but in the Creator?

"Be it so," Paley adds, but he is not out of the intellectual thicket yet. What about instinctive behaviors that impart no immediate gratification, but seem, on the contrary, to mire an animal in pain and distress? How can a bird tolerate days or months of incarceration at the nest for a fleeting moment of carnal pleasure before, for Paley asserts that the female is "often found wasted to skin and bone by sitting upon her eggs." Paley evokes both our empathy and admiration for this sedentary sacrifice:

> Neither ought it . . . to be forgotten, how much the instinct costs the animal which feels it; how much a bird, for example, gives up, by sitting upon her nest; how repugnant it is to her organization, her habits, and her pleasures. . . . An animal delighting in motion, made for motion, . . . is fixed to

her nest, as close as if her limbs were tied down by pins and wires. For my part, I never see a bird in that situation, but I recognize an invisible hand, detaining the contented prisoner from her fields and groves for a purpose, as the event proves, the most worthy of the sacrifice, the most important, the most beneficial.

Paley has cleverly turned the problem to his advantage. Sex can be explained by immediate gratification, though its purpose in the scheme of things be deeper. But incarceration at the nest, by opposing any conceivable motivation of the bird itself, must point more directly to divine intent and imposition. The "invisible hand" that keeps the bird on her nest can only be God himself.

The Reverend William Paley (1743–1805) wrote the most famous and influential entry in a long English tradition with roots at least as far back as John Ray's *Wisdom of God Manifested in Works of the Creation* (1691) and a few twigs persisting even today. Darwin revered Paley's book as a young man, and reminisced to his friend John Lubbock in 1859, just a week before the *Origin of Species* rolled off the presses: "I do not think I hardly ever admired a book more. . . . I could almost formerly have said it by heart." Later in this essay, we shall see that Darwin paid Paley an ultimate debt of gratitude by inverting his former mentor's system to construct his own particular and distinctive version of evolution.

This long tradition bore the name that Paley appropriated for the title of his book—*Natural Theology: or, Evidences of the Existence and Attributes of the Deity, Collected from the Appearances of Nature.* Natural Theology stakes the particular claim that God's nature, as well as his being ("existence and attributes" of Paley's title), can be inferred from the character of objects in the natural world. (Most religious thought today either denies or downplays such a link and does not attempt to validate the idea of divinity from the nature of material objects.)

Natural Theology, first published in 1802, presents five hundred pages of diverse arguments for God's existence, personality, natural attributes, unity, and goodness (in this explicit order of Paley's chapters), all centered upon one primary theme, endlessly hammered: God shows his creating hand in the good design of organisms for their appointed styles of life (wings are optimal for flying, nest behavior for raising offspring). Paley sets forth his theme, in the opening paragraphs of the book, with one of the most famous metaphors in English writing. As the scene opens, the good reverend is walking across a field:

> In crossing a heath, suppose I pitched my foot against a stone, and were asked how the stone came to be there, I might possibly answer, that, for any thing I knew to the contrary, it had lain there forever.

William Paley D.D.

ARCHDEACON of CARLISLE.

Hulton Deutsch Collection Limited

The stone, so rough and formless, can teach us nothing about its origins. "But," Paley continues, "suppose I had found a watch upon the ground, and it should be enquired how the watch happened to be in that place." Now, the answer must be different, for the watch—by its twin properties of complexity and obvious contrivance for a purpose—implies a watchmaker. Complexity and construction for use cannot arise randomly, or even from physical laws of nature (the laws may build something complex, like the chemical structure of a crystal, but not something evidently designed for a purpose, for nature's laws are abstract and impersonal). The watch must have been made on purpose, in order to keep time:

> The inference, we think, is inevitable; that the watch must have had a maker; that there must have existed, at some time and in some place or other, an artificer or artificers who formed it for the purpose which we find it actually to answer; who comprehended its construction, and designed its use.

One additional step completes the argument: Organisms are even more complex, and even more evidently designed for their modes of life, than watches. If the watch implies a watchmaker, then the better design of organisms requires a benevolent, creating God.

> There cannot be design without a designer; contrivance without a contriver. . . . The marks of design are too strong to be got over. Design must have had a designer. That designer must have been a person. That person is GOD.

Paley's argument is scarcely immune from parody, especially since he wrote in such a colorful style. His need to attribute purpose and benevolence to all aspects of our life in this vale of tears does recall his near contemporary, Voltaire's immortal Dr. Pangloss. For example, Paley's earnest resolution to the problem of pain parallels the punch line of an old and feeble joke: "Why did the moron hit himself on the head with a hammer? Because it felt so good when he stopped." (In fairness, Paley also presents the acceptable argument that pain informs the body of danger). Substituting the ills of his age for the hammer of our joke, Paley writes:

> A man resting from a fit of the stone or gout is, for the time, in possession of feelings which undisturbed health cannot impart. . . . I am far from being sure, that a man is not a gainer by suffering a moderate interruption of bodily ease for a couple of hours out of the four-and-twenty.

Nonetheless, I believe that Paley's argument, though quite unacceptable today, deserves our respect as a coherent and subtly defended philosophy from an interesting past—a "fossil worldview" that stretches our mind as we seek to comprehend our own preferences by appreciating the history of alternatives. Self-sustaining arguments are cheap; anyone with half a brain and a reasonable turn of phrase should be able to set forth his own prejudices. The test of a well-constructed defense lies in the identification and disproof of alternatives. If contrary interpretations are fully listed, fairly characterized, and adequately dismissed, then a system can win respect. I admire Paley primarily for his treatment of alternatives to his favored argument.

Paley's central argument includes an assertion—organisms are well designed for definite purpose—and an inference—good and purposeful design implies a designer. We might attack the assertion itself, but the prevalence of good design is an empirical matter not to be settled by a book on philosophy. The assertion, in any case, enjoys wide assent (both in Paley's time and our own). Fish gotta swim and birds gotta fly—and they seem to do so very well indeed. Let us then focus on the validity of the inference. Paley can imagine only two alternatives to his proposition that good and purposeful design implies a designer. Much of his book centers on the dismissal of these competing explanations.

1. Good design exists, but does not imply creation for its current purpose. Paley saw God in the correlation of form with function, specifically, in the divine construction of anatomy for its appointed role: the leg to walk, the hand to write, the mind to glorify God. But suppose that form arises first and function follows. Suppose that form originates for other reasons (direct pro-

duction by physical laws, for example) and then finds a use based on fortuitous fit. Paley grants that such an alternative is conceivable:

> This turn is sometimes attempted to be given, viz. that the parts were not intended for the use, but that the use arose out of the parts. This distinction is intelligible. A cabinet maker rubs his mahogany with fish-skin [I didn't know that the skin of dogfish sharks once served as sandpaper]; yet it would be too much to assert that the skin of the dogfish was made rough and granulated on purpose for the polishing of wood, and the use of cabinet makers.

This argument may work, Paley allows, for simple structures like the skin of a dogfish, but surely not for highly complex contrivances, made of hundreds of parts, all pointing to the same end, and each dependent upon all the others. Nothing so intricate could be made for one purpose and then fortuitously suited for something quite different and entirely unanticipated. Paley writes of the eye:

> Is it possible to believe that the eye was formed without any regard to vision; that it was the animal itself which found out, that, though formed with no such intention, it would serve to see with?

2. Good design exists, and implies production for its current purpose; but adaptations are built naturally, by slow evolution towards desired ends, not by immediate, divine fiat. Evolutionary alternatives were well understood in Paley's time. Darwin provided volumes of evidence and discovered a new and plausible mechanism; he scarcely invented the concept.

Paley could only conceive of evolution as a purposeful sequence of positive steps, building adaptation bit by bit. Thus, he attempted to refute a "Lamarckian" theory of natural change by use and disuse, with inheritance of acquired characters. (Writing in 1802, I doubt that Paley knew Lamarck's work directly, since his French colleague had just begun to publish evolutionary views. But use and disuse represented a common conviction among evolutionists of the time, not an invention made by Lamarck).

Paley provided both empirical and theoretical refutations. He began factually, with an old classic—a good example to be sure, but restricted by modesty to presentation in Latin, lest the unrefined derive some salacious pleasure. Centuries of disuse do not cause organs to disappear, or even to diminish:

> The mammae of the male have not vanished by inusitation; *nec curtorum, per multa saecula, Judaeorum propagini deest praeputium* [nor has the foreskin of Jews become any shorter in offspring through many centuries of circumcision].

(I am reminded of a story told by my father-in-law about life in Saint Louis just before World War I. Underground copies of Krafft-Ebing's *Psychopathia Sexualis* were always in circulation through boys' networks, but all editions then in print retained the author's original device of printing the case studies—and some are doozies, even by today's more permissive standards—in Latin. This fact, he assured me, provided the only impetus for attentive study of a subject then universally taught but otherwise roundly despised.)

Paley also provided some powerful theoretical arguments against evolution by use. If the elephant's short neck implies a great advantage for a long nose, all well and good. But what can a poor proto-elephant do with one-tenth of a trunk:

> If it be suggested, that this proboscis may have been produced in the long course of generations, by the constant endeavor of the elephant to thrust out his nose, (which is the general hypothesis by which it has lately been attempted to account for the forms of animated nature), I would ask, how was the animal to subsist in the meantime, during the process, until this elongation of snout were completed? What was to become of the individual, whilst the species was perfecting?

I accept Paley's arguments and might even be tempted to entertain his conclusions if he had truly accomplished his proper goal of refuting all logically possible alternatives. I believe that he did consider and dismiss all the potential refutations that he could conceive. But now we come to the crux of this essay. True originality is almost always an addition to the realm of the previously conceivable, not a mere permutation of possibilities already in hand. Progress in knowledge is not a tower to heaven built of bricks from the bottom up, but a product of impasse and breakthrough, yielding a bizarre and circuitous structure that ultimately rises nonetheless.

Paley missed a third alternative. We can scarcely blame him. The alternative is weird and crazy, laughable really. No sane person would build anything by such a cruel and indirect route. This third alternative can only work if you have lots of time to spare, and if you are not wedded to the idea that nature must be both efficient and benevolent. The third alternative, like the second, identifies natural evolution as the source of good design, thereby sinking Paley's central conviction that adaptation must imply creation by divine fiat. But instead of viewing evolution as purposeful and positive movement towards the desired goal, this third alternative builds adaptation negatively—by eliminating all creatures that do not vary fortuitously in a favored direction, and preserving but a tiny fraction to pass their lucky legacy into future generations.

As I said, this third alternative is grossly inefficient and defies the logic of a clockwork universe, built by our standards and reasoning. No wonder it never entered Paley's head. The only thing going for this third view—the only reason for even raising such an unpleasant topic here—is the curious fact that nature seems to work this way after all. Nobody ever called this method elegant, but the job gets done. We call this third view "natural selection," or Darwinism. Darwin himself commented most forcefully upon the inefficient and basically unpleasant character of his process, writing to his friend Joseph Hooker in 1856: "What a book a devil's chaplain might write on the clumsy, wasteful, blundering, low, and horribly cruel works of nature!"

The key to understanding Darwin's third alternative lies with a word, unfortunately almost extinct in English, but deserving a revival—hecatomb. A hecatomb is, literally, a massive sacrifice involving the slaughter of one hundred oxen—a reference to ancient Greek and Roman practices. By extension, a hecatomb is any large slaughter perpetrated for a consequent benefit. Natural selection is a long sequence of hecatombs. Individuals vary in no preferred direction about an average form for the population.

Natural selection favors a small portion of this spectrum. Lucky individuals in this portion leave more surviving offspring; the others die without (or with fewer) issue. The average form moves slowly in the favored direction, bit by bit per generation, through massive elimination of less favored forms.

The process might not be so inefficient if the hecatomb only occurred once at the beginning or if the sacrifice diminished from generation to generation. Suppose, for example, that the few survivors of the first hecatomb then automatically produced offspring with tendencies to vary in the favored direction. But Mendelian inheritance doesn't work this way. The few survivors of the first elimination yield offspring that also vary at random about the new average. Thus, the hecatomb in the second generation, and in all subsequent sortings, may be just as intense.

We may use an analogy to symbolize the inefficiency of natural selection by hecatomb. Suppose that a population will be better adapted if it can move from A to B. In direct Lamarckian models, including the only evolutionary scheme that Paley managed to conceptualize, the movement is direct, purposeful, and positive. Members of the population get a push and just walk from A to B. In the clumsy, wasteful, blundering, low, and horribly cruel Darwinian hecatomb, each individual stands at spot A and falls at random. If he happens to fall right along the line to B, he survives to the next trial. All individuals who fall off the line—the vast majority—are summarily shot. After a round of reproduction among the few survivors of this first hecatomb, the second trial begins. Standing now at one body length along the path to B, all individuals fall at random again—and the process continues. The hecatomb is equally pronounced in each round, and the population moves but one body length towards its goal each time. The population will eventually get to point B, but would any engineer favor such a poky and punitive device? Can you blame the divine Paley for not even imagining such a devilish mechanism?

I do not contrast Darwin with Paley as an abstract rhetorical device. Darwin, as quoted above, revered Paley during his youth. In a courageous act of intellectual parricide, he then overthrew his previous mentor—not merely by becoming an evolutionist, but by constructing a particular version of evolutionary theory maximally disruptive of Paley's system and deepest beliefs.

I can imagine two revolutionary ways—one more radical than

the other—to overturn Paley's comfortable and comforting belief that God made us all with shapes and habits beautifully adapted to our modes of life. You might argue that Paley was wrong, that animals are not generally well designed, and that if you insist on seeing God's work in the massive imperfection of nature, then perhaps you ought to revise your notion of divinity. This would be a radical argument, but Darwin devised an even more disturbing version.

Secondly, you might argue (as Darwin actually did) that Paley was quite right: Animals are well adapted to their modes of life. But this good fit is not an emblem of God's benevolence, rather an indirect result of the horrid system of multiple hecatombs known as natural selection. What a bitter pill for Paley—for Darwin allows that Paley described the look of nature correctly, but then argues that the mechanism for this appearance has a mode of action, and an apparent moral force, directly contrary to the intent and benevolence of the God of natural theology.

Where did Darwin get such a radical version of evolution? Surely not from the birds and bees, the twigs and trees. Nature helped, but intellectual revolutions must also have ideological bases. Scholars have debated this question for more than a century, and our current "Darwin industry" of historians has moved this old discussion towards a resolution. The sources were many, various, and exceedingly complex. No two experts would present the same list with the same rankings. But all would agree that two Scottish economists of the generation just before Darwin played a dominant role: Thomas Malthus and the great Adam Smith himself. From Malthus, Darwin received the key insight that growth in population, if unchecked, will outrun any increase in the food supply. A *struggle for existence* must therefore arise, leading by *natural selection* to *survival of the fittest* (to cite all three conventional Darwinian aphorisms in a single sentence). Darwin states that this insight from Malthus supplied the last piece that enabled him to complete the theory of natural selection in 1838 (though he did not publish his views for twenty-one years).

Adam Smith's influence was more indirect, but also more pervasive. We know that the Scottish economists interested Darwin greatly and that, during the crucial months of 1838, while he assembled the pieces soon to be capped by his Malthusian insight, he was studying the thought of Adam Smith. The theory of natural selection is uncannily similar to the chief doctrine of laissez-faire economics. (In our academic jargon, we would say that the two theories are "isomorphic"—that is, structurally similar point for point, even though the subject matter differs). To achieve the goal of a maximally ordered economy in the laissez-faire system, you do not regulate from above by passing explicit laws for order. You do something that, at first glance, seems utterly opposed to your goal: You simply allow individuals to struggle in an unfettered way for personal profit. In this struggle, the inefficient are weeded out and the best balance each other to form an equilibrium to everyone's benefit.

Darwin's system works in exactly the same manner, only more relentlessly. No regulation comes from on high; no divine watchmaker superintends the works of his creation. Individuals are struggling for reproductive success, the natural analog of profit. No other mechanism is at work, nothing "higher" or more exalted. Yet the result is adaptation and balance—and the cost is hecatomb after hecatomb after hecatomb. (I call Darwin's system more relentless than Adam Smith's because human beings, as moral agents, cannot bear these hecatombs. We therefore never let laissez-faire operate without some constraint, some safety net for losers. But nature is not a moral agent, and nature has endless time.)

Adam Smith embodied the guts of his theory—his core insight—in a wonderful metaphor, one of the truly great lines written in the English language. Speaking of an actor in the world of laissez-faire, Adam Smith states:

> He generally indeed neither intends to promote the public interest, nor knows how much he is promoting it. . . . He intends only his own gain, and he is in this, as in many other cases, led by an invisible hand to promote an end which was no part of his intention.

Such a lovely image: The "invisible hand" that produces order, but doesn't really exist at all, at least in any direct way. Darwin's theory uses the same invisible hand, but formed into a fist as a battering ram to eliminate Paley's God from nature. The very features that Paley used to infer not only God's existence, but also his goodness, are, for Darwin, but spin-offs of the only real action in nature—the endless struggle among organisms for reproductive success, and the endless hecatombs of failure.

In this light, we may finally return to poor Paley and feel the poignancy of his inability even to conceptualize Darwin's third alternative—the argument that finally, and permanently, brought his system down. He stood so close, but just didn't have the conceptual tools to put the pieces together. (I do not suggest that Paley would have become a Darwinian if he had recognized the third way. He would surely have rejected evolution by hecatomb, just as he had attacked descent by purposeful step. Yet I remain fascinated by his failure to conceptualize the Darwinian mode at all, for the essence of genius lies in the rare ability to think in new dimensions orthogonal to old schemes, and we must dissect both failures and successes in order to understand this most precious feature of human intellect.)

Darwin received his greatest inspiration from Thomas Malthus and Adam Smith. Paley knew their work as well, yet he didn't draw the implications. For Malthus, Paley actually cites the key line that inspired Darwin's synthesis in 1838 (but in the context of a passage on civil vs. natural evils). Paley writes:

> The order of generation proceeds by something like a geometrical progression. The increase of provision, under circumstances even the most advantageous, can only assume the form of an arithmetic series. Whence it follows, that the population will always overtake the provision, will pass beyond the line of plenty, and will continue to increase till checked by the difficulty of procuring subsistence.

(At this point, Paley adds a footnote: "See this subject stated in a late treatise upon population"—obviously Malthus.)

The influence of Adam Smith is not quite so explicit. But I was powerfully moved (and inspired to write this essay) when I read Adam Smith's great metaphor in Paley's more effusive prose and differing intent. I quoted the line early in this essay: "I never see a bird in that situation, but I recognize an invisible hand, detaining the contented prisoner from her fields and groves for a purpose."

I cite this correspondence as a symbol, not a proof. I realize that it offers no evidence for Paley imbibing the metaphor from Smith. The phrase is obvious enough, and could be independently invented. (Nonetheless, the metaphor of the invisible hand is central to Smith's argument and has always been so recognized. *The Wealth of Nations* was published in 1776—an easy date for Americans to remember—a full generation before *Natural Theology*. So perhaps Paley had caught the rhythm from Smith.) The two usages are diametrically opposed, hence the poignancy of the comparison. Paley's invisible hand is God's explicit intent (though He works, in this case, indirectly through the bird's instinct, and not by a palpable push). Smith's invisible hand is the *impression* of higher power that doesn't actually exist at all. In Darwin's translation, the invisible hand dethrones the God of natural theology.

For some, this tale of shifting usages and ideas may seem a dull exercise in antiquated thought. Yet we have never stopped fighting the same battles, seeking the same solaces, rejecting the same uncomfortable truths. Why are some of us so loath to accept evolution at all, despite overwhelming evidence? Why are so many of us who do accept evolution so unable to grasp the Darwinian argument, or so unwilling, for emotional reasons, to live with it even if we do understand?

This situation may be frustrating for someone like me who has spent a professional lifetime working with the power of Darwinian models and who feels no moral threat in their potential truth (for a fact of nature cannot challenge a precept of morality)—frustrating perhaps, but not hard to comprehend. We leave Paley's world with reluctance because it offered us such comfort, and we enter Darwin's with extreme trepidation because the sources of solace seem stripped away. Consider the happy moral that Paley draws from good design and its divine manufacture:

> The hinges in the wings of an earwig, and the joints of its antennae, are as highly wrought, as if the Creator had nothing else to finish. We see no signs of diminution of care by multiplicity of objects, or of distraction of thought by variety. We have no reason to fear, therefore, our being forgotten, or overlooked, or neglected.

I can offer only two responses—both, I think, powerful and quite conducive to joyous optimism, if this be your fortunate temperament. We may lose a great deal of easy, unthinking, superficial comfort in the rejection of Paley's God. But think what we gain in toughness, in respect for nature by knowledge of our limited place, in appreciation for human uniqueness by recognition that moral inquiry is our struggle, not nature's display. Think also what we gain in increments of real knowledge—and what could be more precious—by knowing that evolution has patterned the history of life and shaped our own origin.

Thomas Henry Huxley faced the same dilemma more than one hundred years ago. Chided by his theologian buddy Charles Kingsley for abandoning the traditional solace of religion in Paley's style, Huxley replied:

> Had I lived a couple of centuries earlier I could have fancied a devil scoffing at me . . . and asking me what profit it was to have stripped myself of the hopes and consolations of the mass of mankind? To which my only reply was and is—Oh devil! truth is better than much profit.

And a gain of such magnitude is no barleycorn.

10 | More Light on Leaves

SOMETIME, in a better world to come, the wolf shall dwell with the lamb on Isaiah's holy mountain. Once, in the better world that was, Leonardo painted exquisite women, Michelangelo rendered the hand of God, and Raphael captured the even greater age of Plato and Aristotle in the *School of Athens*. (I know that these gentlemen have recently mutated into Teenage Ninja Turtles, and so perhaps may we measure the direction of changing excellence in history.)

The myth of a past golden age seems irresistible, but the contrary reality is undeniable. Leonardo built some frightening instruments of war; Michelangelo struggled against the virulent homophobia of his generation; and Raphael died on his thirty-seventh birthday.

A persistent and cardinal legend of this mythology holds that the age of Michelangelo encouraged people of talent to range across all realms of mind and art, to follow that wondrously optimistic motto of Francis Bacon: "I have taken all knowledge for my province." In our present age of narrow specialization, we continue to credit this reverie by referring to a broad-ranging scholar as a "Renaissance man."

But suspicion, deprecation, and narrowness have a pedigree as old as enlightenment. Professionals have always tried to seal the borders of their trade, and to snipe at any outsider with a pretense to amateur enthusiasm (though amateurs who truly love their subject, as the etymology of their status proclaims, often acquire far more expertise than the average time-clock–punching breadwinner). The classic maxim of narrowness—"a cobbler should stick to his last"—dates from the fourth century B.C., the great age of Athens. (A "last," by the way, is a shoemaker's model foot, not an abstract claim about perseverance.)

Nothing—not even acknowledged greatness—can secure a clear passport for distant intellectual migration. Johann Wolfgang von Goethe (1749–1832), who enjoyed the ultimate pleasure of general regard in his own time as the world's greatest poet, complained bitterly of his reception by scientists for considerable labor in their domain (Goethe did serious work, often with

Sketch of Johann Wolfgang Goethe (1817). *Courtesy of Nationale Forchungs- und Gedenkstatten der klassischen deutschen Literatur in Weimar.*

great success, in anatomy, botany, geology, and optics). Near the end of his life, in 1831, Goethe wrote:

> The public was taken aback, for . . . it is expected that a person who has distinguished himself in one field, whose manner and style are generally recognized and esteemed, will not leave his field, much less venture into one entirely unrelated. Should an individual attempt this, no gratitude is shown him; indeed, even when he does his task well, he is given no special praise.

Goethe's spirited defense against this parochialism not only displays his own justifiably expansive ego, but also asserts an intellectual's most precious birthright.

> But a man of lively intellect feels that he exists not for the public's sake, but for his own. He does not care to tire himself out and wear himself down by doing the same thing over and over again. Moreover, every energetic man of talent has something universal in him, causing him to cast about here and there and to select his field of activity according to his own desire.

Six years after Goethe's death, in 1838, the French biologist Isidore Geoffroy Saint-Hilaire devoted an entire article to justifying Goethe's scientific excursions: *Sur les travaux zoologiques et anatomiques de Goethe* (On the zoological and anatomical work of Goethe). (In his very last article, Goethe had defended Isidore's more famous father Etienne Geoffroy in his celebrated confrontation with Georges Cuvier on theories of anatomy, so Isidore's article might be viewed as the repayment of a debt, one generation removed.) Isidore gave incisive expression to the parochial tendencies of scientists:

> Many well-informed people still do not know whether Goethe was limited to propagating ideas already developed in science by reclothing them in the colors of his admirable style, or whether he can claim the greater glory of an originator. Naturalists themselves hesitate to recognize as one of their own a man whom they have been accustomed,

for so long, to admire as a dramatic poet, a novelist, and even as a writer of songs. . . . The more that this distance [between art and science] be viewed as immense, perhaps even unbridgeable, the more we have difficulty in imagining that the same hand that wrote Werther and Faust . . . could hold the anatomical scalpel with skill—and the more we may view this accomplished prodigy as admirable because he was able to combine intellectual qualities that ordinarily exclude each other.

Not that Goethe (or his reputation) needs my defense, but I do devote this essay to fighting the professional narrowness that Isidore Geoffroy identified and deplored—a tendency that has intensified with a vengeance in the 150 years since Isidore's article, for we modern scholars often treat our professions as fortresses and our spokespeople as archers on the parapets, searching the landscape for any incursion from an alien field.

We might mount two kinds of defense for generosity towards incursion from scholars in other fields (beyond the general principle of virtue in ecumenicism and variety). The weaker claim—I shall call it universalism—holds that all good thinkers operate in pretty much the same way, and that benefits offered by outsiders are largely quantitative. Intellectual progress is tough, and we need all the help we can get, so why bar access to brilliant people in other disciplines? Although I will offer a different defense for Goethe, universalism is often a good argument. For example, I would not struggle against sexism (as many scientists have) by claiming that women tend to reason in a different but equally valuable way. I regard such a claim as both false and demeaning. We need to open all fields to women because it is simply absurd, given the rarity of genius, to recruit from only half the potential pool.

The second and stronger claim—I shall call it *special insight*—holds that we should value outsiders not simply as more bodies, but as potentially applying to their extracurricular concerns a fresh and different mode of thinking imported from their central profession. (I call this claim stronger because narrow professionals might accept the argument of more bodies, but usually rebel against special insight with the rallying cry of all parochialisms—

how dare they come into my garden and tell me how to cultivate my tomatoes!)

In the case of Goethe and science, I advance this second claim of special insight for two reasons. First, I feel that characteristic ways of thinking in the arts—the role of the imagination, holistic vs. reductionistic approaches, for example—might enlighten science (not because scientists never think in this "artful" manner, but because the unpopularity of these styles among professionals greatly limits their fruitful use, and an infusion from outside might therefore help). Second, Goethe himself viewed his treatment of biological problems as different from that of most full-time scientists, and he attributed his unconventional approach to his training and practice in the arts.

In particular, Goethe argued that his artist's perspective led him to view nature as a unity, to search for integration among disparate parts, to find some law of inherent concord. Goethe wrote:

> What is all intercourse with nature if by the analytic method we merely occupy ourselves with individual material parts, and do not feel the breath of the spirit, which prescribes every part its direction, and orders, or sanctions, every deviation, by means of an inherent law!

I may then pose the cardinal question of this essay: Did Goethe get any mileage for his unconventional "artist's" approach in science? Did it work? The answer, I think, is undoubtedly "yes." We might hold that Goethe's general brilliance allowed him to succeed whatever cockamamie method he happened to use—and that his artist's vision of integration and imagination didn't really help after all. But we might also take him at his word, admit the efficacy of his approach, and try to appreciate the message of pluralism and the artificiality of conventional boundaries among disciplines.

Goethe had a taste of success early in his career, in 1784, when he discovered a new bone in the human upper jaw. He called this bone the *intermaxillary*; others dubbed it *Goethe's bone.* He insisted that such a bone must exist in humans (prior to any evidence and in the face of general denial) because other terrestrial vertebrates possessed it—and such a bone must therefore belong to the archetype, or abstract generating plan, of all reptiles, birds, and mammals. (This bone is called the premaxillary in other vertebrates; it generally holds the upper incisor teeth in mammals.) In humans, the small intermaxillary fuses with other bones of the upper jaw and cannot be recognized in skeletons after birth; Goethe ascertained its existence by noting the sutures that form before fusion in embryos. In an essay written in 1832, the year of his death, Goethe recalled this discovery as "the first battle and the first triumph of my youth." He attributed his success to his artist's vision of necessary unity: The bone must exist in humans if all terrestrial vertebrates share a common and abstract plan of development.

Goethe published his most important biological work in 1790—*Versuch die Metamorphose der Pflanzen zu erklären* (An attempt to explain the metamorphosis of plants). This work, a pamphlet devoid of illustrations or charts and consisting of 123 numbered, largely aphoristic passages, can scarcely be called a document of conventional science. It embodies the two principles that Goethe attributed to his artistic predilections—bold hypotheses based on assumptions of inherent unity. And yet, though Goethe's central notion cannot be sustained, this curious little work is full of insight and has exerted a strong influence over the history of morphology (a word coined by Goethe). We may accept the assessment of Goethe's scientific champion Etienne Geoffrey Saint-Hilaire, written in 1831:

> The professional knowledge of the naturalist—what one might call the mechanical part—appears nowhere; no descriptions of flowers are given; no experiments are noted. It is the book of a scientist for its fund of ideas but, in its format, it is the book of a philosopher who expresses himself as a poet. Nonetheless, we must accept it as an excellent treatise in natural history.

Most subtle arguments can be ridiculed by oversimplification and stereotype. Goethe's theory of plant form has been particularly subject to such unfair treatment because his drive to find unifying themes did lead him to a bold hypothesis, all too easily caricatured. Goethe worked within a developing morphological tradition generally called *unity of type*. He longed to find an archetype—an abstract generating form—to which all the parts of plants might be related as diversified products.

Many of Goethe's colleagues sought the archetype of animal skeletons in the vertebra. (Etienne Geoffroy tried to homologize the external carapace of arthropods with the internal skeleton of vertebrates, and to identify the abstract vertebra as archetype of both. He actually claimed that insects therefore lived within their own vertebrae, and that insect legs represented the same structure as vertebrate ribs.) Goethe, following the same approach in another kingdom, held that the archetypal form for all plant parts—from cotyledons, to stem leaves, to sepals, petals, pistils, stamens, and fruit—could be found in the leaf.

The caricature of Goethe's theory therefore proclaims "all plant parts are leaves" (with the implied corollary, "ha, ha, ha, what nonsense"). But Goethe said no such thing. First of all, the archetype is not an actual maple leaf or pine needle, but an abstract generating form, from which stem leaves depart least in actual expression. Goethe defended his use of the common word *leaf* in describing his archetypal idea:

> We ought to have a general term with which to designate this diversely metamorphosed organ and with which to compare all manifestations of its form. . . . We might equally well say that a stamen is a contracted petal, as that a petal is a stamen in a state of expansion; or that a sepal is a contracted stem leaf approaching a certain state of refinement, as that a stem leaf is an expanded sepal.

Secondly, Goethe's theory applies only to the lateral and terminal organs of ordinary plants, not to the supporting roots and stems. Goethe's defensive reaction to criticism on this point does not rank among his best arguments, but I cannot fault him for omitting stems and roots; after all, a theory for all appended parts is no mean thing, even if the underlying superstructure goes unaddressed. Goethe says that he copped out because roots are such lowly objects!

> My critics have taken me to task for not considering the root in my treatment of plant metamorphosis. . . . I was not con-

cerned with it at all, for what had I to do with an organ which takes the form of strings, ropes, bulbs and knots . . . , an organ where endless varieties make their appearance and where none advances. And it is advance solely that could attract me, hold me, and sweep me along on my course.

Goethe's theory does propose a leaf archetype, but his full account of plant form is a subtle interplay among three great general forces in nature: the universal and inherent archetype, and the impact upon it of both directional and cyclical factors. The interaction of these three principles—stability, direction, and recurrence—produces the natural object that we call a plant.

Within the 1790 essay, Goethe expressed the central principle of his system in measured tone: "The organs of the vegetating and flowering plant, though seemingly dissimilar, all originate from a single organ, namely, the leaf." In a private document, written in 1831, he became more effusive: "[I have traced] the manifold specific phenomena in the magnificent garden of the universe back to one simple general principle." To friends, as to the great philosopher J. G. Herder in 1787, he became positively effusive (dare I say florid?):

The archetypal plant as I see it will be the most wonderful creation in the whole world, and nature herself will envy me for it. With this model and the key to it, one will be able to invent plants without limit to conform, that is to say, plants which even if they do not actually exist nevertheless might exist and which are not merely picturesque or poetic visions and illusions, but have inner truth and logic. The same law will permit itself to be applied to everything that is living.

Goethe dissects and compares, trying to find the leaflike basis of apparently diversified and disparate structures. The fused sepals, for example, forming the *calyx* (cup) at the base of a flower, are leaves that fail to separate when a cutoff of nutriment stops expansion of the stem: "If the flowering were retarded by the infiltration of superfluous nutriment, the leaves would be separated and would assume their original shape. Thus, in the calyx, nature forms no new organ but merely combines and modifies organs already known to us."

When parts are too modified to show connection and reduction to the leaf archetype in one type of plant, Goethe uses the comparative approach to find sufficiently similar shapes in other species. Even the most disparate *cotyledons* (first growths from the seed) eventually attain a tolerably leaflike form in some species:

> They are often misshapen, crammed, as it were, with crude matter, and as much expanded in thickness as in breadth; their vessels are unrecognizable and scarcely distinguishable from the mass as a whole. They bear almost no resemblance to a leaf, and we might be misled into regarding them as special organs. Yet in many plants the cotyledons approach leaf form: they flatten out; exposed to light and air, they assume a deeper shade of green; their vessels become distinct and begin to resemble veins.

If Goethe's system were, as often portrayed, no more than a theory of leaf-as-archetype, it would have no claim to interesting completeness, for it would not explain systematic variation in form up the stem, and would therefore not stand as a full attempt to explain both similarities and characteristic differences in the parts of plants. But, in his most fascinating intellectual move, Goethe produces a complete account by grafting two additional principles onto the underlying notion of leaf-as-archetype: the progressive refinement of sap, and cycles of expansion and contraction. We may regard these principles as ad hoc or incorrect today, but the power of their conjunction with the archetypal idea can still be grasped and appreciated with much profit.

These two additional principles embody both necessary sides of the grandest Western metaphor for intelligibility in any growing, or historically advancing, system—arrows of direction and cycles of repeatability (I called them *time's arrow* and *time's cycle* in my 1987 book on the discovery of geological time—see bibliography). We must, in any scientific process unfolding through time, be able both to identify vectors of change (lest time have no history, defined as distinctness of moments) and underlying constant, or cyclical principles (lest temporal sequences be nothing but uniqueness after uniqueness, with nothing general to identify at all). Goethe, faced with observations of both directionality and repeatability up the stem, recognized the need for both sides of this primal dichotomy.

1. *Refinement of sap as a directional principle.* Up and down; heaven and hell; brain and psyche vs. bowels and excrement; tuberculosis as a noble disease of airy lungs vs. cancer as the unspeakable malady of nether parts (see Susan Sontag's important book, *Illness as Metaphor*): This major metaphorical apparatus of Western culture almost irresistibly applies itself to plants as well, with gnarly roots and tubers as things of the ground and fragrant, noble flowers as topmost parts, straining towards heaven. Goethe, by no means immune to such thinking in a romantic age, viewed a plant as progressing towards refinement from cotyledon to flower. He explained this directionality by postulating that each successive "leaf" progressively filters an initially crude sap. Flowering is prevented by these impurities and cannot occur until they have been removed. The cotyledons begin both with minimal organization and refinement, and with maximum crudity of sap:

> We have found that the cotyledons, which are produced in the enclosed seed coat and are filled to the brim, as it were, with a very crude sap, are scarcely organized and developed at all, or at best roughly so.

The plant moves towards its floral goal, but too much nutriment delays the process of filtering sap, as material rushes in and more stem leaves must be produced for drainage. A decline in nutriment allows filtering to attain the upper hand, producing sufficient purification of sap for flowering:

> As long as cruder saps remain in the plant, all possible plant organs are compelled to become instruments for draining them off. If excessive nutriment forces its way in, the draining operation must be repeated again and again, rendering inflorescence almost impossible. If the plant is deprived of nourishment, this operation of nature is facilitated.

Finally, the plant achieves its topmost goal:

> While the cruder fluids are in this manner continually drained off and replaced by pure ones, the plant, step by

step, achieves the status prescribed by nature. We see the leaves finally reach their fullest expansion and elaboration, and soon thereafter we become aware of a new aspect, apprising us that the epoch we have been studying has drawn to a close and that a second is approaching—the epoch of the flower.

2. *Cycles of expansion and contraction.* If the directional force worked alone, then a plant's morphology would be a smooth continuum of progressive refinement up the stem. Since, manifestly, plants display no such pattern, some other force must be working as well. Goethe specifies this second force as cyclical, in opposition to the directional principle of refining sap. He envisages three full cycles of contraction and expansion during growth. The cotyledons begin in a retracted state. The main leaves, and their substantial spacing on the stem, represent the first expansion. The bunching of leaves to form the sepals at the base of the flower marks the second contraction, and the subsequent elaboration of petals the second expansion. Narrowing of the archetypal leaf to form pistils and stamens identifies the third contraction, and the formation of fruit the last and most exuberant expansion. The contracted seed within the fruit then starts the cycle again in the next generation. Put these three formative principles together—the archetypal leaf, progressive refinement up the stem, and three expansion-contraction cycles of vegetation, blossoming, and bearing fruit—and the vast botanical diversity of our planet yields to Goethe's vision of unity:

> Whether the plant vegetates, blossoms, or bears fruit, it nevertheless is always the same organs with varying functions and with frequent changes in form, that fulfill the dictates of nature. The same organ which expanded on the stem as a leaf and assumed a highly diverse form, will contract in the calyx, expand again in the petal, contract in the reproductive organs, and expand for the last time as fruit.

How shall we judge Goethe's botanical theory today? In one immediate sense, of course, it is false: Sap is not refined up the stem, and nothing expands and contracts in regular waves during growth. But falsity is not a foolproof criterion for judging impor-

tance or capacity for suggesting insight. Many false ideas have
been immensely useful, if only because the process of disproof so
often leads to greater knowledge and integration. Consider two
famous statements, both cited before in these essays, but worth
repeating as one of the most important (if slightly paradoxical)
truths of intellectual life. First, the economist Vilfredo Pareto (I
certainly appreciate the botanical metaphor in this context):

> Give me a fruitful error any time, full of seeds, bursting with
> its own corrections. You can keep your sterile truth for
> yourself.

Second, Charles Darwin:

> False facts are highly injurious to the progress of science,
> for they often endure long; but false views, if supported by
> some evidence, do little harm, for every one takes a salutary
> pleasure in proving their falseness.

Or, for that matter, consider Goethe's own words (from a posthumous essay published in 1833):

> A false hypothesis is better than none at all. The fact that it
> is false does not matter so much. However, if it takes root
> [another botanical metaphor!], if it is generally assumed, if
> it becomes a kind of credo admitting no doubt or scrutiny—
> that is the real evil, one which has endured through the
> centuries.

If ever an idea qualified as a "fruitful error," as a "salutary"
"false view . . . supported by some evidence," then place Goethe's theory of plant form at the head of the list—as the loveliest
and most refined fruit of the last expansion. First of all, Goethe's
particular claims do record many elements of empirical truth.
Leaves may not provide a basis for all form, but many plant parts
are modifications of leaf primordia—take a close look at a flower
petal. Sap is not filtered and refined up the stem, and no simple
force expands and contracts the archetypal pattern in regular
cycles, but directional and repetitive trends do shape a plant during growth, even if Goethe misconstrued the actual causes.

But Goethe's vision provides a second and more important reason for treating his theory with respect and for appreciating the "artistic" aspects of his presentation. Great ideas, whether true or false, do drive our research forward by focusing our thoughts and suggesting new pathways of exploration. Goethe's theory has the great virtue of reducing an enormously complex issue, otherwise chaotic and confusing, to three important and expansive principles. The theory, in this sense, is both inspiring and beautiful—two words rarely granted much status in scientific discourse, but worthy of our attention, at least as prods to action, if not as criteria of truth. Goethe's three principles are basic and true, important components of any comprehensive account of nature—whatever the limits of his particular application to plants. We must search for underlying rules and principles to generate the otherwise uncoordinated variety of related objects. And we cannot make historical sequences intelligible (including the growth of plants) unless we identify both directional and repetitive aspects, for we need both uniqueness and underlying lawlike structure to make any sense of processes that develop in time. The history of life, for example, is both a tale of genealogical unfolding from one special object to the next (time's arrow), and of recurring processes (mass extinction, rules of ecological order, break-up and joining of continents, transgression and regression of seas) that graft some broad predictability upon the string of unique events (time's cycle).

Great theories are expansive; failures mire us in dogmatism and tunnel vision. I do not know the actual context of Goethe's famous dying words. Perhaps he was only asking for another candle, all the better to view the faces of his beloved one last time. But perhaps he was begging the Almighty for the greatest gift that fruitful theories can provide—*Mehr Licht!* (more light!).

Time in Newton's Century

11 | On Rereading Edmund Halley

MY DEAR COLLEAGUE Sewall Wright saw the comet in 1910 while working on the railroads in South Dakota—and then, as he so strongly wished, lived to witness its return in 1986. Mark Twain born under the comet's waning light in 1835, died at its next passage seventy-six years later. But the vast majority of us get only one chance—or none at all. So we celebrate and, if intellectually inclined, we also cerebrate. If any natural happening ever received more than its merited share of written attention, we can only nominate the return of Halley's comet in 1986, especially since that miserable iceball mocked our long anticipation by putting on such a poor show. I therefore fully intended to ignore both Mr. Halley and his cursed comet in the monthly essays that form the basis for this series of books.

But, as good intentions so often succumb to pervasive temptation, I confess that I was drawn to the man by a curious omission or underplaying that I detected in the flood of articles written about Halley in the light (hmm!) of his namesake's return. We are all parochial at heart and tend to view wide-ranging geniuses like Halley as members of our own fraternity, even for limited contributions. I insist that Thomas Jefferson was primarily a paleontologist, and insurance salesmen surely view Charles Ives as a compatriot who occasionally dabbled in composition (as they, no doubt, appropriate Edmund Halley, who devised some of the first actuarial tables).

In my parish, Edmund Halley ranks as a geologist who occasionally looked upward. We claim him for a five-page article published in the *Philosophical Transactions of the Royal Society of London.*

Its title exhausts a good part of page one: "A short account of the cause of the saltness of the ocean, and of the several lakes that emit no rivers; with a proposal by help thereof, to discover the age of the world." In short, Halley wrote one of the finest and most influential papers with a testable proposal for that *primum desideratum* of our discipline—the earth's age. Moreover, his method, though ultimately flawed, engendered much fruitful research and was, before the development of radioactive dating, among the two or three leading contenders for addressing this fundamental question.

Yet this paper and its interesting idea, so central to my profession and its history, got lost in the popular articles on Halley. *Discover* magazine abandoned its tradition of contemporaneity and named Halley their scientist of the year—but gave his geological work less than a paragraph. Carl Sagan and Ann Druyan's *Comet* awarded just a few lines more. John Noble Wilford's long article in the *New York Times* science section (October 29, 1985) omitted this work entirely from a long listing of Halley's achievements.

This veil of silence forced me back to my old Xerox of Halley's original, lovingly made in my graduate student days at the American Museum of Natural History, from a beautifully bound set of this oldest scientific journal in English. I had often assigned its offspring to students but had not read the original for several years. Perhaps if I encountered something intriguing, I might actually find a chink of difference and be able to add, rather than merely reiterate, if I chose to write about Halley after all. I plunged in and read with mounting disappointment. Nothing unremembered, nothing unusual. Oh, I did manage a tiny contribution to the great issue that seemed to obsess press commentators—the spelling and pronunciation of the man's name. We were instructed so often to say "Edmond" (not "Edmund"), that "o" became an insider's badge of sophistication. But he is Edmund in this article (and the indifferent spelling of Halley's time probably made either quite acceptable to him). As to the other vexatious mystery—why Americans, flagrantly disregarding one of the few decent guides that English spelling offers for pronunciation, insist on calling the man "Hayley"—I can only conjecture (as many others have) that our minds were poisoned by a certain Haley (properly "Hayley" by virtue of the single "l") who made a lot of noise when I was a teenager and also called his group the Comets.

Reading long after midnight, I finally came to the last paragraph, having no fun at all as the clock struck one (and unable to get that wretched tune out of my head). Then I had a moment of discovery, that one instant in ten thousand that makes a scholar's life so exciting, and that justifies the tedium and discipline accompanying any serious intellectual work (Edison's old allocation of effort between perspiration and inspiration is just about right—ninety-nine to one). I realized, in short, that I, and every other comment I have ever read about Halley's proposal for the earth's age, have interpreted him precisely the wrong way round. I also think I know why. The difficulty lies not with anything Halley wrote. His meaning could not have been stated more explicitly. Rather, for concerns of our own, and by a traditional misreading of the history of science, we have simply passed over Halley's own construction and imposed our preferences upon his reasoning. It's a damned shame, too—for his intent is both interesting and instructive for us today.

Halley's article makes a simple and elegant proposal. (The bare bones of the method have not been misinterpreted, only Halley's view of its meaning.) He assumes that the oceans were originally fresh and have become progressively more salty through an influx of dissolved materials transported by rivers. Since rivers flow in, but nothing flows out, salt must accumulate steadily. (Evaporation returns ocean waters to rivers in the earth's hydrologic cycle, but evaporated waters are fresh and leave their salts behind.) We usually regard river water as fresh, but Halley recognized correctly that streams carry tiny (and untasteable) amounts of dissolved salts:

> But the rivers in their long passage over the earth do imbibe some of the saline particles thereof, though in so small a quantity as not to be perceived, unless in these their depositories [that is, lakes and oceans] over a long tract of time.

Halley recognized that this argument for the source of oceanic salt suffered one grave methodological defect. The world ocean is a sample of one. How can the ocean, by itself, prove the general proposition that basins with riverine inlets, but no outlets, become progressively more salty with time? Perhaps the ocean is just a special case proving nothing but its individuality, not the largest representative of a general process.

The cleverness of Halley's argument lies in his recognition that lakes, properly classified and divided, serve as smaller systems representing the same process he proposed for oceans. So he sorted large lakes into those—most of them—that have both inlets and outlets, and the few that, like the oceans, receive waters from rivers but provide no exit beyond evaporation.

He could find only four in this second category comparable with oceans—the Caspian Sea, the "Mare Mortuum" (Dead Sea), "the lake on which stands the City of Mexico," and Titicaca in Peru. All are salt to varying degrees, while all freshwater lakes fell into his first category. Halley's taxonomy of lakes had confirmed his theory for the origin of salt in oceans—a fine example of the methodological principle that sample sizes can often be increased only by recognizing proper analogues in other classes of objects.

Halley now felt ready to advance his argument for the age of the earth (or at least for its oceans):

> Now if this be the true reason of the saltness of these lakes,
> 'tis not improbable but that the ocean it self is become salt
> from the same cause, and we are thereby furnished with an
> argument for estimating the duration of all things, from an
> observation of the increment of saltness in their waters.

If we could measure the salinity of modern oceans, then determine the amount of salt brought in by rivers each year, we could extrapolate back to an initial time of no salt at all and estimate the age of the oceans. Halley recognized that his method required a set of simplifying assumptions that might not be strictly true—rough constancy of annual influx, no appreciable loss of salt in buried sediments, for example. But he felt that his method might give a reasonable first-order estimate.

Halley realized that he could not hope to measure the annual influx accurately—too much variation among too many rivers and probably too small a total compared with the amount of salt now in the oceans. But we could, for the benefit of posterity, make accurate measurements of salt now in oceans and lakes; for, a few centuries hence, the total increase should be palpable enough to permit a good estimate of average annual increment. Halley advised:

> This argument can be of no use to ourselves, it requiring very great intervals of time to come to our conclusion. . . . I recommend it therefore to the Society, as opportunity shall offer, to procure the experiments to be made of the present degree of saltness of the ocean, and of as many of these lakes as can be come at, that they may stand upon record for the benefit of future ages.

The geological literature contains a "standard" interpretation of Halley's contribution. It points out, first of all and quite correctly, that Halley's method was wrong—a good try to be sure, but ultimately based on a false premise. Halley assumed that since salt entered the oceans every year, yet the sea was not saturated (as the greater salinity of the Dead Sea attested), newly entering salt must be added to the amount already present. But many of nature's cycles are maintained in dynamic balance between influxes and outflows, long before most components reach some theoretically maximal level. Our atmosphere could maintain a lot more carbon dioxide, but until we began messing with an old balance by burning massive amounts of fossil fuel, carbon dioxide had remained relatively steady at percentages much smaller than the atmosphere can hold (as we may discover to our great sorrow if current rises lead to a runaway greenhouse effect).

Most components of the atmosphere and ocean are in such dynamic balance on our ancient earth. (In a sense, such equilibria must exist, for the earth is so old that any directional increment, however small, would lead to saturation in a fraction of historical time.) Oceanic salt persists at its current level in a dynamic balance, or steady state, between influx from rivers and numerous processes, including burial in sediments and biological uses, that constantly remove about the same amount that enters. Perhaps, right at the beginning of things, an originally fresh ocean accumulated salt in Halley's manner. But that process of initial increase ended long ago, and Halley's method cannot reach so far back into the abyss of time.

This usual presentation of Halley's crucial error is then balanced in traditional accounts by warm praise—for two main reasons. First, Halley wins kudos for making the first serious quantitative proposal to determine the earth's age. Moreover, though Halley felt that he could not apply the method himself, his suggestions were followed by later scientists, particularly toward the end of the nineteenth century by the great Irish geologist John Joly who used the accumulation of salt to propose a date of 100 million years for the earth. Although Joly's estimate was vastly too small—the earth is some 4.5 billion years old—his work represented a great advance on previous speculative traditions that had led to little but hot air and had rarely dared to imagine dates even so old as Joly's.

Secondly, Halley has been proclaimed a hero in the false view of history that sees light and truth locked in perpetual warfare with religion. Halley does begin his article by rejecting a literal interpretation of Genesis for the earth's age. He accepts, because scripture so states, that humans have lived on earth for some 6,000 years but denies a creation of all things just five days before:

> Whereas we are there told that the formation of man was the last act of the creator, 'tis no where revealed in scripture how long the earth had existed before this last creation, nor how long those five days that proceeded it may be to be accounted; since we are elsewhere told, that in respect of the almighty a thousand years is as one day, being equally no part of eternity; nor can it well be conceived how those days should be to be understood of natural days, since they are mentioned as measures of time before the creation of the sun, which was not till the fourth day.

But Halley writes these lines as a liberal theist, not as a scientist engaged in a conscious battle with religion. The word *scientist* didn't exist in Halley's day, and close ties between rational science and sensible religion were sought by most scholars engaged in work that we would now call scientific. Halley, in other words, speaks here for the liberal tradition of nonliteral interpretation.

The traditional literature usually unites these themes of praise in a single phrase: Halley ranks with the heroes of geology because his method, though flawed, does provide a *minimal* estimate of the earth's age (the time of accumulation before oceans reached steady-state). The *Discover* man-of-the-year article concludes: "His effort was useful because, in arriving at a minimum age for the earth, it inspired others to look for better geological clocks."

And so I have always read, and taught, Halley's argument for the earth's age—as a *minimal* estimate proposed to burst the bonds of a biblical literalism that made science impossible because ordinary causes could not produce the geological work required in only a few thousand years. And so I read it again last week—until I came to the last paragraph. Here Halley conveys one of the most important and subtle points of good scientific methodology—a lesson that ranks above all others in rules of procedure that must be purveyed to advanced students beginning their own careers in independent research.

In complex historical sciences like geology, few situations can be as well controlled as ideal laboratory experiments. Biases are unavoidably and intrinsically contained within available data. Since these biases cannot always be removed, researchers must follow one cardinal rule—they must be sure that recognized biases fall in a direction that will make confirmation of their hypothesis *less* likely (for if sources of bias tend to support favored views, how can you know whether a positive result records a preferred explanation or simply the inherent bias).

Halley begins the last paragraph by admitting a bias that he cannot correct:

> If it be objected that the water of the ocean, and perhaps some of these lakes, might at the first beginning of things, in some measure contain salt, so as to disturb the proportionality of the encrease [*sic*] of saltness in them, I will not dispute it . . .

How lovely, I said to myself. He is about to state the principle that bias must lie against a favored outcome. I must note this and point it out to students. I read on and was not disappointed. Halley makes the argument with fine precision (continuing from the quotation above):

. . . but shall observe that such a supposition would by so much contract the age of the world within the date to be derived from the foregoing argument. . . .

I smiled benignly, began to read on, and then experienced the moment of truth so like that classic scene of cartoon or comedy when the policeman sees a flying mouse or a walking snowman, smiles happily with a "how-nice, a-flying-mouse" look, then does a double take, drops whistle from mouth in astonishment, and says, "A flying mouse!" Wait a minute! Halley is supposed to be arguing for an age *longer* than tradition, an expansion of biblical limitations. But he actually says that a bias he can't remove will make the earth seem *older* than it really is, the very kind of bias—one favoring your hypothesis—that *must* be avoided. (If a bag contains one hundred beans and you observe that one new bean is added each day, you will assume, by analogy with Halley's argument about originally saltless oceans, that beans have been added for one hundred days. But if the bag started with twenty-five beans, the process will only be seventy-five days old, not one hundred as you estimated. In the same manner, if the oceans started with some salt, but you assume they began saltless, your age by Halley's method will be greater than the true age.)

I then became seriously puzzled. Does Halley have his methodology backward? What is going on? How can he be trying to expand the earth's age and then admit a bias that will make it appear older? Either Halley was a methodological dunce or something must be desperately wrong with the traditional view that he was trying to set a minimal age for the earth. In fact, Halley is telling us that he has set a *maximal* age, and he is clearly damned pleased with himself. I read further (again continuing the last quotation):

. . . the foregoing argument, which is chiefly presented to refute the ancient notion, some have of late entertained, of the eternity of all things.

So Halley thought he was doing the exact opposite of what all posterity has attributed to him. He contends that he was establishing a maximal age to refute a notion of eternity. We say that he was seeking a minimal age to expand the literal biblical chronology. Why—given the admirable clarity of Halley's words—have we so grievously misstated his intent?

Eternity is no longer an issue for us (in discussing the earth's duration). We all assume that our planet had a determinable beginning. Since we have not entertained this alternative for several centuries, we lack a context for grasping Halley's last paragraph. We read right through his words because they make no sense to us. We, as veterans of several creationist waves from Scopes to Arkansas, fully understand the threat to science of biblical literalism. We are therefore led to read Halley falsely in our light and see him as a fighter for expanded time rather than, as he insists, a measurer who would fix an actual date in order to eliminate the possibility of infinite existence.

Fortunately, I chose to dust off Halley's article while I was busy reading the protogeologies of late seventeenth-century British savants (primarily Thomas Burnet's *Sacred Theory of the Earth*) for another project (see my book *Time's Arrow, Time's Cycle*). I was therefore predisposed to read Halley as he intended.

We can grasp why eternity seemed an even greater danger than biblical literalism only when we understand Halley and his generation as struggling to find the basis for a science of historical events (both Burnet and Halley were friends of Newton and members of a scientific generation with common goals). Burnet, for example, rails on and on through 400 pages against the idea of eternity, and he fights this great battle, or so he says, because eternity precludes the possibility of meaningful history defined as a sequence of distinct and recognizable events linked by ordinary ties of cause and effect. Burnet identifies, as his main enemy:

> . . . this Aristotelian doctrine, that makes the present form of the earth to have been from eternity, for the truth is, this whole book is one continued argument against that opinion.

If the earth is eternal, then no event can be distinctive. All must occur again and again, and we fall into incomprehension, for our struggles and dreams lose any meaning as unique events in finite time. Eternity destroys history; it "takes away the subject of our discourse," as Burnet writes.

Jorge Luis Borges, in his uniquely exquisite way, expressed the incomprehensibility of infinity in *The Book of Sand*. In this story, Borges procures an infinite book. He cannot find its end, for no matter how furiously he turns the pages, as many remain between him and the back cover. The book contains small illustrations, spaced 2,000 pages apart. None is ever repeated, and Borges fills a notebook with their sequence, never coming any closer to a termination. Finally, he understands that this precious book is actually monstrous and obscene, and he loses it permanently on a shelf deep in the stacks of the Argentine National Library. He has understood the dilemma of eternity: "If space is infinite, we may be at any point in space. If time is infinite, we may be at any point in time."

Halley sought to disprove this most unthinkable of systems by resolving Borges's dilemma and setting a definite point—an actual age in years—for the earth's beginning. Halley, in short, was fighting for history. If we view him as a great historical scientist who advanced his proposal for dating the earth as a blow for rationality itself, then we can understand the last paragraph and its message for us.

Halley fights for history from both sides; his short article is both a specific proposal for measuring the earth's age and a beautifully crafted defense of historical science in general. He does argue against biblical literalism—to gain enough time so that ordinary causes may shape geological history. Without time in abundance, we will need miracles to cram such richness into just a few thousand years. The traditional reading of Halley stops here.

But Halley insists that his struggle for a comprehensible history proceeds primarily from the other end, by stealing time from eternity—for the bias of his method can only justify its employment against a claim for *greater* ages than he might measure. I think that we must take Halley at his word, for he was too astute a methodologist to misuse the primary criterion of bias. Halley believed that he had set a maximal age for our planet. Surely, the earth could not be eternal, for biases in Halley's method could only make it younger than his own measured maximum.

When we understand what Halley really sought, we can also grasp the reason for his failure. He was trying to establish a rational science of history. To do this, he needed a criterion that would change constantly in a recognizable way through time—so that each moment would be distinctly different from every other, thus avoiding Borges's dilemma. He thought that the accumulation of salt in oceans and lakes, linearly increasing through time, would provide such a criterion. The primary struggle of historical science ever since Halley has centered upon the search for phenomena that change constantly and therefore mark the passage of time. Halley knew exactly what history needed, but he chose the wrong criterion for interesting reasons.

Halley may have burst the bonds of biblical literalism, but he had no inkling whatever of time's true immensity. (The 100-million-year age so often attributed to him is Joly's nineteenth-century date using Halley's method. Halley himself never dared to think in more than thousands.) The clearest evidence for Halley's limited perspective, a notion shared by all contemporaries who tried to date the earth (see next essay), lies within his argument about salt, when he laments that ancient Romans and Greeks did not measure the salinity of oceans:

> It were to be wished that the ancient Greek and Latin authors had delivered down to us the degree of the saltness of the sea, as it was about 2,000 years ago; for then it cannot be doubted but that the difference between what is now found and what then was, would become very sensible.

If Halley had recognized how infinitesimally tiny a fraction of earth history these 2,000 years actually represented, he would not have been so confident that the increment between then and now could set a metric for determining the beginning itself. Two thousand years was an appreciable part of the tens of thousands that his wildest fancies could conceive.

An earth as young as Halley imagined might have provided criteria for history in such simple physical processes as the influx of salt from rivers. But, as I argued above, simple systems generally equilibrate or reach some completed state over truly great durations. Components of atmospheres and oceans reach equilibrium; they do not change steadily over billions of years. Unless we can find something truly big (fuel of a star) or numerous (number of atoms subject to radioactive decay) relative to time available, physical objects make poor criteria of history.

The best signs of history are objects so complex and so bound in webs of unpredictable contingency that no state, once lost, can ever arise again in precisely the same way. Life, through evolution, possesses this unrepeatable complexity more decisively than any other phenomenon on our planet. Scientists did not develop a geological time scale—the measuring rod of history—until they realized that fossils provided such a sequence of uniquely nonrepeating events.

When I began these essays in 1974, I chose for my general title a phrase from the last paragraph of Darwin's *Origin of Species*—"this view of life." I selected this passage because I love the science of history. Darwin used this phrase to contrast the richness of life's history with the timeless cycling of simpler physical systems, in particular planets in their orbits. Halley knew what a science of history required, but he could not grasp why simple systems did not provide good criteria because he dared not even imagine how old the earth might really be. Darwin sensed the scope of time and knew that only life's complexity could map its richness:

> There is grandeur in this view of life. . . . Whilst this planet has gone cycling on according to the fixed law of gravity, from so simple a beginning endless forms most beautiful and most wonderful have been, and are being, evolved.

Postscript

Although this is one of the few articles on Halley written during the season of his celebrated eponymous object, but not treating the subject of comets at all, I must nonetheless report briefly on my viewing experience, because I saw something so touching and learned something important thereby.

I had been waiting all my conscious life, and I wasn't going to miss it. Views from Boston were especially lousy (low on the horizon and therefore invisible from just about everywhere), and the comet was putting on a crummy show anyway (and anywhere). Further south meant higher in the sky and a better shot at seeing something. I had to visit the Smithsonian Tropical Research Station in Panama anyway, so I timed my trip for an optimal view. A staff scientist at the Institute picked me up at 3:00 A.M., as the best time for sighting came just before dawn. We drove far way from the city of Panama and set up a telescope in a dark field.

That night and morning provided two special pleasures. First, Halley's comet made quite a surprising impression on me. I had been told many times to expect nothing interesting or exciting, so I came prepared for disappointment (and only because a lifetime's promise to oneself cannot be easily canceled in the mind). I hoped for little, but saw quite a bit. The comet looked so different from everything else up there—from all the pinpoints of light that I know so well as a lifelong stargazer. The comet was faint, but fuzzy rather than concentrated. Everything else, however bright, was a point; Halley's comet formed a broad line subtending nearly ten degrees of celestial arc. If you know the night sky as a friend, something so different amidst all your buddies can be awesome, if smallish.

Second, I was moved far more by a human scene. As we drove back to the city of Panama along the causeway by the sea just as the sun began to rise, I noticed crowds of Panamanian people, ever denser the closer we got to town (for the majority had to walk, even though viewing improved with distance). Most were family groups. Fathers and mothers pointed to the sky, showing the comet to their young children telling them perhaps that they might live to see it again. As the dawn broke, these people appeared as silhouettes against the brightening sky—parents pointing upward towards their once-in-a-lifetime view. Hundreds of poor and carless citizens had bustled their children out long before dawn and walked away from the city lights to line the causeway with human curiosity. Who will dare to say that people do not have a sense of wonder, or do not care about science and nature, if a little fuzzy line, properly publicized, can make us all citizens of the universe, at least for a short morning?

12 | Fall in the House of Ussher

I AM UNCOMFORTABLE ENOUGH in a standard four-in-hand tie; pity the poor seventeenth-century businessmen and divines, so often depicted in their constraining neck ruffs. The formidable gentleman in the accompanying engraving commands the Latin title *Jacobus Usserius, Archiepiscopus Armachanus, Totius Hiberniae Primas,* or James Ussher, Archbishop of Armagh, and Primate of All Ireland. He is known to us today almost entirely in ridicule—as the man who fixed the time of creation at 4004 B.C., and even had the audacity to name the date and hour: October 23 at midday.

Let me begin with a personal gloss on the caption to this engraving, for my misreading embodies, in microcosm, the entire theme of this essay. I confess that I have always been greatly amused by the term *primate,* used in its ecclesiastical sense as "an archbishop . . . holding the first place among the bishops of a province." My merriment must be shared by all zoologists, for primates, to us, are monkeys and apes—members of the order Primates. Thus, when I see a man described as a "primate," I can't help thinking of a big gorilla. (Humans, of course, are also members of the order Primates, but zoologists, in using the term, almost always refer to nearly 200 other species of the group—that is, to lemurs, monkeys, and apes.)

But my amusement must be labeled as silly, parochial, and misguided. The title comes from the Latin *primas,* meaning "chief" or "first." In the mid-eighteenth century, Linnaeus introduced the word to zoology as a designation for the "highest" order of mammals—the group including humans. But the ecclesiastical

JACOBUS USSERIUS, ARCHIEPISCOPUS ARMACHANUS, TOTIUS HIBERNIÆ PRIMAS

usage has an equally obvious claim to proper etymology and sub-stantial precedence in usage (the *Oxford English Dictionary* traces this meaning to 1205). Thus, we zoologists are the usurpers, not the guardians of a standard. (I wonder if preachers laugh when they see the term in a zoological book and think of a baboon running about in a neck ruff.) In any case, the archbishop of Ar-magh is titular head, hence primate, of the Anglo-Irish church, just as the archbishop of Canterbury is primate of all England.

This little tale mimics the forthcoming essay in miniature for two reasons:

1. I shall be defending Ussher's chronology as an honorable effort for its time and arguing that our usual ridicule only records a lamentable small-mindedness based on mistaken use of present criteria to judge a distant and different past—just as our current amusement in picturing a primate of the church as a garbed ape inverts the history of usage, for the zoological definition is derivative and the ecclesiastical primary.

2. The mental picture of a prelate as a garbed ape reinforces the worst parochialism that scientists often invoke in interpreting their history—the notion that progress in knowledge arises from victory in battle between science and religion, with religion defined as unthinking allegiance to dogma and obedience to authority, and science as objective searching for truth.

James Ussher (1581–1656) lived through the most turbulent of English centuries. He was born in the midst of Elizabeth's reign and died under Cromwell (who gave him a state funeral in Westminster Abbey, despite Ussher's royalist sentiments and his previous support for the executed Charles I). As a precocious scholar with a special aptitude for languages, Ussher entered Trinity College, Dublin, at its founding in 1594, when he was only thirteen years old. He was ordained a priest in 1601 and became a professor at Trinity (1607) and then vice chancellor on two occasions in 1614 and 1617. With his appointment as Archbishop of Armagh in 1625, he became head (or primate) of the Anglo-Irish church—a tough row to hoe in this preeminently Catholic land ("Romish" or "papist" as Ussher always said in the standard deprecations of his day). Ussher was vehement and unrelenting in his verbal assaults on Roman Catholicism (he wasn't too keen on Jews and other "infidels" either, but the issue rarely came up). His 1626 "Judgement of the Arch-Bishops and Bishops of Ireland" begins, for example:

> The religion of the papists is superstitious and idolatrous; their faith and doctrine erroneous and heretical; their church . . . apostatical; to give them therefore a toleration, or to consent that they may freely exercise their religion . . . is a grievous sin.

One may cringe at the words (and no one can take Ussher as a model of toleration), but he was, in fact, regarded as a force for moderation and compromise at a time of fierce invective (read Milton's anti-Catholic pamphlets sometime if you want to get a feel for the rhetoric of those troubled years). Despite his opinions, Ussher continued to espouse debate, discussion, and negotiation. He preached to Catholics and delighted in meeting their champions in formal disputations. His own words were harsh, but he believed in triumph by force of argument, not by banishment, fines, imprisonment, and executions. In fact, even the hagiographical biographies, written soon after Ussher's death, criticize him for lack of enthusiasm in the daily politics of ecclesiastical affairs and for general unwillingness to carry out policies of intolerance. He was a scholar by temperament and, at best, a desultory administrator. He was in England at the outbreak of the civil war in 1642 and never returned again to Ireland. He spent most of his last decade engaged in study and publication—including, in 1650, the source of his current infamy: *Annales veteris testamenti, a prima mundi origine deducti,* "Annals of the Old Testament, deduced from the first origin of the world."

Ussher became the symbol of ancient and benighted authoritarianism for a reason quite beyond his own intention. Starting about fifty years after his death, most editions of the "authorized," or King James, translation of the Bible began to carry his chronology in the thin column of annotations and cross-references usually placed between the two columns of text on each page. (The Gideon Society persisted in placing this edition in nearly every hotel room in America until about fifteen years ago; they now use a more modern translation and have omitted the column of annotations, including the chronology.) There, emblazoned on the first page of Genesis, stands the telltale date: 4004 B.C. Ussher's chronology therefore acquired an almost canonical status in English Bibles—hence his current infamy as a symbol of fundamentalism.

To this day, one can scarcely find a textbook in introductory geology that does not take a swipe at Ussher's date as the opening comment in an obligatory page or two on older concepts of the earth's age (before radioactive dating allowed us to get it right). Other worthies are praised for good tries in a scientific spirit (even if their ages are way off—see previous essay on Halley), but Ussher is excoriated for biblical idolatry and just plain foolishness. How could anyone look at a hill, a lake, or a rock pile and not know that the earth must be ancient?

One text discusses Ussher under the heading "Rule of Authority" and later proposals under "Advent of the Scientific Method." We learn—although the statement is absolute nonsense—that Ussher's "date of 4004 B.C. came to be venerated as much as the sacred text itself." Another text places Ussher under "Early Speculation" and later writes under "Scientific Approach." These authors tell us that Ussher's date of 4004 B.C. "thus was incorporated into the dogma of the Christian Church" (an odd comment, given the tradition of Catholics, and of many Protestants as well, for allegorical interpretation of the "days" of Genesis). They continue: "For more than a century thereafter it was considered heretical to assume more than 6,000 years for the formation of the earth."

Even the verbs used to describe Ussher's efforts reek with disdain. In one text, Ussher "pronounced" his date; in a second, he "decreed" it; in a third, he "announced with great certainty that . . . the world had been created in the year 4004 B.C. on the 26th of October at nine o'clock in the morning!" (Ussher actually said October 23 at noon—but I found three texts with the same error of October 26 at nine, so they must be copying from each other.) This third text then continues: "Ussher's judgment of the age of the earth was gospel for fully 200 years."

Many statements drip with satire. Yet another textbook—and this makes six, so I am not merely taking potshots at rare silliness—regards Ussher's work as a direct "reaction against the scientific explorations of the Renaissance." We then hear about "the pronouncement by Archbishop Ussher of Ireland in 1664 that the Earth was created at 9:00 A.M., October 26, 4004 B.C. (presumably Greenwich mean time!)" Well, Ussher was then eight years dead, and his date for the earth's origin is again misreported. (I'll pass on the feeble joke about Greenwich time, except to note that such issues hardly arose in an age before rapid travel made the times of different places a matter of importance.)

Needless to say, in combating the illiberality of this textbook tradition, I will not defend the substance of Ussher's conclusion—for one claim of the standard critique is undeniably justified: A 6,000-year-old earth did make a scientific geology impossible because any attempt to cram the empirical record of miles of strata and life's elaborate fossil history into such a moment requires a belief in miracles as causal agents.

Fair enough, but what sense can be made of blaming one age for impeding a much later system that worked by entirely different principles? To accuse Ussher of delaying the establishment of an empirical geology is much like blaming dinosaurs for holding back the later success of mammals. The proper criterion must be worthiness by honorable standards of one's own time. By this correct judgment, Ussher wins our respect just as dinosaurs now seem admirable and interesting in their own right (and not as imperfect harbingers of superior mammals in the inexorable progress of life). Models of inevitable progress, whether for the panorama of life or the history of ideas, are the enemy of sympathetic understanding, for they excoriate the past merely for being old (and therefore both primitive and benighted).

Of course Ussher could hardly have been more wrong about 4004 B.C., but his work was both honorable and interesting—therefore instructive for us today—for at least four reasons.

1. The excoriating textbook tradition depicts Ussher as a single misguided dose of darkness and dogma thrown into an otherwise more enlightened pot of knowledge—as if he alone, representing the church in an explicit rearguard action against science and scholarship, raised the issue of the earth's age to recapture lost ground. No idea about the state of chronological thinking in the seventeenth century could be more false.

Ussher represented a major style of scholarship in his time (see previous essay on Halley for discussion of another contemporary style—one more congenial to our current views, but no more popular than Ussher's mode in a seventeenth-century context). Ussher worked within a substantial tradition of research, a large community of intellectuals striving toward a common goal under an accepted methodology—Ussher's shared "house" if you will pardon my irresistible title pun. Today we rightly reject a cardinal premise of that methodology—belief in biblical inerrancy—and we recognize that this false assumption allowed such a great error in estimating the age of the earth. But what intellectual phenomenon can be older, or more oft repeated, than the story of a large research program that impaled itself upon a false central assumption accepted by all practitioners? Do we regard all people who worked within such traditions as dishonorable fools? What of the scientists who assumed that continents were stable, that the hereditary material was protein, or that all other galaxies lay within the Milky Way? These false and abandoned efforts were pursued with passion by brilliant and honorable scientists. How many current efforts, now commanding millions of research dollars and the full attention of many of our best scientists, will later be exposed as full failures based on false premises?

The textbook writers do not know that attempts to establish a full chronology for all human history (not only to date the creation as a starting point) represented a major effort in seventeenth-century thought. These studies did not slavishly use the Bible, but tried to coordinate the records of all peoples. Moreover, the assumption of biblical inerrancy doesn't provide an immediate and dogmatic answer—for many alternative readings and texts of the Bible exist, and scholars must struggle to a basis for choice among them. As a primary example, different datings for key events are given in the Septuagint (or Greek Bible, first translated by the Jewish community of Egypt in the third to second centuries B.C. and still used by the Eastern churches) and in the standard Hebrew Bible favored by the Western churches.

Moreover, within shared assumptions of the methodology, this research tradition had considerable success. Even the extreme values were not very discordant—ranging from a minimum, for the creation of the earth, of 3761 B.C. in the Jewish calendar (still in use) to a maximum of just over 5500 B.C. for the Septuagint. Most calculators had reached a figure very close to Ussher's 4004. The Venerable Bede had estimated 3952 B.C. several centuries before, while J. J. Scaliger, the greatest scholar of the generation just before Ussher, had placed creation at 3950 B.C. Thus, Ussher's 4004 was neither idiosyncratic nor at all unusual; it was, in fact, a fairly conventional estimate developed within a large and active community of scholars. The textbook tradition of Ussher's unique benightedness arises from ignorance of this world, for only Ussher's name survived in the marginal annotations of modern Bibles.

2. The textbook detractors assume that Ussher's effort involved little more than adding up ages and dates given directly in the Old Testament—thus implying that his work was only an accountant's act of simple, thoughtless piety. Another text-

book—we are now up to seven—states that Ussher's 4004 was "a date reconstructed from adding up the ages of people named in the lineages of the scripture." But even a cursory look at the Bible clearly shows that no such easy solution is available, even under the assumption of inerrancy. You can add the early times, from creation up to the reign of Solomon—for the requisite information is provided by an unbroken male lineage supplying the key datum of father's age at the birth of a first son. But this easy route cannot be carried forward into the several hundred years of the kingdom, from Solomon's reign to the destruction of the Temple and the Babylonian captivity—for here we are only given the lengths of rule for kings, and several frustrating ambiguities (including overlaps or co-regencies of a king and his successor) were widely acknowledged but not easily resolved. Finally, how can you use the Old Testament to reach the crucial birthday of Christ and thus connect the older narrative to the present? For the Old Testament stops in the period of Ezra and Nehemiah, the fifth century B.C. in Ussher's chronology.

James Barr explains the problems and complexities in an excellent article, "Why the World Was Created in 4004 B.C.: Archbishop Ussher and Biblical Chronology" (see bibliography). He divides the chronological enterprise into three periods, each with characteristic problems, as mentioned above. You can add up during the first period (creation to Solomon), but which text do you use? The ages in the Septuagint* are substantially longer and add more than 1,000 years to the date of creation. Ussher solved this dilemma by using the Hebrew Bible and ignoring the alternatives.

In the second period, you really have to struggle to establish a coherent time line through the period of the kings. You feint and shift, try to correlate the dates given for the two kingdoms of Israel and Judah, then attempt to link in the few ages given for events other than beginnings and ends of reigns. The result, with

---

*The name Septuagint derives from the legend that seventy-two translators (close to the Latin *septuaginta,* or "seventy")—six from each of the twelve tribes of Israel—worked in separate rooms and made their own translations. When they compared their results, all were identical. If the linguistic mishmash seems odd— Jews translating The Hebrew Bible into Greek in Egypt—remember that Alexander the Great conquered Egypt and established the Ptolemies as a Greek ruling family.

luck and adjustment, is a coherent network of mutually supporting times.

For the third period of more than 400 years from Ezra and Nehemiah to the birth of Jesus you cannot use the Bible at all—for no information exists. Ussher and all other chronologists therefore tried to link a known event in the period of kings with a datable episode in another culture—and then to use the timetables of other peoples until another lateral feint could be made back into the New Testament. Ussher proceeded by correlating the death of the Chaldean king Nebuchadnezzar II with the thirty-seventh year of the exile of Jehoiachin (as stated in 2 Kings 25:27). (Nebuchadnezzar was, of course, prominent in Jewish history for conquering Jerusalem in 586 B.C. and deporting its prominent citizens—the so-called Babylonian captivity.) Ussher could then calculate through the Chaldean and the subsequent Persian records, eventually reaching the period of Roman rule and the birth of Jesus.

3. But where did Ussher get October 23, 4004? Surely, neither the Bible nor any other source gives a specific date, even if you can estimate the year. Was this date, at least, a bow to dogma, even if the rest of the chronology has more scholarly roots?

No, not dogma, but a different style of interpretive argument—one based on symbol and eschatology rather than listed chronology. (This style cannot be labeled as dogma, if only because each point became a subject of lively disagreement and fierce debate among scholars. No resolution was ever obtained, so the church obviously imposed no answer ex cathedra.)

First of all, the date 4004 coordinates comfortably with the most important of chronological metaphors—the common comparison of the six days of God's creation with 6,000 years for the earth's potential duration: "But, beloved, be not ignorant of this one thing, that one day is with the Lord as a thousand years, and a thousand years as one day" (2 Peter 3:8). Under this widely accepted scheme, the earth was created 4,000 years before the birth of Christ and could endure as much as 2,000 years thereafter (a proposition soon to be tested empirically and, we all hope, roundly disproved!).

But why 4004 and not an even 4000 B.C.? By Ussher's time, chronologists had established an error in the B.C. to A.D. transition, for Herod died in 4 B.C.—and if he truly talked to the Magi, feared the star, and ordered the slaying of the innocents, then Jesus could not have been born after 4 B.C. (an oxymoronic statement, but acceptable as a testimony to increasing knowledge).

Thus, if Jesus was born in 4 B.C., eschatological tradition should fix the date of creation at 4004 B.C., without any need for complex, sequential calculation of genealogies. This situation must inspire a nasty suspicion that Ussher "knew" the necessity of 4004 B.C. right from the start and then jiggered the figures around to make everything come out right. Barr, of course, considers this possibility seriously but rejects it for two reasons. First, Ussher's chronology extends out to several volumes and 2,000 pages of text and seems carefully done, without substantial special pleading. Second, the death of Herod in 4 B.C. doesn't establish the birth of Jesus in the same year. Herod became king of Judea (Roman puppet would be more accurate) in 37 B.C.—and Jesus might have been born at other times in this thirty-three-year interval. Moreover, other traditions argued that the 4,000 years would run from creation to Christ's crucifixion, not to his birth—thus extending the possibilities to A.D. 33. By these flexibilities, creation could have been anywhere between 4037 B.C. (4,000 years to the beginning of Herod's reign) and 3967 B.C. (4,000 years to the Crucifixion). Four thousand four is in the right range, but certainly not ordained by symbolic tradition. You still have to calculate.

But what about October 23? Here, chronology cannot help. Many scholars, from the Venerable Bede to the great astronomer Johannes Kepler, argued for spring as an appropriate season for birth and the chosen time of Babylonian, Chaldean, and other ancient chronologies. Others, including Jerome, Josephus, and Ussher, favored fall, largely because the Jewish year began then, and Hebrew scriptures formed the basis of chronology.

Now an additional problem must be faced. The Jewish chronology is based on lunar months and therefore very hard to correlate with a standard solar calendar. Ussher, recognizing no basis for a firm calibration, therefore decided to establish creation as the first Sunday following the autumnal equinox. (Sunday was an obvious choice, for God created in six days and rested on the seventh, and the Jewish Sabbath falls on Saturday.)

But if creation occurred near the autumnal equinox, why October 23, more than a month from the current date? For this final piece of the puzzle, we need only recognize that Ussher was still using the old Julian (Roman) calendar. The Julian system was very similar to our own, but for one apparently tiny difference—it did not suppress leap years at the century boundaries. (Not everyone knows that our present system—which keeps more accurate time than the Julian—omits leap years at all century transitions not divisible by 400. Thus, 1700, 1800, and 1900 were not leap years, but 1600 was and 2000 will be.) This difference seems tiny, but errors accumulate over millennia. By 1582, the discrepancy had become sufficiently serious that Pope Gregory XIII proclaimed a reform and established the system that we still live by— called, in his honor, the Gregorian calendar. He dropped the ten days that had accumulated from the "extra" leap years at century boundaries in the Julian system (this was done by the clever device of allowing Friday, October 15, to follow Thursday, October 4, in 1582).

We now enter the religious tensions of the time. Recall Ussher's fulminations against popery, an attitude shared by his Anglican brethren in charge. The Gregorian reform smelled like a Romish plot, and Ussher's contemporaries would be damned if they would accept it. (England and the American colonies finally succumbed to rationality and instituted the Gregorian reform in 1752. This delay, by the way, is responsible for the ambiguity in George Washington's birth, sometimes given as February 11 and sometimes as February 22, 1732. He was born under the Julian calendar, and eleven days, rather than ten, had to be dropped by this later time.) In any case, if the Julian discrepancy accounted for ten extra days in the 1,600 or so years between its institution and the Gregorian reform, Ussher realized that the disparity would amount to just over thirty days for the additional time from 4004 B.C.—thus fixing the creation at October 23, rather than about two-thirds through September, as by our present calendar.

One final point. Why high noon on the day of creation? The inception of Genesis reads:

> In the beginning God created the heaven and the earth. And the earth was without form, and void; and darkness was upon the face of the deep. And the spirit of God moved upon the face of the waters. And God said, Let there be light. . . .

Now you cannot have days without alternations of light and darkness, so Ussher began chronology with the creation of light, which he fixed, for no given reason, at high noon. He wrote, "In ipse primi diei medio creata est lux" (In the middle of the first day, light was created).

But what about the phrases in Genesis that precede the creation of light? Here we encounter an old exegetical problem: Does the text present an epitome of the whole process in these lines, or does it say that God made matter before creating light? Ussher accepted the latter reading and argued that a creation of matter "without form and void" took place during the night before the creation of light. Thus, a precreation, a slipping of material into place, occurred on the night of October 22—yielding several "temporary hours" (Ussher's words) before the overt creation of light on October 23.

4. Ussher's chronology is a work within the generous and liberal tradition of humanistic scholarship, not a restrictive document written to impose authority. As Barr notes, Ussher's *Annales* presents a chronology for all human history (meaning Western history, for he knew no other well enough), from the creation—and you must remember that humans were made five days thereafter, so earthly history is, essentially, human history—to the fall of Jerusalem in A.D. 70. Barr writes:

> It is a great mistake, therefore, to suppose that Ussher was simply concerned with working out the date of creation: this can be supposed only by those who have never looked into its pages. . . . The *Annales* are an attempt at a comprehensive chronological synthesis of all known historical knowledge, biblical and classical. . . . Of its volume only perhaps one sixth or less is biblical material.

Socrates told us to know ourselves, and no datum can be more important for humanism than an accurate chronology serving as a framework for the epic of our cultures, our strivings, our failures, and our hopes.

The figure of Ussher that begins this article comes from the only work of his that I own—a comprehensive catechism prepared for children and their families, entitled *A body of divinity: or, the sum and substance of Christian religion.* Catechisms may simplify, but they have the virtue of laying basic belief right on the line, without the hemming and hedging so intrinsic to academic texts.

I was delighted by Ussher's defense of his chronology in this catechism—simple words that illustrate the basic humanism of his enterprise. How do we know about creation? he asks—and responds: "Not only by the plain and manifold testimonies of Holy Scripture, but also by light of reason well directed." His main quarrel, we note, is not with other timings of the human epic, but with Aristotle's ahistorical notion of eternity (see previous essay for discussion of Halley's similar primary concern). "What say you then to Aristotle, accounted of so many the Prince of Philosophers; who laboreth to prove that the world is eternal." Ussher answers his own question by defending God's majesty against a mere unmoved mover of eternal matter, for Aristotle "spoileth God of the glory of his Creation, but also assigneth him to no higher office than is the moving of the spheres, whereunto he bindeth him more like to a servant than a lord."

I close with a final plea for judging people by their own criteria, not by later standards that they couldn't possibly know or assess. We castigate Ussher for making the creation so short—a mere six days, where we reckon billions for evolution. But Ussher fears that six days might seem too long in the opinion of his contemporaries, for why should God, who could do all in an instant, so spread out his work? "Why was he creating so long, seeing he could have perfected all the creatures at once and in a moment?" Ussher gives a list of answers, but one caught my attention both for its charm and for its incisive statement about the need for sequential order in teaching—as good a rationale as one could ever devise for working out a chronology in the first place! "To teach us the better to understand their workmanship; even as a man which will teach a child in the frame of a letter, will first teach him one line of the letter, and not the whole letter together."

4 | Musings

Clouds of Memory

13 | Muller Bros.
Moving & Storage

I OWN MANY OLD and beautiful books, classics of natural history bound in leather and illustrated with hand-colored plates. But no item in my collection comes close in personal value to a modest volume, bound in gray cloth and published in 1892: *Studies of English Grammar,* by J. M. Greenwood, Superintendent of Schools in Kansas City. The book belonged to my grandfather, a Hungarian immigrant. He wrote on the title page, in an elegant European hand: "Prop. of Joseph A. Rosenberg, New York." Just underneath, he added in pencil the most eloquent of all possible lines: "I have landed. Sept. 11, 1901."

Papa Joe died when I was thirteen, before I could properly distill his deepest experiences, but long enough into my own ontogeny for the precious gifts of extensive memory and lasting influence. He was a man of great artistic sensibility and limited opportunity for expression. I am told that he sang beautifully as a young man, though increasing deafness and a pledge to the memory of his mother (never to sing again after her death) stilled his voice long before my birth. He never used his remarkable talent for drawing in any effort of fine arts, though he marshaled these skills to rise from cloth-cutting in the sweatshops to middle-class life as a brassiere and corset designer. (The content of his chosen expression titillated me as a child, but I now appreciate the primary theme of economic emancipation through the practical application of artistic talent). Yet, above all, he expressed his artistic sensibilities in his personal bearing—in elegance of dress (a bit on the foppish side, perhaps), grace of movement, beauty of handwriting, ease of mannerism.

Sadly, I have no snapshot of Papa Joe and me. But here he is with my cousin Adele, impeccably dressed as always.

I well remember one manifestation of this rise above the ordinary, both because we repeated the act every week and because the junction of locale and action seemed so incongruous, even to a small child of five or six. Every Sunday morning, Papa Joe and I would take a stroll to the corner store on Queens Boulevard to buy the paper and a half-dozen bagels. We then walked to the great world-class tennis stadium of Forest Hills, where McEnroe and his ilk still cavort. A decrepit and disused side entrance sported a rusty staircase of three or four steps. With his unfailing deftness, Papa Joe would take a section of the paper that we never read and neatly spread several sheets over the lowermost step (for the thought of a rust flake or speck of dust in contact with his trousers filled him with horror). We would then sit down and have the most wonderful man-to-man talk about the latest base-ball scores, the rules of poker, or the results of the Friday night fights.

I retain a beautiful vision of this scene: The camera pans back and we see a tiny staircase, increasingly dwarfed by the great stadium. Two little figures sit on the bottom step—a well-dressed elderly man gesturing earnestly, a little boy listening with adoration.

Certainty is both a blessing and a danger. Certainty provides warmth, solace, security, an anchor in the unambiguously factual events of personal observation and experience. I know that I sat on those steps with my grandfather because I was there, and no external power of suggestion has ever played havoc with this most deeply personal and private experience.

But certainty is also a great danger, given the notorious fallibility—and unrivaled power—of the human mind. How often have we killed on vast scales for the "certainties" of nationhood and religion? How often have we condemned the innocent because the most prestigious form of supposed certainty—eyewitness testimony—bears all the flaws of our ordinary fallibility?

Primates are visual animals par excellence, and we therefore grant special status to personal observation, to being there and seeing directly. But all sights must be registered in the brain and stored somehow in its intricate memory. And the human mind is both the greatest marvel of nature and the most perverse of all tricksters: Einstein and Loge inextricably combined.

This special (but unwarranted) prestige accorded to direct observation has led to a serious popular misunderstanding about science. Since science is often regarded as the most objective and truth-directed of human enterprises, and since direct observation is supposed to be the favored route to factuality, many people equate respectable science with visual scrutiny—just the facts ma'am, and palpably before my eyes. But science is a battery of observational and inferential methods, all directed to the testing of propositions that can, in principle, be definitely proven false. A restriction of compass to matters of direct observation would stymie the profession intolerably. Science must often transcend sight to win insight. At all scales, from smallest to largest, quickest to slowest, many well-documented conclusions of science lie beyond the strictly limited domain of direct observation. No one

has ever seen an electron or a black hole, the events of a picosec-
ond or a geological eon.

One of the phoniest arguments raised for rhetorical effect by
"creation scientists" tried to deny scientific status to evolution
because its results take so much time to unfold and therefore
can't be seen directly. But if science required such immediate
vision, we could draw no conclusions about any subject that studies the past—no geology, no cosmology, no human history (including the strength and influence of religion) for that matter.
We can, after all, be reasonably sure that Henry V prevailed at
Agincourt even though no photos exist and no one has survived
more than five hundred years to tell the tale. And dinosaurs really
did snuff it tens of millions of years before any conscious ob-
server inhabited our planet. Evolution suffers no special infirmity
as a science because its grandest events took so long to unfold
during an unobservable past.

Moreover, eyewitness accounts do not deserve their conventional status as ultimate arbiters even when testimony of direct
observation can be marshaled in abundance. In her sobering
book, *Eyewitness Testimony* (1979), Elizabeth Loftus debunks,
largely in a legal context, the notion that visual observation confers some special claim for veracity. She identifies three levels of
potential error in supposedly direct and objective vision: misperception of the event itself, and the two great tricksters of passage
through memory before later disgorgement—retention and re-
trieval.

In one experiment, for example, Loftus showed 40 students a
three-minute videotape of a classroom lecture disrupted by 8
demonstrators (a relevant subject for a study from the early
1970s!). She gave the students a questionnaire and asked half of
them, "Was the leader of the 12 demonstrators . . . a male?"; and
the other half, "Was the leader of the 4 demonstrators . . . a
male?" One week later, in a follow-up questionnaire, she asked
all the students, "How many demonstrators did you see entering
the classroom?" Those who had previously received the question
about 12 demonstrators reported seeing an average of 8.9 people; those told of 4 demonstrators claimed an average of 6.4. All
had actually seen 8, but formed a later judgment as a compromise
between their actual observation and the largely subliminal
power of suggestion in the first questionnaire.

People can even be induced to "see" totally illusory objects. In another experiment, Loftus showed a film of an accident, followed by a misleading question: "How fast was the white sports car going when it passed the barn while traveling along the country road?" (The film showed no barn, and a control group received a more accurate question: "How fast was the white sports car going while traveling along the country road?") A week later, 17 percent of students in the first group stated that they had seen the nonexistent barn; only 3 percent of the controls reported a barn.

Thus, we are easily fooled on all fronts of both eye and mind: seeing, storing, and recalling. The eye tricks us badly enough; the mind is even more perverse. What remedy can we possibly suggest but constant humility, and eternal vigilance and scrutiny? Trust your memory as you would your poker buddy (one of my grandfather's mottos from the steps).

With this principle in mind, I went searching for those steps last year after more than thirty years of absence from my natal turf. I exited the subway at 67th Avenue, walked to my first apartment at 98-50, and then set off on my grandfather's route for Queens Boulevard and the tennis stadium.

I was walking in the right direction, but soon realized that I had made a serious mistake. The tennis stadium stood at least a mile down the road, too far for those short strolls with a bag of bagels in one hand and a five-year-old boy attached to the other. In increasing puzzlement, I walked down the street and, at the very next corner, saw the steps and felt the jolt and flood of memory that drives our *recherches des temps perdus*.

My recall of the steps was entirely accurate—three modest flagstone rungs, bordered by rusty iron railings. But the steps are not attached to the tennis stadium; they form the side entrance to a modest brick building, now crumbling, padlocked, and abandoned, but still announcing its former use with a commercial sign, painted directly on the brick in the old industrial style— "Muller Bros. Moving & Storage"—with a telephone number below from the age before all-digit dialing: ILlinois 9-9200.

Obviously, I had conflated the most prominent symbol of my old neighborhood, the tennis stadium, with an important personal place, and had constructed a juxtaposed hybrid for my mental image. Yet my memory of the tennis stadium soaring

The side wall of Muller Bros. as it appears today with its painting in the old industrial style. *Photograph by Eleanor Gould.*

above the steps remains strong, even now in the face of conclusive correction.

I might ask indulgence on the grounds of inexperience and relative youth, for my failure as an eyewitness at the Muller Bros. steps. After all, I was only an impressionable lad of five or so, when even a modest six-story warehouse might be perceived as big enough to conflate with something truly important.

But I have no excuses for a second story. Ten years later, at a trustable age of fifteen, my family made a western trip by automobile: I have specially vivid memories of an observation at Devil's Tower, Wyoming (the volcanic plug made most famous as a landing site for aliens in *Close Encounters of the Third Kind*). We approach from the east. My father tells us to look for the tower from tens of miles away, for he has read in a guidebook that it rises, with an awesome near-verticality, from the dead-flat Great Plains, and that pioneer families used the tower as a landmark and bea-

con on their westward trek. We see the tower, first as a tiny projection, almost square in outline, at the horizon. It gets larger and
larger as we approach, assuming its distinctive form and finally
revealing its structure as a conjoined mat of hexagonal basalt
columns. I have never forgotten the two features that inspired my
rapt attention: the maximal rise of verticality from flatness, forming a perpendicular junction, and the steady increase in size from
a bump on the horizon to a looming, almost fearful giant of a rock
pile.

Now I know, I absolutely *know* that I saw this visual drama, as
described. The picture in my mind of that distinctive profile,
growing in size, is as strong as any memory I possess. I *see* the
tower as a little dot in the distance, as a midsized monument, as a
full field of view. I have told the story to scores of people, comparing this natural reality with a sight of Chartres as a tiny toy
tower twenty miles from Paris, growing to the overarching symbol and skyline of its medieval city.

In 1987, I revisited Devil's Tower with my family—the only
return since my first close encounter thirty years before. I
planned the trip to approach from the east, so that they would see
the awesome effect—and I told them my story, of course.

In the context of this essay, my dénouement will be anticlimac-
tic in its predictability, however acute my personal embarrassment. The terrain around Devil's Tower is mountainous; the
monument cannot be seen from more than a few miles away in
any direction. I bought a booklet on pioneer trails westward, and
none passed anywhere near Devil's Tower. We enjoyed our visit,
but I felt like a perfect fool. Later, I checked my old log book for
that high school trip. The monument that rises from the plain,
the beacon of the pioneers, is Scottsbluff, Nebraska—not nearly
so impressive a pile of stone as Devil's Tower.

And yet I still *see* Devil's Tower in my mind when I think of that
growing dot on the horizon. I see it as clearly and as surely as
ever, though I now know that the memory is false.

This has been a long story for a simple moral. Papa Joe, the
wise old peasant in a natty and elegant business suit, told me on
those steps to be wary of all blandishments and to trust nothing
that cannot be proved. We must extend his good council to our
own interior certainties, particularly those that we never question
because we regard eyewitnessing as paramount in veracity.

Yours truly on the fateful steps. *Photograph by Eleanor Gould.*

Of course we must treat the human mind with respect, for nature has fashioned no more admirable instrument. But we must also struggle to stand back and to scrutinize our own mental certainties. This last line poses an obvious paradox, if not an outright contradiction, and I have no resolution to offer. Yes, step back and scrutinize your own mind. But with what?

14 | Shoemaker and Morning Star

KOKO, the obsequious tailor promoted to public executioner in Gilbert and Sullivan's *Mikado*, maintains "a little list of society offenders who might well be underground"—and he means dead and buried, not romantically in hiding. He places into the lengthy ledger of those "who never would be missed," a variety of miscreants, including nearly all lawyers and politicians, and even, in a bow to his Victorian prejudices (the true setting beneath the Japanese exterior), "that singular anomaly, the lady novelist." But the most deserving character in Koko's compendium, for he haunts all times and places, is "the idiot who praises, with enthusiastic tone, all centuries but this, and every country but his own."

Admittedly, we do live in a conceptual trough that encourages such yearning for unknown and romanticized greener pastures of other times. The future doesn't seem promising, if only because we can extrapolate some disquieting present trends into further deterioration: pollution, nationalism, environmental destruction, and aluminum bats. Therefore, we tend to take refuge in a rose-colored past—lemonade and cookies in a rocking chair on the porch of a warm summer's evening. (I actually participated in all these lovely anachronisms, after a lecture last year, in the intellectually dynamic but architecturally frozen Victorian village of Chautauqua, and I was thoroughly charmed until I remembered that, at the actual time recaptured à la Rockwell, my ancestors worked in sweatshops and lived in tenements, while all black people in town probably dwelled in shacks, literally on the other side of the railroad tracks.)

Foolish romanticism about the past feeds on our selective memory, our fundamental human ability—the only rescue from madness in this world of substantial woe—to discard the ugly and reconstruct our former lives and surroundings to our liking. I do not doubt the salutary, even the essential, properties of this curiously adaptive human trait, but we must also record the down side. Legends of past golden ages become impediments when we try to negotiate our current dilemmas. Blubbery nostalgia clouds any hope for rational understanding. (I don't remember the fifties as a wonderfully pleasant and carefree time, and nostalgia for that particular decade of McCarthyism and the cold war seems strongest in young people who weren't even alive then!) Mythology about a happy and simpler past also presents a seemingly limitless arena for commercial exploitation.

I encountered an interesting, though minimally disturbing, example of this commercial side during a recent family visit last month to the Amana Colonies in Iowa. The seven Amana villages were founded in the mid-nineteenth century by German pietists, members of the Society of True Inspirationists who, like so many religious minorities, left a scene of Old World persecution for a new life in America. They first settled near Buffalo, New York, and then, in 1855, moved to Iowa, spurred westward by cheap, abundant, and fertile land.

Utopian communities in America had variable success; few lasted for very long, and those that survived usually held their membership tight (and unrebellious) by strong, shared religious bonds. Amana was a truly communistic society; members ate in communal kitchens and used no money (common foodstuffs could be taken from supply bins "according to need," while the colonies issued scrip for purchase of rarer items in company stores). They endured in this admirable and decidedly un-American fashion until 1932, when a variety of inevitabilities, from economic woes of the depression to a "youth revolt" spurred by desire for personal ownership of standard consumer goods, provoked what residents still call the "Great Change." Amana split its major affairs of church and economy (the former has been declining ever since, the latter booming)—and the work of the fields, shops and industries transmogrified, in the good old American way, into a joint stock company.

The villages remain small and pleasant, displaying an architecture both simple and elegant in Shakeresque fashion. But the main street of Amana (the central village) is abuzz with businesses, all designed to separate tourists from dollars by promoting the bucolic and agrarian simplicity of a romanticized past. Some, like the Amana Furniture Shop, at least feature indigenous (if remarkably pricy) crafts of the original inhabitants; others offer utilitarian, and more economically accessible, products of local bakeries and vineyards. But many ply the objects of other states and nations, forging their link to Amana only in the "product image" of nostalgia and bucolia—and producing a dispiriting sameness that struck me as a country counterpart to the identical Crabtree and Evelyn soap store found in every yuppie boutique mall of urban America.

The best evidence of conscious intent in this well-crafted commercial image lies in the near invisibility imposed upon the largest building and biggest employer in the region—Amana Refrigeration, Inc., covering 1.2 million square feet in the territorial heart of Middle Amana. The company is not mentioned in the official brochure, and its location goes unannounced on the official map. No leaflet or flyer can be found at the official visitors center, although every tiny shop and country product receives copious notice and advertisement. Yet, surely, most Americans know the name Amana through the fine refrigerators, air conditioners, and microwave ovens manufactured by this exemplary company.

We might attribute this strange silence to justice or oversight if Amana Refrigeration bore no relationship to the villages, or if the factory sought some form of local anonymity, but neither argument holds. The company was founded in 1934, by George C. Foerstner, an Amana resident freed to indulge his commercial skills by the Great Change. Foerstner's dubious and personal use of the Amana name created tension with the Amana Society, the joint stock company formed to manage village businesses after the Great Change. This tension ended creatively in 1936, when the Amana Society bought the plant and made Foerstner its principal manager. The society ran the factory with outstanding success until 1965, when Raytheon purchased the name and works to the great benefit of the villages (a deeper source of current prosperity, I would guess, than apricot bread or rhubarb wine).

Moreover, the factory does not hide itself behind a facade of corn stalks. Hourly tours are offered to the public from a spacious and well-appointed visitors' center (though no notice of the tours can be found in any standard tourist literature available everywhere else within miles). You will not, I trust, charge me with unwarranted cynicism if I conclude that the villages are trying their damnedest to sequester the most prominent bearer of their name in the interest of a bucolic vision that has become eminently profitable itself.

In any case, I confess that the new image of the old is entirely infectious. I was having a wonderful time reading old German hymnals and samplers in the museum, watching the inevitable blacksmith at work, even copping a free sip of that rhubarb wine. I almost began to picture myself in this better and innocent world, supping freely with my fellows and bringing in the sheaves: no more essay deadlines, and no more suffering with the Boston Red Sox; no nukes, no seatbelts, no sweat but by the honest brow.

Then I came upon the Great Reminder (make that capital G, capital R) so freely available in any town as the ultimate antidote to waves of romantic nostalgia for a simpler past—the gravestones of dead children. In 1834, as the True Inspirationists began to contemplate their move to America, Friedrich Rückert wrote the set of poems that Gustav Mahler would later use for his searing song cycle of 1905—*Kindertotenlieder,* or "songs for dead children." Rich or poor, city or country, all nineteenth-century parents knew that many of their children would never enter the adult world. All my Victorian heroes, Darwin and Huxley in particular, lost beloved children in heartrending circumstances. I cannot believe that the raw pain could ever be much relieved by a previous, abstract knowledge of statistical inevitability—and, on this powerful basis alone, I would never trade even the New York subways for a life behind John Deere's plow that broke the plains. Imagine the mourning, or just the constant anxiety:

> In diesem Wetter, in diesem Braus
> nie hätt' ich gesendet die Kinder hinaus!

(In this weather, in this rainstorm, I would never have let the children outside.)

The graveyard of Middle Amana is spartan in its simplicity.

The identical, small white stones are laid out in rows, by strict sequence of death date, starting in the upper left corner and proceeding in book order. The German names are a panoply of objects, professions, descriptions, and moral states—Salome Kunstler (artist), Frau Geiger (violin), and Herr Rind (cow). The longer names do not fit across the small stone, and inscriptions must depart from ultimate simplicity by arching the many letters between the severe borders—Herr Schmiedehammer (sledge-hammer), Morgenstern (morning star), and Schuhmacher (shoe-maker).

I was looking for more of the arching names when I came upon a particularly stark example of the Great Reminder. The stone read, simply: "Emil Neckwinder, died 23 Nov. 1897, 1 day olt" (a conflation of the German "alt" and the English "old"—an example of languages and cultures in transition). Emil's twin sister Emma lies just beside him—"died 11 Dez. [again the German spelling], 3 weeks olt."

The loss of twins, though tragic, would not mark an unusual event. But the neighboring disruption of symmetry caught my attention, for two stones broke the severe geometrical pattern of even arrays. They stand in the space between two rows, directly in front of Emil and Emma's last resting place. In 1904, Frau Neckwinder bore another set of twins, and named them once more with E. Again, they both died—Evaline on May 23 "0 week alt" (fully in German this time), Eva on September 27, "4 months olt." The geometry itself is so eloquent; what more need be said?—the exception in an otherwise unvarying order of even rows, the intercalation into a linear sequence, permitted so that both pairs of infant twins might lie together in death.

Nearer my home in Lexington, in the graveyard just behind the Commons where our nation began in blood on April 19, 1775, a larger stone marks another kind of dying during our Revolutionary War: "This monument is erected to the memory of 6 children of Mr. Abijah Childs and Mrs. Sarah his wife." All died between August 19 and September 6 of 1778, presumably in an epidemic of infectious disease now quickly and eminently curable: Sarah at age thirteen (on August 28), Eunice at age twelve (on August 23), Abijah, Jr., at age eleven (on September 6), Abigail at age seven (on August 29), Benjamin at age four (on August 24), and Moses at "3 wanting 8 days" (on August 19.)

The gravestone of the infant Emil Neckwinder who died in his first day of life. *Photograph by Deborah Gould.*

Abijah, Sr., and Sarah lie behind, the husband dead at age seventy on August 30, 1808, his wife at age seventy-eight on March 3, 1812, as another war began. Sarah had to endure the death of at least one more child—Isaac, who must have been but a year old when a plague swept six siblings away, and who died on November 20, 1811, at age thirty-four. Isaac's grave bears one of the four-line doggerels so common on headstones of the time:

> Death like an overflowing flood
> Doth sweep us all away.

My hands are on the gravestones of one pair of Neckwinder twins.
Note the markers of the second set of twins in the foreground.
Photograph by Deborah Gould.

> The young, the old, the middle aged
> All to death become a prey.

These inscriptions are particularly poignant on the gravestones of children and young adults. Most state a rote acceptance
of the Lord's inscrutable will and read like a mantra copied from a
pattern book (the source, I suspect, for most inscriptions, given
their incessant repetition). Good psychology for mourners perhaps, but forgive a modernism if I doubt the sincerity of stated
calm and understanding. Sometimes, a lament of sadness strikes
closer to immediate reactions—as in this verse for three-month-
old Nathan, on a stone for another family, but standing right next
to the grave of Abijah Childs, Sr.

> This lovely babe so young and fair
> Call'd hence by early doom

The common gravestone for six children of Abijah and Sarah Childs. All died in an epidemic within one month. *Photograph by Deborah Gould.*

Just came to show how sweet a flow'r
In Paradise would bloom.

But bitterness sometimes breaks through. In a tiny cemetery on a windswept hill in Lower Island Cove, Newfoundland—a plot that also contains a monument for the four La Shana brothers lost at sea on May 25, 1883—I read of William Garland, who died in 1849 at age twenty-five:

Wherefore should I make my moan
Now the darling child is dead
He to rest is early gone
He to paradise is fled.
I shall go to him, but he
Never shall return to me.

The message is so simple, so commonplace, so often made—yet infinitely worth repetition in light of the curious human psy-

Tombstone of Moses Childs, youngest child to die. *Photograph by Deborah Gould.*

chology that paints our past rosy by selective memory of the good. Koko burst this bubble in his "little list," while another Gilbertian character, the sham-sensitive poet Reginald Bunthorne, understood the path of exploitation:

> Of course you will pooh-pooh whatever's fresh and new,
> And declare it's crude and mean,
> For art stopped short in the cultivated court
> Of the Empress Josephine.

A foolish or self-serving man like Bunthorne may make such an argument for realms of taste that admit no objective standard. But the directional, even the progressive, character of human knowledge and technology cannot be denied. Medicine, properly called the "youngest science" by Lewis Thomas, has not been among the most outstandingly successful of human institutions. Most improvements in longevity can be traced to a better understanding of nutrition and sanitation, not to any "cure" of disease. The germ theory of disease provided our one conspicuous triumph under conventional models of cure based on causal understanding, but more lives have been saved, even here, by prevention due to better sanitation, than by direct battle against bacteria. As for chronic conditions of aging and self-derailment—including most heart disease, strokes, and cancers—our success has been limited. Even so, and with all these strictures, modern medicine allows our children to grow up. The death of a child is now an unexpected tragedy, not a grim prediction. For this one transcendent reason alone, what sane person would choose any earlier time as a favored age for raising a family?

Technological progress is often less ambiguous and more linear. (I need hardly say that I define progress, in this sense, by internal standards of design and efficiency, not by resulting benefit to human life or planetary health. Technological progress will as likely do us in as raise us up.) If you need to get somewhere fast, airplanes beat horses, and if you need to rise, elevators are more pleasant than shank's mare.

I have only one reason for taking up this old subject within a series of essays devoted to evolutionary biology. We evolutionists do hold a key to appreciating the universal (or at least the planetary) significance of this progressive potential in human technology. At least we know how bizarre and unusual such short-scale linearity must be in the history of our part of the cosmos. Human culture has introduced a new style of change to our planet, a form that Lamarck mistakenly advocated for biological evolution, but that does truly regulate cultural change—inheritance of acquired characters. Whatever we devise or improve in our lives, we pass directly to our offspring as machines and written instructions. Each generation can add, ameliorate, and pass on, thus imparting a progressive character to our technological artifacts.

Nature, being Darwinian, does not work in this progressive way with our bodies. Whatever we do by dint of strength to improve our minds and physiques—from the blacksmith's big right arm in Lamarck's Amanaesque metaphor to the accumulated knowledge of a modern computer wonk—confers no genetic advantage upon our offspring, who must learn these skills from scratch using the tools of cultural transmission.

This fundamental difference between Lamarckian and Darwinian styles of change explains why cultural transformation can be rapid and linear, while biological evolution has no intrinsic directionality and follows instead, and ever so much more slowly, the vagaries of adaptation to changing local environments.

Cultural transformation, in its Lamarckian mode, therefore unleashed a powerful new force upon the earth—producing all the ills of our current environmental crisis, and all the joys of our confidently growing children. But we should not scoff at poky, old, biological evolution, for this Darwinian style of change also placed a potent source of novelty into the cosmos.

By contrast, passage of time in the physical universe either lacks directionality (therefore excluding the essence of history, defined as a pattern of distinctive change imparting uniqueness to moments) or possesses only the longest-scale linearity of stellar burn-out or universal expansion from a big-bang dot. Our planet did not know the full richness of history until Darwinian change bowed in with the evolution of life.

I thought of this key distinction between physical and biological time as I searched for long, curving names in the cemetery of Middle Amana: Schuhmacher und Morgenstern—shoemaker and morning star; the human technologist vs. the planet Venus, bright in the sky just before dawn. And I remembered that Charles Darwin had drawn the very same contrast in the final lines of the *Origin of Species*. When asking himself, in one climactic paragraph, to define the essence of the difference between life and the inanimate cosmos, Darwin chose the directional character of evolution vs. the cyclic repeatability of our clockwork solar system:

There is grandeur in this view of life. . . . Whilst this planet has gone cycling on according to the fixed law of gravity, from so simple a beginning endless forms most beautiful and most wonderful have been, and are being, evolved.

Shoemaker vs. morning star.

Authenticity

15 | In Touch with Walcott

I GREW UP in New York and, beyond a ferry ride or
two to Hoboken (scarcely qualifying as high adventure or rural
solitude), never left the city before age ten. But I read about
distant places of beauty and quiet, and longed to visit the American West. I fulfilled my dream during a family automobile trip at
age fifteen. I remember my first views of Yellowstone, the Grand
Canyon, Carlsbad Caverns. Yet, for reasons that I have never
fathomed, my strongest memories of awe are reserved for the
vast flatness of the Great Plains, for hundreds of miles of wheat
and corn in Kansas, Nebraska, South Dakota, and Minnesota. I
loved the symmetry of the fields, the endless flatness broken by
Victorian farmhouses and their windbreaks of trees, the adjacent
silos, the small towns marked by their signatures of water towers
and grain elevators (the analog of church steeples for navigation
in any old European village).

I feel no differently today. Two summers ago, I drove west with
my family from Minneapolis, not on Interstate 90 (the enemy of
regionalism), but on Route 14, with sidetrips to the lovely Victorian mainstreet of Pipestone, built of beautiful, soft red Sioux
Falls quartzite, and to the Corn Palace in Mitchell, South Dakota.
Only one town breaks the pattern of solitude and timelessness in
350 miles between Mankato, Minnesota, and the state capital of
Pierre, South Dakota. Right on Route 14, midway between Sleepy
Eye, Minnesota, and Blunt, South Dakota, stands De Smet,
little town on the prairie, childhood home of Laura Ingalls
Wilder.

The success of Wilder's wonderful books about her pioneer

childhood (I have read them all to my kids), not to mention the inevitable TV spinoff series, has converted De Smet into a commercial island (either desert or oasis according to your values) of hagiography. The main gift shop for Wilder paraphernalia sells an amazing array of items, from expensive furniture down to tiny bits of memorabilia at two bits a pop (bookmarks, pencils, cottonwood twigs from Pa's trees). As testimony to our odd desire to possess, even in replica, some tangible property of a heroine or her bloodline, I was most amused by one of the twenty-five-cent items—the calling card of Laura's daughter Rose (who became an ultraconservative journalist, an opponent of income taxes, social security, and all New Deal programs).

As something of a squirrel myself, I dare not be too critical. I confess that I also own some calling cards, but only two and both genuine. For what item could better symbolize the continual presence of an intellectual hero than this most overt testimony of personal presence from an age without telephones? I proudly own cards associated with two remarkable men who shared both a name and a calling: Charles Darwin and Charles Doolittle Walcott.*

---

*I received Charles Darwin's personal card from an anthropological colleague who found it in his museum, amidst some items that Darwin had collected. He most kindly sent it to me with the following note: "I assume that Mr. Darwin meant to leave this card for you; you must have been out when he called." The Walcott card belonged to his wife, the formidable Mary Vaux Walcott (note T. H. Clark's description later in this essay). In the original version of this piece, I missed the little squiggle of an *s* at the end of the title—Mrs. rather than Mr.— and attributed the card to C. D. Walcott himself. (Having companions in error is no excuse, but the editor of *Natural History* told me that several levels of editors and compositors had scrutinized the article and its illustrations, and no one had noticed the misattribution.)

Attentive readers, of course, noted my error; the resulting correspondence was fascinating and, I suppose, predictable. Most mentioned the error gently, acknowledging the easy confusion if names of husbands and wives be distinguished only by a tiny wavy line at the end of a superscript. A few angry feminists accused me of just one more lamentable example in ignoring women, to the point of making them invisible. My response here must be quite contrary: I had, of course, not forgotten the old custom, particularly among wives of social rank, for submerging individuality by self-identification as Mrs. plus a husband's full name. But, as this custom is now happily passing away, thanks in large part to increasing sensitivity produced by feminist critiques, I simply hadn't been alert enough to remember and suspect. When I saw "Charles Doolittle Walcott," I assumed the man himself and didn't scrutinize the preceding title. My error, in other words,

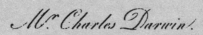

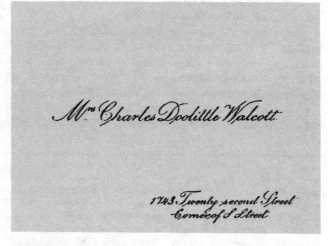

The calling cards of Mr. Charles Darwin and Mrs. Charles Doolittle Walcott.

The subject of calling cards, with its overt theme of personal greeting, inevitably raises the question of intellectual genealogy. If I actually own his card, how far back must I go to touch Charles Darwin? Since Darwin died more than one hundred years ago in 1882, a metaphorical handshake might seem so distant as to be

---

arose from the successes of feminism and not as a rearguard action against the proper acknowledgment of women.

This change to correct attribution does not compromise the essay which, after all, began with an anecdote about honoring Laura Ingalls Wilder through purchasing replicas of her daughter's calling card.

uninteresting in contemplation. But intellectual genealogies tend to be surprisingly short since people of importance touch so many lives, and at least a few of the anointed will be blessed with great longevity. (True bloodlines, by contrast, tend to pass through many more generations in a given length of time both because children arrive early in life and because connections must pass through small numbers of progeny, often including no one with a long lifespan.)

In fact, I can touch Darwin through only two or three intermediaries. I studied vertebrate paleontology with Ned Colbert. Ned, as a young man, was the personal research assistant of Henry Fairfield Osborn, president of the American Museum of Natural History. I now touch Darwin either directly or by one more step, depending on which version of the most famous Osborn legend you endorse.

Osborn was a very smart man, but his immodesty greatly outran his considerable intelligence. He once published an entire book, dedicated to listing his publications and photographing his medals and degree certificates. (He cites, as a specious rationale in his forward, a simple and selfless desire to encourage young scientists by demonstrating the potential rewards of diligence.) Tales of Osborn's smugness and arrogance continue to permeate the profession, more than fifty years after his death. The most famous story begins with W. K. Gregory, who took over Osborn's course in vertebrate paleontology after the great man retired. Once a year, Gregory would take his students to visit the haughty professor emeritus. At one such meeting, Osborn rose from his desk and stiffly shook each student's hand. An interlude of increasingly uncomfortable silence followed, for no one knew how to address a person of such eminence. Osborn himself finally broke the silence, saying: "When I was a young man about your age, I worked for a year in the laboratory of E. Ray Lankester in London—and one day Charles Darwin [T. H. Huxley, in the other version] walked in—and I shook his hand, so I know how you all feel now." Thus, my short linkage to Darwin needs only two or three steps—either Colbert–Osborn–Darwin or Colbert–Osborn–Huxley–Darwin.

Walcott (1850–1927), my other Caroline hero, may not be so well known, but he ranks as a giant within my profession of paleontology. Charles Doolittle Walcott was the world's greatest expert on rocks and fossils of the Cambrian period—the crucial time, beginning some 550 million years ago, when modern multicellular life first arose in a geological whoosh (a few million years) called the "Cambrian explosion." Walcott, a master administrator, was also the most powerful man in American science during his prime. In 1907, he moved from head of the United States Geological Survey to secretary (their name for boss) of the Smithsonian Institution, where he died with his boots on in 1927. He persuaded Andrew Carnegie to found the Carnegie Institute of Washington and encouraged Woodrow Wilson to establish the National Research Council. He was an intimate of every president from Teddy Roosevelt to Calvin Coolidge. In 1920, at age seventy, he made the following entry in the yearly summation at the end of his diary:

> I am now Secretary of Smithsonian Institution, President National Academy of Sciences, Vice Chairman National Research Council, Chairman Executive Committee Carnegie Institute of Washington, Chairman National Advisory Committee for Aeronautics. . . . Too much but it is difficult to get out when once thoroughly immersed in the work of any organization.

As a deeply traditional and conservative doer and thinker, Walcott suffered the unkind fate of many people who hold immense power in their times, but do not pass innovative ideas along to posterity—erasure from explicit memory, despite endurance of unrecognized influence. Few outside the profession of paleontology may remember Walcott's name today, but no scientist has exceeded his power and prestige.

Since discoveries tend to outlive personalities, Walcott's name has survived best in attachment to his most enduring single accomplishment—his remarkable find, lucky in Pasteur's sense of fortune favoring the prepared mind, of the world's most important fossils, the animals of the Burgess Shale (see my book, *Wonderful Life*). In 1909, high in the Canadian Rockies, Walcott discovered the closest object to a holy grail in paleontology—a fauna blessed with complete representation, thanks to the rare preservation of soft anatomy, from the most crucial of all times, right after the Cambrian explosion.

Snatching defeat from the jaws of victory, Walcott then proceeded to misinterpret these magnificent fossils in the deepest possible way. He managed to shoehorn every single Burgess species into a modern group, calling some worms, others arthropods, still others jellyfish. The Burgess animals became a small group of simple, primitive precursors for later, successful lineages. Under such an interpretation, life began in primordial simplicity and moved inexorably, predictably onward to more and better.

Walcott's reading, originally published in 1911 and 1912, went virtually unchallenged for more than half a century. But during the past twenty years, an elegant and comprehensive recollection and restudy by Harry Whittington and his students has entirely reversed Walcott's interpretation. Burgess diversity—in range of anatomical designs, not number of species—exceeded the scope of all organisms living today. The history of life is therefore a story of decimation and limited survival (with enormous success to a few of the victors, insects for example), not a tale of steady progress and expansion. Moreover, we have no evidence that survivors prevailed for any conventional cause rooted in anatomical superiority or ecological adaptation. We must entertain the strong suspicion that this early decimation worked more as a grand-scale lottery than a race with victory to the swift and powerful. If so, then any rerun of life's tape would yield an entirely different set of survivors. Since *Pikaia,* the first recorded member of our own lineage (Chordata), lived as a rare component of the Burgess fauna, most replays would not include the survival of our ancestry—and we would be wiped out of history. Conscious life on earth is this tenuous, this accidental.

I regard this reinterpretation of the Burgess Shale as the most important paleontological conclusion of my lifetime. If you accept my judgment on the importance of the Burgess Shale, then Charles Doolittle Walcott must join the roster of great scientists who achieved both power in their own time and immortality later. Whatever his interpretation, no one can take away his discovery.

But why did Walcott so thoroughly misinterpret his most important fossils? I suggested two basic reasons in a long chapter of my recent book, *Wonderful Life.* First, and however mundane the point, Walcott was so insanely busy with administrative tasks that he never had time for adequate study of the Burgess animals. His papers of 1911 and 1912 were only meant as preliminary accounts, but he never wrote the main act. (He had zealously—and jealously—guarded the Burgess fossils for a grand retirement project, but died in office at age seventy-seven.) Even if Walcott had been intellectually inclined to let the Burgess fossils tell him their amazingly unconventional story, he never found time for a proper conversation. Testimonies to Walcott's lifestyle abound in the Smithsonian archives, but I found no document more poignant than the following statement submitted to his bank:

> I enclose herewith the affidavit that you wish. I used to sign my name Chas. D. Walcott. I now use only the initials, as I find it takes too much time to add in the extra letters when there is a large number of papers or letters to be signed.

Second, and more importantly, I do not think that Walcott was inclined to free and open conversation with the Burgess fossils. I don't accuse him of biases stronger than those of most scientists; I merely say that we all live within our own constraining world of concepts, and that the historical record grants us insight into the scope and power of Walcott's preconceptions. Walcott, from the depths of his traditionalism, was fiercely committed to a view of life's history as predictably progressive and culminating in the ordained appearance of human intelligence. As a Christian evolutionist, he believed that God had established the law of natural selection expressly to produce this intended result in the long run. At the height of controversy with fundamentalism in the days of Bryan and the Scopes trial, Walcott rejected the usual view (held by both scientists and theologians then and now) that science and religion occupy separate intellectual spheres demanding equal respect. He held instead that science must validate a "correct" view of divine guidance through the immensity of geological time in order to avoid accusations of atheism. Walcott therefore authored an appeal, signed and published in 1923, a year before Scopes was indicted in Tennessee. Endorsed by "a number of conservative scientific men and clergymen" (including Herbert Hoover and Henry Fairfield Osborn of my introduction), this statement held, in part:

It is a sublime conception of God which is furnished by science, and one wholly consonant with the highest ideals of religion, when it represents Him as revealing Himself through countless ages in the development of the earth as an abode for man and in the age-long inbreathing of life into its constituent matter, culminating in man with his spiritual nature and all his God-like power.

How could a man committed to this view of life ever interpret his own most precious discovery as the maximally varied source of a grand lottery that just happened to include human ancestors among the lucky survivors? Walcott had to view the Burgess animals as a limited array of primitive precursors for a history of life that would progress and diverge predictably.

As I was writing *Wonderful Life,* learning more and more about Walcott and coming to understand his extraordinary energy and influence, I developed a strong urge, following the game of intellectual genealogy, to establish personal contact with him. Darwin died too long ago for any hope of a shortest link with the deceased—via one intermediary who knew the hero personally. But Walcott died in 1927, and someone who interacted with him as a colleague (not merely as a child who once shook his hand) might still be living. I thus longed to "touch" Walcott as I prepared his biography. (I do not know why we wish to add this personal dimension to a scholarly endeavor; some would even consider such a quest either foolishly trivial or actually harmful in its potential for prejudicial intrusion into hopes for objectivity. But I am content that fellowship can thus reach across time.)

All my attempts failed. I tried my colleague Bill Schevill, who collected at the Burgess quarry with Percy Raymond in 1930. No luck. I anticipated success with G. Evelyn Hutchinson, the world's greatest ecologist, who died in 1991 in his late eighties. In 1931, Hutchinson had written an important article on the anatomy and mode of life for two key Burgess creatures. But he had begun his work in 1929 and had never met Walcott. I moved my quest to the back burner and proceeded to other things.

Then, out of the blue on February 9, 1990, I received a three-page letter from T. H. Clark, Logan Professor of Paleontology Emeritus at McGill University. I knew of Clark by reputation of course, for all geologists respect his classic textbook (written with my colleague Colin Stearn) on *The Geological Evolution of North America.* But I did not know that he is ninety-seven years old (and still professionally active, with a technical paper now in press in the *Journal of Paleontology*), and I was quite unprepared to encounter one of life's frequent (and most pleasant) ironies—to quest hard for something and fail, and then to have it drop into your lap by pure good fortune, without the slightest effort. Clark began:

Dear Professor Gould:

I have just finished reading your latest book—*Wonderful Life*—much to my pleasure and edification. . . . We have in the Redpath Museum [of McGill University] two lots of Burgess Shale fossils. . . . The second lot consists of nearly fifty specimens . . . collected by myself, in 1924, when I was second in command of the Harvard Summer School located in the southern Canadian Rockies under Dr. Percy Raymond.

Note: I entered Harvard in 1913, approaching Dr. Raymond to introduce myself as a fledgling freshman who wanted nothing less than a degree in invertebrate paleontology and was greeted with this apparent pessimistic reception and advice and the words, "Very well. But I warn you that if you continue in such foolishness you'll be the last paleontologist alive by the time you retire. There's no future in it. . . ."

In 1924 Raymond asked me to take charge of the Harvard Summer School, headquarters at Banff, while he had complete tooth removal. We ended the six weeks tour at Field [Canadian Pacific Railway town near the Burgess Shale]. Henry Stetson and I hired horses and crossed over the Kicking Horse Pass to climb to Walcott's Quarry, staying there two whole days and collecting what I thought was a wonderful lot.

So far, so good. Sweet reminiscences of a favorite place, and bright vindication for my profession, not large, but numbering a healthy few thousand, and full of vibrant debate, as Clark continues to work on the way to his centenary. Percy Raymond, by the way, was a great expert on trilobites who taught at Harvard from the office I now occupy. Raymond made an important collection of Burgess Shale fossils in 1930. Harry Whittington, genius of the contemporary Burgess revision, was Raymond's successor, and lived in the same office just before me. Thus, my own interest in the Burgess Shale is, at least in part, dynastic. But I was still unprepared for the joy of Clark's last paragraph:

> I have an incident to record which may interest you. During the 1924 trip we were visited by Dr. Walcott and his wife who spent all afternoon with us. Walcott and Raymond stayed together while Mrs. Walcott cruised around camp criticizing everything down to the tent pegs. . . . with apologies for causing you to lose so much time reading these notes.
>
> I remain, yours sincerely,
>
> T. H. Clark

Clark's letter came with a note from his research associate Ingrid Birker, stating that, since the publication of my book, "we have been inundated with requests from all sorts of local biologists, geologists, historians, and students of natural science to see the Burgess Shale material. . . . So we have decided to give a small presentation at the end of March using Clark's photos from 1924 and a selection of specimens. If the fates allow this will be a delightful occasion to hear someone who has indeed had a wonderful life and can still share the enthusiasm and humor of it. You are most welcome to join our little conversazione."

Some invitations for travel engender considerable mulling; a few, like a child's first invitation to a circus or a ballgame, need not be made twice. I went to Montreal on March 28, to meet Dr. Clark—and to touch Charles Doolittle Walcott.

Call me a foolish romantic, but we paleontologists take connection seriously; nothing but fragile continuity, by the thinnest genealogical lines, brings the reward of persistence. Humans are here today because our particular line never fractured—never once at any of the billion points that could have erased us from history. The tangible establishment of an intellectual genealogy may be only symbolic, but symbols count, and my handshake with

Photograph of C. D. Walcott (left) and Percy Raymond
taken in camp by T. H. Clark in 1924. *Courtesy of Redpath Museum:
McGill University.*

Clark (combined with a memory of his with Walcott) will be my
personal metaphor for the deeper genetic connections that make
life possible and paleontology practical.

Clark recalls that when they met in 1924 he gave Walcott a
reprint of his first paper—on strata of the Beekmantown Formation in Quebec. He then took a photo of Walcott and Percy Raymond—the younger man, now toothless, slumped, and pear-
shaped (much like my demeanor in the field), but Walcott, at age
seventy-four, erect as could be, with military bearing. Now, sixty-
six years later, after the longest professorial tenure in the history

T. H. Clark on horseback in 1924. *Courtesy of Redpath Museum: McGill University.*

of McGill University (for Clark started there in 1924 and remains on the rolls as Logan Professor of Paleontology Emeritus), T. H. Clark gave me copies of his latest reprints, and I reciprocated in kind. We need such continuity in a crazy world that wrenches and uproots our bearings, often on a monthly or yearly schedule (see Essay 16).

I thought that my day in Montreal would be sweet and memorable, but I had not anticipated the greatest pleasure of all—intellectual challenge. I thought that Clark would give an anecdotal talk, seated before his fossils, telling us tales of C. D. Walcott's passion for photography, Mary Walcott's love of wildflowers, the great days of railroads and pack trains, and the heroic age of fossil collecting.

No such thing: We met in the Redpath Museum, a wonderful Victorian building opened in 1882. The lecture hall is an architectural gem, built like a Renaissance dissecting theater à la Vesalius (but ornamented throughout with Victorian filigree). The lecturer therefore stands on the floor at the bottom of a pit, with seats for spectators rising sharply in several concentric, semicircular tiers. T. H. Clark, I soon realized, had not come for a pleasant chat. His Burgess fossils were laid out on a table in the middle of the pit, but Clark stood at the end of the table and spoke for forty minutes without notes on the life and works of Charles Doolittle Walcott—a beautiful presentation chock full of information and organized, as all good talks must be, around a distinctive and integrative point of view. (How I wish that academics would grasp this most basic requirement of all successful public speaking or any intellectual discourse at all!)

As I listened to Clark, something slowly dawned upon me, and my simple pleasure at being in Montreal to "touch" Walcott blossomed into the deepest possible delight. Clark was not merely presenting a chronological account of Walcott's life and works. Rather, he had read my book carefully, had studied my chapter on Walcott, and *did not like it* (in the best sense of treating the same factual material from an entirely different perspective). In the gentlest, kindest possible way, Clark had organized his talk to defend the living hero of his professional youth, a man he once met, from my "modernist" historiographical approach.

Wonderful Life has been a source of great satisfaction for me. I wrote the book with real passion, for the work of my colleagues (described therein) is so elegant and so important. The book has done well in the overt sense that publishers cherish: it paid leisurely and friendly visits to the best-seller lists of three countries, and got its share of rave reviews (and some real first-class stinkers for contrast). But nothing gave me nearly so much pleasure as the thought that my views had so engaged a man who knew Walcott, that he chose, at age 97, to prepare and deliver his first public talk in many years to set the record straight from his point of view.

Scientists of my generation, and back at least to anyone who

T. H. Clark lecturing in 1990. *Courtesy of Redpath Museum: McGill University.*

began a career after World War II, tend to be cynical about science as a public institution (however idealistically we may think or act in any personal quest for knowledge and discovery). After all, we have grown up with Hiroshima, the perjured testimony of tobacco company statisticians, the rush for biotech profits in the corporate world, and, above all, the debasement of our own world by measurement of all accomplishment in grant dollars. But I have met several older scientists who not only mouthed—as we all can do—the ideals of science as objective and truth-directed by a quiet but overwhelming moral force (with scientists as humble and privileged participants in this inexorable process), but who also actually seemed to believe and live by this lovely vision. (The vision strikes me as so false—as a statistical statement about norms of science, not as an accessible position for admirable individuals—that I confess to great trouble in comprehension. But I cannot deny the power and beauty of the image, nor do I fail to appreciate its salutary role as an impetus to productive work and psychic benevolence, whatever its truth value).

From my own point of view, I had treated Walcott with fairness and abiding respect (I will even confess to a feeling of awe when I contemplate his accomplishments). I had tried to ferret out his biases and prejudices, but these, to me, mark his humanity, not his failing as a scientist. (Moreover, a key part of the Burgess drama—Walcott's great and interesting error of the shoehorn—must be explicated, for basic reasons of historical truth, in terms of his constraining biases). But from the standpoint of Clark's generation, and from the heart of his beautiful vision, such talk can only degrade the work of a great scientist—for science gathers information in objective respect for nature's factuality, and Walcott was a great and hyperproductive empiricist.

Clark unified his wonderful talk with a metaphor that he wove skillfully in and around his narrative of Walcott's remarkable accomplishments. When Walcott was young, and limited by restricted resources to local travel for collection of trilobites around his native town of Utica, New York, he rode (in Clark's metaphor) on a bicycle. When he joined the Geological Survey, he could collect his data "and put them on a rail, on a freight car." And then, "when he became director of the survey, he was given a locomotive and as many freight cars as he wanted. He filled them up, one after the other, bringing them back to Washington."

Clark continually referred to his metaphor as "a fantasy," but I cannot imagine a more appropriate summary for his noble vision of science. Think of all the components. A railway line runs in a definite direction, as science moves towards an external truth. Freight cars bear the factual goods that form the content of the enterprise. The locomotive, propelled by its own steam, represents the proper scientist hauling his booty from nature down the highway of truth.

Clark's metaphor even provided an image for the aspect of Walcott's career that he could not depict in a thoroughly positive light—restricted time for scientific work, and no time at all for the Burgess Shale, as a consequence of administrative burdens.

> On one of his railroad trips, fairly late in life, he didn't notice an open switch in front of him. The locomotive tripped the open switch; the cars all behind followed off the main line into the wilderness. And there they stayed for a number of years. . . . Nothing more was done on the wonderful collection that he found.

Walcott's locomotive (his drive and persona) continued forward as his administrative activities, but he had left his freight cars, full of scientific facts, on the sidings.

Science, for Clark, does move inexorably forward towards truth, but progress can be stalled by externalities. The freight cars remained on the sideline for nearly fifty years, until a "young fellow" came along and said: "I know what's wrong with the fossils from the Burgess Shale, and I think I will have a crack at them." (Harry Whittington began his Burgess work in his late fifties and continued into his seventies. As I approach that magic mid-century mark, I note with delight that some colleagues might still call me a "young fellow" even ten years from now.) So Whittington went back to pick up Walcott's freight cars and to continue down the straight and narrow road to truth: "He said, well, we have to get those cars on their way; so he jumped aboard the locomotive, backed up, and got the whole train onto the main track."

What a remarkable antithesis of interpretation for the same basic story. What I had read as a deep (and fascinating) conceptual error, permanently constraining and born of ideological commitments, was, for Clark, merely a pause (however unfortunate) in the inexorable progress of scientific knowledge.

I could simply end by defending my own reading, while expressing respect for Clark, but I would not dare such a conclusion for two reasons. First, we need mythology, in the admirable sense of universal and inspirational legend (not the pejorative connotation of falsehood)—and Clark's vision of science is mythology in the best and noblest meaning. Second, who can say that my historiography is "right" and his "wrong." They are both faithful to the factual record, and they both offer insight. I will defend my account as a best telling for themes of motivation, social context, and interplay of fact and theory in the history of ideas.

But my themes do not exhaust the content of science; and, as I listened to Clark, I realized that I had left something out in my choice of focus. Clark kept emphasizing the immensity of Walcott's empirical work—his massive multivolumed monographs on fossils and strata. (At the conclusion of his talk, Clark held up Walcott's weighty, two-volume treatise on Cambrian brachiopods as an epitome of his excellence and zeal. Clark then ended his talk with a stunning one liner: "Believe it or not; [Walcott's] middle name was Doolittle.") And I realized that I had virtually excluded this part of Walcott from my account. Oh, I mentioned his voluminous empiricism, but almost as a side comment. How could I do such a thing—a side comment for work that, by sheer volume of time and effort, had consumed most of Walcott's professional life (even if his Burgess error will stand as his paramount mark upon intellectual history).

Lives are too rich, too multifaceted for encompassing under any one perspective (thank goodness). I am no relativist in my attitude towards truth; but I am a pluralist in my views on optimal strategies for seeking this most elusive prize. I have been instructed by T. H. Clark and his maximally different vision. There may be no final answer to Pilate's inquiry of Jesus (John 18:37), "What is truth?"—and Jesus did remain silent following the question. But wisdom, which does increase with age, probes from many sides—and she is truly "a tree of life to them that lay hold upon her."

Touching Walcott via the minimum of one intermediary.
Courtesy of Redpath Museum: McGill University.

16 | Counters and Cable Cars

SAN FRANCISCO, *October 11, 1989:*

In a distinctive linguistic regionalism, New Yorkers like me stand "on line," while the rest of the nation waits patiently "in line." Actually, I spend a good part of my life trying to avoid that particular activity altogether, no matter what preposition it may bear. I am a firm supporter of the Yogi Berra principle regarding once fashionable restaurants: "No one goes there anymore; it's too crowded."

Consequently, in San Francisco this morning, I awoke before sunrise in order to get my breakfast of Sears's famous eighteen pancakes (marvel not, they're very small) before the morning crush of more amenable hours rendered the restaurant uninhabitable on Berra's maxim. Then out the door by 7:30 to the cable car stop at Union Square for a ride that thrills me no less in middle life than on my first trip as a boy. What moment in public transportation could possibly surpass that final steep descent down Russian Hill? (For a distant second and third in America, I nominate the Saint Charles streetcar of New Orleans, last of the old-time trolley lines, as it passes by the antebellum houses of the Garden District, and the Staten Island Ferry, only a nickel in my youth and the world's most distinguished cheap date, as it skirts the Statue of Liberty by moonlight.) I travel during the last minutes of comfort and accessibility. By 9:00 A.M., long lines of tourists will form and no one will want to ride anymore.

We paleontologists are driven, almost by professional definition, to an abiding respect for items and institutions that have prevailed and prospered with integrity in an unending sea of

238

change (although I trust that we can also welcome, even foster, intellectual innovation). I love Sears restaurant with its familiar, uniformly excellent and utterly nonyuppie breakfast menu. And I adore those Victorian cars with their wooden seats and their distinctive sounds—the two-clang signal to move, the hum of the cable perpetually running underground, the grasp of the grip as it takes hold to pull the passive car along.

As I ride, I ponder a psychological puzzle that has long intrigued me: Why does authenticity—as a purely conceptual theme—exert such a hold upon us? An identical restaurant with the same food, newly built in the San Francisco segment of a Great Cities Theme Park, would supply me with nothing but calories; a perfect replica of a cable car, following an even hillier route in Disneyland, would be a silly bauble.

Authenticity comes in many guises, each contributing something essential to our calm satisfaction with the truly genuine. Authenticity of *object* fascinates me most deeply because its pull is entirely abstract and conceptual. The art of replica making has reached such sophistication that only the most astute professional can now tell the difference between, say, a genuine dinosaur skeleton and a well-made cast. The real and the replica are effectively alike in all but our abstract knowledge of authenticity, yet we feel awe in the presence of bone once truly clothed in dinosaur flesh and mere interest in fiberglass of identical appearance.

If I may repeat, because it touched me so deeply, a story on this subject told once before in these volumes (Essay 12 in *The Flamingo's Smile*): A group of blind visitors met with the director of the Air and Space Museum in Washington to discuss greater accessibility, especially for the large objects hanging from the ceiling of the great atrium and perceptible only by sight. The director asked his guests whether a scale model of Lindbergh's *Spirit of St. Louis,* mounted and fully touchable, might alleviate the frustration of nonaccess to the real McCoy. The visitors replied that such a solution would be most welcome, but only if the model were placed directly beneath the invisible original. Simple knowledge of the imperceptible presence of authenticity can move us to tears.

We also respect an authenticity of *place.* Genuine objects out of context and milieu may foster intrigue, but rarely inspiration.

London Bridge dismantled and reassembled in America becomes a mere curiosity. I love to watch giraffes in zoo cages, but their jerky, yet somehow graceful, progress over the African veld provokes a more satisfying feeling of awe.

Yet, until today, I had not appreciated the power of a third authenticity of *use*. Genuine objects in their proper place can be devalued by altered use—particularly when our avid appetite for casual and ephemeral leisure overwhelms an original use in the honorable world of daily work.

Lord knows, being one myself, I have no right to complain about tourists mobbing cable cars. Visitors have an inalienable right to reach Fisherman's Wharf and Ghirardelli Square by any legal means sanctioned and maintained by the city of San Francisco. Still, I love to ride incognito at 7:30 A.M. with native San Franciscans using the cable car as a public conveyance to their place of work—Asian students embarking on their way to school as the car skirts by Chinatown, smartly dressed executives with their monthly transit passes.

But I write this essay because I experienced a different, unanticipated, and most pleasant example of authenticity of use in Sears this morning. (I could not have asked for a better context. The Bay Area, this week, is experiencing a bonanza in authenticity of place—as the Oakland A's and the San Francisco Giants prepare for the first single-area World Series since 1956, when the seventh and last "subway series" of ten glorious childhood years in New York, 1947 to 1956, produced Don Larsen's perfect game and the revenge of my beloved Yankees for their only defeat, the year before, by the Dodgers in their true home in Brooklyn. Think what we would lose if, in deference to October weather and a misplaced sense of even opportunity, the World Series moved from the home cities of full season drama to some neutral turf in balmy Miami or New Orleans.)

I have always gone to Sears with other people and sat at a table. This time I went alone and ate at the counter. I had not known that the counter is a domain of regulars, native San Franciscans on their way to work. One man gets up and says to the waitress, "Real good, maybe I'll come back again sometime." "He's in here every morning," whispers the waitress to me. Another man takes the empty seat, saying "Hi, honey" to the woman on the next stool. "You're pretty early today," she replies. "The works!"

he says, as the waitress passes by. "You got it," she replies. A few minutes later, she returns with a plate of pancakes and a dish of scrambled eggs. But first she slides the eggs off the plate onto a napkin, blotting away the butter. "No good for him," she explains. He begins a discussion on the relative merits of cloth napkins and paper towels in such an enterprise. Good fellowship in authenticity of use; people taking care of each other in small ways of enduring significance.

As I present talks on evolutionary subjects all around America, I can be sure of certain questions following any speech: Where is human evolution going? What about genetic engineering? Are blacks really better at basketball? (Both the dumb and the profound share this character of inevitability.) I must rank the ecological question high on the list of perennial inquiries. It is usually asked with compassion, but sometimes with pugnacity: Why do we need to save all these species anyway?

I know the conventional answers rooted in practicality. I even believe them: You never know what medical or agricultural use might emerge from species currently unknown or ignored; beneficial diversity of gene pools in cultivated species can often be fostered by interbreeding with wild relatives; interconnectedness of ecological webs may lead to dire and unintended consequences for "valued" species when "insignificant" creatures are rubbed out. Still, I prefer to answer with an ethical, more accurately a viscerally aesthetic, statement espoused by nearly all evolutionary biologists as a virtual psychic necessity for choosing to enter the field in the first place: We relish diversity; we love every slightly different way, every nuance of form and behavior—and we know that the loss of a significant fraction of this gorgeous variety will quench our senses and our satisfactions in any future meaningfully defined in human terms (potential recovery of diversity several million years down the road is too abstract and conjectural for this legitimately selfish argument—see Essay 2). What in the world could possibly be more magnificent than the fact that beetle anatomy presents itself in more than half a million separate packages called species?

I have always been especially wary of "soft" and overly pat analogies between biological evolution and human cultural change. (Some comparisons are apt and informative, for all modes of change must hold features in common; but the mechanisms of biological evolution and cultural change are so different that close analogies usually confuse far more than they enlighten.) Nonetheless, aesthetic statements may claim a more legitimate universality, especially when an overt form rather than the underlying mechanism of production becomes the subject of consideration. If you feel aesthetic pleasure in proportions set by the "golden section," then you may gain similar satisfaction from a nautilus shell or a Greek building despite their maximally different methods and modes of construction. I do, therefore, feel justified in writing an essay on the moral and aesthetic value of diversity both in natural and in human works—and in trying to link the genesis and defense of diversity with various meanings of authenticity.

(Also, if I may make a terrible confession for a working biologist and a natural historian: I grew up on the streets of New York, and I suppose that one never loses a primary affection for things first familiar—call it authenticity of place if you wish. I do think that America's southwestern desert, in the four corners region around Monument Valley, is the most sublime spot on earth. But when I crave diversity rather than majesty, I choose cities and the products of human labor, as they resist conformity and embody authenticity of object, place, and use. My motto must be the couplet of Milton's "L'Allegro ed Il Penseroso"—from the happy rather than the pensive side:

> Towered cities please us then
> And the busy hum of men.

Several years ago I visited India on a trip sponsored by Harvard's Natural History Museum. My colleagues delighted in arising at 4:00 A.M., piling into a bus, driving to a nature reserve, and trying to spot the dot of a tiger at some absurd distance, rendered only slightly more interesting by binoculars. I yearned to be let off the bus alone in the middle of any bazaar in any town.)

Natural diversity exists at several levels. Variety permeates any nonclonal population from within. Even our tightest genealogical groups contain fat people and thin people, tall and short. The primal folk wisdom of the ages proclaims enormous differences in temperament among siblings of a single family. But the greatest dollop of natural diversity arises from our geographical divisions—the differences from place to place as we adapt to varying environments and accumulate our distinctiveness by limited contact with other regions. If all species, like rats and pigeons, lived all over the world, our planet would contain but a tiny fraction of its actual diversity.

I therefore tend to revel most in the distinctive diversity of geographical regions when I contemplate the aesthetic pleasure of differences. Since I am most drawn to human works, I find my greatest joy in learning to recognize local accents, regional customs of greeting and dining, styles of architecture linked to distinctive times and places. I also, at least in my head if not often enough in overt action, think of myself as a watchdog for the preservation of this fragile variety and an implacable foe of standardization and homogenization.

I recognize, of course, that official programs of urban layout and road building must produce more elements of commonality than a strict aesthetic of maximal diversity might welcome. After all, criteria of design have a universality that becomes more and more pressing at upper limits of size and speed. If you have to move a certain number of cars through a given region at a stated speed, the road can't meander along the riverbanks or run through the main streets of old market towns. Public buildings and city street grids beg for an optimal efficiency that imposes some acceptable degree of uniformity.

But the sacred task of regionalism must be to fill in the spaces between with a riotous diversity of distinctive local traditions—preferably of productive work, not only of leisure. With this model of a potentially standardized framework for roads and public spaces filled in, softened, and humanized by local products made by local people for local purposes—authenticity of object, place, and use—I think that I can finally articulate why I love the Sears counter and the cable cars in the early morning. They embody all the authenticities, but they also welcome the respectful stranger. (Again, nature and human life jibe in obedience to basic principles of structural organization. Ecological rules and principles—flow of energy across trophic levels, webs of interaction that define the "balance of nature"—have a generality corresponding to permissible uniformity in the framework of public space. But local diversity prevails because different organisms embody the rules from place to place—lions or tigers or bears as predictable carnivores of three separate continents—just as uniquely local businesses should fill the slots within a more uniform framework.)

I also now understand, with an intellectual argument to back a previous feeling, what I find so troubling about the drive for standardization, on either vernacular (McDonald's) or boutique levels (Ghirardelli Square or Harborside or Quincy Market or how can you tell which is where when all have their gourmet chocolate chip cookie cart and their Godiva chocolate emporium?). I cannot object to homogenization per se, for I accept such uniformity in the essential framework of public spaces. But McDonald's introduces standardization at the wrong level by usurping the smaller spaces of immediate and daily use, the places that cry out for local distinction and an attendant sense of community. McDonald's is a flock of pigeons ordering all endemic birds to the block, a horde of rats wiping out all the mice, gerbils, hamsters, chinchillas, squirrels, beavers, and capybaras. The Mom-and-Pop chain stores of Phoenix and Tucson are almost a cruel joke, a contradiction in terms.

I grew up in Queens, next to a fine establishment called the T-Bone Diner (it is still there, *mirabile dictu*). The contrast between railroad-car–style diners of my youth and McDonald's of my midlife brings us to the heart of the dilemma. Diners were manufactured in a few standardized sizes and shapes—many by the Worcester Car Company in my adopted state—and then shipped to their prospective homes. Owners then took their standard issue and proceeded to cultivate the distinctness that defines this precious item of American culture: menus abounding with local products and suited to the skills and tastes of owners; waiters and waitresses with a flair for uniqueness, even eccentricity, of verve, sassiness, or simple friendliness; above all, a regular clientele forged into a community of common care. McDonald's works in precisely the opposite way and becomes perverse in its incongruity. It enters the small-scale domain of appropriate uniqueness within the interstices of an allowable uniform framework. It even occupies spaces of widely differing designs, placements, and previous uses. It then forges this diversity into a crushing uniformity that permits not a millimeter of variation in the width of a fry from Oakland to Ogunquit.

But we are not defeated. Uniqueness has a habit of crawling

back in and around the uniformities of central planning. Uniqueness also has staying power against all the practical odds of commercial culture because authenticities speak to the human soul. Many of those old diners are still flourishing in New England. I am at least a semiregular at one of the finest. On my last visit, the counter lady pointed to a jar with dollar bills. A regular customer, she told me, had a sick child in need of an operation—and everyone was kicking in, if only as a symbol of support and community. No one even mentioned the jar to casual customers on that particular morning, but I was simply told to contribute: no pleas, no harangues, no explanations beyond the simple facts of the case. Our communities are many, overlapping, and of various strengths. I am proud to be part of this aggregate, forged to a coherent species by a common place of local integrity. So long as these tiny communities continue to form in the interstices of conformity, I will remain optimistic about the power of diversity. And I will remember Elijah's discovery during his flight from Jezebel (1 Kings 19:11–12): "After the wind an earthquake. . . . And after the earthquake a fire. . . . And after the fire a still, small voice."

Postscript

As the dateline indicates, I wrote this essay just a week before the great San Francisco earthquake of October 17. This violently altered circumstance converted my closing line into an utterance that, if intended after the fact rather than written unwittingly before, might seem overly pointed, if not verging on cruel. In using Elijah to reemphasize my central contrast between small-scale, local, and distinctive diversity (the "still, small voice") and global effects (well represented by general catastrophes), I was, I freely confess, also trying to make a small joke about San Francisco as the location of my essay—for the 1906 earthquake did wreak destruction with a tremor followed by fire.

Little did I know that my attempt at humor would soon be turned so sour by nature. I could, of course, just change the ending, sink this postscript, and fudge a fine fit with history. But I would rather show what I wrote originally—appropriate to its moment, but not a week later—as a testimony to nature's continuing power over our fortunes, and as a working example of another theme so often addressed in these essays: the quirky role of unique historical events both in nature and in human life.

The earthquake has also illuminated several other points that I raised about authenticity and local diversity. The World Series, although delayed, was not moved to neutral turf but honored baseball's powerful tradition for authenticity of place, despite the practical difficulties. My line about "people taking care of each other in small ways of enduring significance," although meant only as a comment about the Sears counter, soon extended to the whole region. Every fire or flood provokes endless rumination and pious commentary on why we need disaster to bring out the best in us. But clichés are hackneyed because they are true, and the framework of this essay does put a different twist upon a commonplace. Just as McDonald's marks the dark side by bringing the allowable conformity of large-scale public space into the inappropriate arena of local distinctiveness, human kindness after disaster, on the bright side, has a precisely opposite effect, for pervasive trouble promotes the usual caring of small and local communities to the large and overt domain of anonymity and callousness. Now how can this still, small voice be heard and felt at all scales all the time?

5 | Human Nature

17 | Mozart and Modularity

DAINES BARRINGTON (1727–1800), a lawyer and wealthy member of the lesser nobility, published so many short articles on such a variety of subjects that he could scarcely avoid a reputation as a dilettante. In numerous communications to the Royal Society of London, he discussed the landing place of Caesar in Britain, the merely local nature of Noah's flood, the antiquity of playing cards, and the death of Dolly Pentreath, the last native speaker of Cornish (an extinct branch of the Celtic languages). Some of his colleagues considered him superficial and overly credulous. One detractor even composed a heroic couplet in his dishonor:

> Pray then, what think ye of our famous Daines?
> Think of a man denied by Nature brains!

Then, in 1764, Barrington happened onto something truly important. But, stung by rebukes for his previous carelessness and hyperbole, Barrington proceeded cautiously. He waited six years before publishing his observations as a note in the *Philosophical Transactions of the Royal Society of London,* Britain's leading scientific journal both then and now. And he began his article by invoking the classical literary form for understatement—*litotes.* (These Greek terms for parts of speech and forms of rhetoric have paralyzed generations of schoolkids who can't remember the difference between a dactyl and a synecdoche. Monty Python got back at professorial pedants by making great merriment with "litotes" and its improbable pronunciation.) Litotes (from the Greek *litos,* meaning "small" or "meager") is a form of understatement that expresses an affirmative by the negative of its contrary—as in "not bad" for "good." In his opening paragraph, Barrington used litotes in a near apology to readers for taking their time:

> If I was to send you a well attested account of a boy who measured seven feet in height when he was not more than eight years of age, it might be considered as not undeserving the notice of the Royal Society.

In the second paragraph, Barrington sneaked up a bit further upon his actual discovery:

> The instance which I now desire you will communicate to that learned body, of as early an exertion of most extraordinary musical talents, seems perhaps equally to claim their attention.

The third paragraph, though only in historical retrospect, drops the bombshell:

> Joannes Chrysostomus Wolfgangus Theophilus Mozart, was born at Saltzbourg in Bavaria, on the 17th of January, 1756. . . . Upon leaving Paris [in 1764 at age eight] he came to England, where he continued more than a year. As during this time I was witness to his most extraordinary abilities as a musician, both at some publick concerts, and likewise by having been alone with him for a considerable time at his father's house; I send you the following account, amazing and incredible almost as it may appear.*

---

*I originally thought that this paragraph contained three errors—and said so in the original version, with a snide (if unstated) implication that Barrington, perhaps, deserved his reputation as dilettante. But Joseph B. Russell wrote to inform me that Salzburg was, in 1756 (before Austria existed), subject to the Duchy of Bavaria, and that Saltzbourg is an acceptable Anglicization. Mozart was, however, born on January 27, not January 17. This may be a simple error of transcription or typesetting—or, perhaps, Barrington got confused by some aspect of the approximately ten-day difference imposed by the Gregorian calendric reform, then so recently accepted in England (see Essay 12 on this point). In any case, my respect for Barrington, evident throughout the essay, increases again.

Litotes had ceded to overt wonderment.

Mozart's skills were so astounding that Barrington even doubted his extreme youth; could father Leopold's game be an elaborate ruse, passing off a well-trained adult midget as a young son? Barrington delayed publication for six years until he could obtain proof in the form of Mozart's birth certificate from the register of Salzburg, "procured from his excellence Count Haslang, envoy extraordinary and minister plenipotentiary of the electors of Bavaria and Palatine" (you just gotta believe somebody with a title like that).

Leopold Mozart made quite a business of showing off his precocious son. Barrington, graced with a private visit, proceeded as any intellectual would: He tested eight-year-old Wolfgang for a variety of musical skills in reading, memory, and improvisation, and his letter to the Royal Society is a report of his impressions. (I learned about this publication at a special exhibit on Mozart at the British Museum. Barrington's article, entitled "Account of a

Young Mozart at the piano. *Courtesy of Photographie Bulloz, Paris.*

very remarkable young musician," appeared in 1770, in volume 60 of the *Philosophical Transactions.* The notion that young Mozart had served as subject for a scientific paper in England's leading journal was too much to resist as a topic for this series. What better symbol could we possibly advance for the fruitful interaction of art and science?)

One issue, above all, fascinated Barrington as he observed Mozart and affirmed in spades all the reports he had heard about the young child's precocity (for Barrington sought, in this article, to plumb the nature of genius itself, not merely to explicate Mozart who, remember, was then just a remarkable little boy, not yet an icon of Western achievement): Apparent "wholeness" must be decomposable into separate modules, each subject to independent development. How else could a mere child be so transcendent in one particular arena, but ordinary in most other ways? This idea of dissociability must provide a key to understanding human talents: Genius is not integral, but must result from a hypertrophy of particular modules.

Barrington cites two examples of dissociation in grasping the nature of genius. First, he marvels at Mozart's musical sophistication in an otherwise ordinary and rambunctious eight-year-old boy. If young Wolfgang had been a miniature adult, as adept in manners as in music, then genius might be portrayed as integral, but he acted like an ordinary kid in all domains outside his special talent:

> I must own that I could not help suspecting his father imposed with regard to the real age of the boy, though he had not only a most childish appearance, but likewise had all the actions of that stage of life. For example, whilst he was playing to me, a favorite cat came in, upon which he immediately left his harpsichord, nor could we bring him back for a considerable time. He would also sometimes run about the room with a stick between his legs by way of horse.

Second, Barrington gained some insight about the dissociability of basic emotions. He asked Mozart to improvise songs expressing particular emotions—a song of love and a song of anger. Again, Barrington took refuge in litotes to describe the successful result:

[The love song] had a first and a second part, which . . . was of the length that opera songs generally last: if this extemporary composition was not amazingly capital, yet it was really above mediocrity, and showed most extraordinary readiness of invention.

The song of rage was even more dramatically successful:

This lasted also about the same time with the Song of Love; and in the middle of it, he had worked himself up to such a pitch, that he beat his harpsichord like a person possessed, rising sometimes in his chair.

But how could an eight-year-old boy, with presumably limited experience, at least of sexual love, so abstract and distill these basic modules of our emotional repertoire? This could only be possible, Barrington reasoned, if the fundamental emotions reside in our behavioral storehouses as dissociable packages. Our totality must be an amalgam of separable components.

We have, before and ever since, been fascinated with such "splinter skills"—extraordinary talents in otherwise undistinguished or even severely handicapped people—for the same reason that so intrigued Barrington: Such dissociation seems to argue for a separate origin and causation of talents that we would prefer to view, but cannot on this evidence, as expressions of a more general genius. We all know the standard examples of chess grandmasters who cannot balance their check books, and mentally handicapped people with prodigious skills in apparently instantaneous numerical calculation or reckoning the day of the week for any date over centuries or millennia.

For all the criticism that Barrington received as an injudicious dilettante, this time he chose well—both in subject and argument. For the principle of dissociation, and construction from separable modules, is central to our understanding of any complex system that arises by natural evolution. Barrington identified the right issue for his wonderment, and the breadth of application extends well beyond divine Mozart to the evolution of any complex organism and the structure of mind. Integral wholeness may sound warm, fuzzy, and romantic, but dissociability is the necessary way of the world.

Since principles are often best illustrated by exposing the fallacy of their contraries, I present the most important, and probably most intelligent, argument ever raised against evolution by a great scientist in the turbulent generation before Darwin. In the *Discours préliminaire* to his four-volume work on fossil vertebrates, published in 1812, Georges Cuvier denied the possibility of evolution by affirming the doctrine of intrinsic and nondissociable wholeness.

Cuvier designated his principle as "the correlation of parts," maintaining that all features of an organism are intricately designed and coordinated to function in a certain optimal way. No part can change by itself. Any conceivable alteration in one organ would require the redesign of every other feature, for optimal function requires complete integration:

> Every organized individual forms an entire system of its own, all the parts of which mutually correspond, and concur to produce a certain definite purpose, by reciprocal reaction, or by combining towards the same end. Hence none of these separate parts can change their forms without a corresponding change in the other parts of the same animal, and consequently each of these parts, taken separately, indicates all the other parts to which it belonged.

Cuvier used this principle primarily to argue that he could reconstruct entire organisms from fossil fragments, because one bone implied a necessary shape for all others. But Cuvier had a second, even grander motive—the denial of evolution. How can transmutation occur if parts cannot alter separately, or at least with some degree of independence? If each tiny modification requires a redesign of absolutely every other feature, then inertia itself must debar evolution. How can we imagine a coordinated change of all parts every time some minute advantage might attend a slight alteration in one feature? Cuvier continued:

> Animals have certain fixed and natural characters, which resist the effects of every kind of influence, whether proceeding from natural causes or human interference; and we have not the smallest reason to suspect that time has any more effect upon them than climate.

The logic of this argument is impeccable. If parts are not dissociable, then evolution cannot occur. "All for one" might be good morality for a Musketeer but cannot describe the pathway of natural change in complex systems. Yet logical arguments are only as good as their premises. The chain of inference may be irrefutable, but if the premise be false, then the conclusion will probably fail as well. To cite the harsh motto of our computer age: GIGO, or garbage in, garbage out (no matter how phenomenal the inner workings of the machine).

Cuvier's logic was correct, but his premise of total integrity is false. Evolution does proceed (as it must) by dissociating complex systems into parts, or modules made of a few correlated features, and by altering the various units at differing rates and times. Biologists refer to this principle as "mosaic evolution," and we need look no farther than the history of our own species. Human ancestors, like Lucy and her early australopithecine cousins, evolved an upright posture of nearly modern design before any substantial enlargement of the brain had occurred.

This cardinal principle of dissociability works just as well for the mental complexities of emotions and intelligence as for designs of entire bodies. As he began to compile the notes that would lead to his evolutionary theory, Charles Darwin recognized that he could not give an evolutionary account of human emotions without the principles of modularity and dissociation.

He wished, for example, to trace facial gestures to antecedent states in ancestral animals. But if the human complement forms an integrated array, locked together by our unique consciousness, then a historical origin from simpler systems becomes impossible. Darwin recognized that two principles must underlie the possibility of evolution. First, gestures cannot be subject to fully conscious control; some, at least, must represent automatic, evolved responses. As evidence for ancestral states, Darwin cited several gestures that make no sense without modern morphology, but must have served our forebears well. In sneering, we tighten our upper lips and raise them in the region of our canine teeth. This motion once exposed the fighting weapons of our ancestors (as it continues to uncover the long and sharp canines in many modern mammals that perform the same gesture), but human canines are no bigger than our other teeth and this inherited reaction has lost its original function.

Second, just as young Mozart could separate and abstract single emotions, Darwin realized that standard facial gestures must be modules of largely independent action—and that the human emotional repertoire must be more like the separate items in a shopper's bag than the facets of an unbreakable totality. Evolution can mix, match, and modify independently. Otherwise we face Cuvier's dilemma: If all emotions are inextricably bound by their status as interacting, optimal expressions, then how can anything ever change?

Late in his life, Darwin wrote an entire book on this subject: *The Expression of the Emotions in Man and Animals* (1872). But his youthful jottings in the so-called M Notebook of 1838, hastily scribbled in the months before he codified his theory of natural selection in September of the same year, are even more compelling for their telegraphic expression of excitement in discovery and novel explanation. Darwin later labeled this notebook as "full of Metaphysics on Morals and Speculations on Expression." His fascinating notes on emotional gestures center on modularity and its importance for evolutionary explanation. Each feeling is linked to a gesture; we have limited control over the form of a gesture, and its evolutionary meaning must often be sought in a lost ancestral function. Darwin wrote:

> He may despise a man and say nothing, but without a most distinct will, he will find it hard to keep his lip from stiffening over his canine teeth.—He may feel satisfied with himself, and though dreading to say so, his step will grow erect and stiff like that of turkey.... With respect to sneering, the very essence of an habitual movement is continuing it when useless,—therefore it is here continued when uncovering the canine useless.

Darwin then speculates on the further evolution of emotions treated as separate entities. He argues that sighing is still directly useful in humans "to relieve circulation after stillness." Yet we might retain the gesture as a sign for an accompanying emotion even if the physiological benefit disappeared: "If organization were changed, I conceive sighing might yet remain just like sneering does."

I received my clearest insight into the modularity of facial expressions not from any scientific writing, but from viewing the world's greatest sculpture: Michelangelo's Moses in the church of San Pietro in Vincoli (St. Peter in Chains) in Rome. Moses, bearing the tablets of the ten commandments, has just come down from Mount Sinai. Suffused with holiness, and with joy at the gift he may now bestow upon his people, he looks around only to see the Israelites worshipping the golden calf. His face is a maelstrom of emotions: zeal and ardor for what he has witnessed on the mountain, rage at his people for their transgression, deep sorrow for human weaknesses. The sublimity of the statue lies in the richness of this mixture upon one face—as if Moses has become everyman (in every major state of feeling).

I visited this statue several times and felt its power but could not grasp how Michelangelo had put so much into one face. On my last trip, and largely by chance, I think that I found a guide to the solution. Michelangelo understood—whether viscerally or explicitly, I do not know—the principle of modularity. When I focused on one feature and covered the rest of the face, I saw only one emotion each time. The eyebrows speak one message, the nose another, the lips a third. The rich face can be decomposed into modules of feeling, but the totality stuns us by integration.

Many of the most famous experiments in animal ethology affirm and extend the principle of modularity. Consider Niko Tinbergen's classic work on begging for food in newly hatched gulls (so beautifully described in his charming book, *The Herring Gull's World*). The newborns peck vigorously at their parents' beak, apparently aiming for a red spot near the tip of the mandible. If an infant makes proper contact, the parent regurgitates a parcel of food and the baby gull gets its first meal.

But what inspires the pecking behavior? The baby gull has no conscious understanding of a reward to be gained. It has never eaten before and cannot know what a knock on a parent's bill will provide. The behavior must be innate and unlearned.

At what, then, does the baby bird direct its pecks? At first consideration, one might conjecture that the entire form or gestalt of the parent would provide an optimal target. After all, what could be more appealing than the parent's totality—a full, three-dimensional image with the right movements and odors. But consider the issue a bit more deeply: The hatchling has never seen a bird. Can the complexities of the entire parental form be engrafted innately upon its untested brain? Wouldn't the goal be more readily achieved—easier to program if you will—if the hatchling responded to one or a few abstract particulars, that is, to modules extracted from the total form?

In an exhaustive series of experiments, Tinbergen showed that hatchling gulls do respond to modules and abstractions. They peck preferentially at long and skinny objects, red things, and regions of markedly contrasting colors. As an effect of this simplified modularity, they hit the spot at the tip of the parental bill— the only red region at the end of a long object, in an area of contrasting color with surrounding yellow. Complex totality may be beyond the cognitive capacity of a hatchling gull, but any rich object can be broken down to simpler components and then built up. Any developing complexity—whether in the cognitive growth of an individual or the evolution of a lineage—may require this principle of construction from modules.

If hatchling gulls favor abstractions (and don't perceive parental totality), then Tinbergen reasoned that he might construct a "super-gull"—a model exaggerating the key modules. This "improved" version might elicit more attention than the actual parents themselves. This idea bore fruit as hatchlings preferred several remarkably artificial dummies to real birds. For example, narrow sticks longer than real bills, and color patches more starkly contrasted with surroundings than the red spot on an actual bill, elicited more pecks from hatchlings than did an accurately modeled head.

Tinbergen then generalized these observations to the important concept of a *super-normal stimulus*— an artificial exaggeration that elicits more favor or response than the feature itself. (In his book, Tinbergen includes an amusing discussion of his struggle to find a good name for this phenomenon. He first spoke of a *super-optimal* stimulus, but finally rejected the term as oxymoronic—for optima, by definition, cannot be exceeded. He then remarks that 'supernatural' would be a good term, if it were not used already in another sense." Finally, he settled upon super-normal).

Many animals exploit this modular principle of super-normal stimuli to gain advantage over others. In the classical example, cuckoos subvert the propensities of their hosts to feed any chick in the nest that rises higher, squawks louder, or opens its beak wider. The mother cuckoo lays an egg in another bird's nest. The egg itself is an accurate mimic and often can't be distinguished from the host's own products. But the cuckoo hatchling quickly outstrips its nestmates in growth and may even toss them out to their death. The unwitting adult hosts, fooled by the super-normal stimuli arising from the large, loud cuckoo chick, continue to feed the usurper and murderer.

Obviously, if modularity didn't often rule over accurate perception of totalities, super-normal stimuli would not exist. Host parents would know their own children and reject the cuckoo. Hatchling gulls would peck at parental beaks (that might feed them), rather than at cardboard dummies with exaggerated features.

Modularity pervades all neurological organization, right up to what Darwin called "the citadel itself"—human cognition. This principle of breaking complexity into dissociable units does not disappear at the apex of known organization. Humans might not be fooled *in toto* by the analog of a cuckoo chick, but the fashion industry knows how well, and how sheepishly, we respond to a plethora of super-normal stimuli.

Interestingly, Darwin accompanied his M-notebook jottings on the modularity of emotional gestures with similar statements about cognitive items and units:

> People who can multiply large numbers in their head must have this high faculty, yet not clever people. . . . The great calculators, from the confined nature of their associations (is it not so in punning) are people of very limited intellects, and in the same way are chess players. . . . The son of a fruiterer in Bond St. was so great a fool that his father only left him a guinea a week, yet he was inimitable chess player.

The concept of modularity, explicitly so called, lies at the heart of much innovative research in cognitive science. The brain does a great deal of work by complex coordination among its parts, but we have also known for a long time that highly particular aptitudes and behaviors map to specific portions of the cerebral cortex. The modules are often stunningly precise and particular, as illustrated by unusual losses and misperceptions of people who have suffered damage to highly localized regions of cortex (see

Oliver Sacks' wonderful book, *The Man Who Mistook His Wife for a Hat*). The September 19, 1991, issue of *Nature* tells the remarkable story of two men who suffered localized strokes that seriously impaired their ability to use and recognize vowels but not consonants. Surely, we would have regarded our separation of sounds into vowels and consonants as an artificial division of a totality—yet this distinction may record a deeper mapping of cerebral modules.

Mozart was not yet Mozart when Daines Barrington witnessed his incredible performance. He was just a bratty kid at the acme of precocity. In fact, Barrington even speculated on his potential for future contributions. He spoke of another prodigy named John Barratier who knew Latin at four, Hebrew at six, and who translated the travels of Rabbi Benjamin, complete with learned notes and glosses, at eleven. But we know little of Barratier today because he died before the age of twenty.

Barrington notes the unhappy tendency of geniuses to die young, and he expresses his hope for Mozart by comparing him with England's greatest musical guest, the German emigré Handel. Young Handel may not have been quite so precocious as Mozart, but he did live a long and remarkably productive life, from 1685 to 1759, and Barrington took comfort:

> I am the more glad to state this short comparison between these two early prodigies in music, as it may be hoped that little Mozart may possibly attain to the same advanced years as Handel, contrary to the common observation that such *ingenia praecocia* are generally short lived.

Barrington got half his wish. Mozart lived long enough to become Mozart, but died so young, at thirty-five, that his early demise has become the canonical example of a genre—the tragic and uncertain lives led by so many artists. (I wrote this essay on the very day of the two hundredth anniversary of Mozart's death. I wish that I had been able to compose this piece at the proper Handelian distance of forty years hence, which would have been good for Mozart and good for me too).

Daines Barrington thought that he was writing a scientific article about the modularity of human abilities. The later exaltation of Mozart makes us view his work in a more particular light—as a testimony about the early life of everyone's favorite musical prodigy. Is Barrington's article part of science or of art? Perhaps, for once, these are truly false modules, and our intellectual life would benefit by more integration. If Mozart had died before *Mitridate* (a teenage opera), Barrington's article would endure as a respectable scientific account of a generic musical prodigy. I thank God for *Don Giovanni* (and I promise to tolerate every Musak rendition of *Eine Kleine Nachtmusik* in exchange). But even if Mozart had died in childhood, in the frosts of an English winter (in their damned buildings without central heating), his contribution to our understanding of the human mind would still be no mean thing, no small potatoes.

18 | The Moral State of Tahiti—and of Darwin

CHILDHOOD PRECOCITY is an eerie and fascinating phenomenon. But let us not forget the limits; age and experience confer some blessing. The compositions that Mozart wrote at four and five are not enduring masterpieces, however sweet. We even have a word for such "literary or artistic works produced in the author's youth" (*Oxford English Dictionary*)—*juvenilia*. The term has always borne a derogatory tinge; artists certainly hope for substantial ontogenetic improvement! John Donne, in the second recorded use of the word (1633) entitled his early works: "Iuuenilia: or certaine paradoxes and problemes."

I shouldn't place myself in such august company, but I do feel the need to confess. My first work was a poem about dinosaurs, written at age eight. I cringe to remember its first verse:

> Once there was a *Triceratops*
> With his horns he gave big bops
> He gave them to an allosaur
> Who went away without a roar.

(I cringe even more to recall its eventual disposition. I sent the poem to my boyhood hero, Ned Colbert, curator of dinosaurs at the American Museum of Natural History. Fifteen years later, when I was taking his course as a graduate student, Colbert happened to clean out his old files, found the poem, and gleefully shared it with all my classmates one afternoon.)

Now, a trivia question on the same theme: What was Charles Darwin's first published work? A speculation on evolution? Perhaps a narrative of scientific discovery on the Beagle? No, this greatest and most revolutionary of all biologists, this inverter of the established order, published his first work in the *South African Christian Recorder* for 1836—a joint article with *Beagle* skipper Robert FitzRoy on "The Moral State of Tahiti." (The standard catalogue of Darwin's publications lists one prior item—a booklet of *Beagle* letters addressed to Professor Henslow and printed by the Cambridge Philosophical Society in 1835. But this pamphlet was issued only for private distribution among members—the equivalent of an informal modern Xeroxing. "The Moral State of Tahiti" represents Darwin's first public appearance in print, and biographers record it as his first publication—even though the article is mostly FitzRoy's, with long excerpts from Darwin's diaries patched in and properly acknowledged.)

The great Russian explorer Otto von Kotzebue had poured fuel on an old and worldwide dispute by arguing that Christian missionaries had perpetrated far more harm than good in destroying native cultures (and often cynically fronting for colonial power) under the guise of "improvement." FitzRoy and Darwin wrote their article to attack Kotzebue and to defend the good work of English missionaries in Tahiti and New Zealand.

The two shipmates began by noting with sorrow the strong anti-missionary sentiments that they had encountered when the *Beagle* called at Capetown:

> A very short stay at the Cape of Good Hope is sufficient to convince even a passing stranger, that a strong feeling against the Missionaries in South Africa is there very prevalent. From what cause a feeling so much to be lamented has arisen, is probably well known to residents at the Cape. We can only notice the fact: and feel sorrow.

Following a general defense of missionary activity, FitzRoy and Darwin move to specific cases of their own prior observation, particularly to the improved "moral state" of Tahiti:

> Quitting opinions . . . it may be desirable to see what has been doing at Otaheite (now called Tahiti) and at New Zealand, towards reclaiming the 'barbarians.' . . . The *Beagle* passed a part of last November at Otaheite or Tahiti. A

more orderly, quiet, inoffensive community I have not seen in any other part of the world. Every one of the Tahitians appeared anxious to oblige, and naturally good tempered and cheerful. They showed great respect for, and a thorough good will towards, the missionaries; . . . and most deserving of such a feeling did those persons appear to be.

FitzRoy and Darwin were, obviously, attentive to a possible counterargument—that the Tahitians have always been so decent, and that missionary activity had been irrelevant to their good qualities by European taste. The article is largely an argument against this interpretation and a defense for direct and substantial "improvement" by missionaries. Darwin, in particular, presents two arguments, both quoted directly from his journals. First, Tahitian Christianity seems deep and genuine, not "for show" and only in the presence of missionaries. Darwin cites an incident from his travels with native Tahitians into the island's interior, far from scrutiny. (This incident must have impressed Darwin powerfully, for he told the tale in several letters to family members back home and included an account in his *Voyage of the Beagle*):

Before we laid ourselves down to sleep, the elder Tahitian fell on his knees, and repeated a long prayer. He seemed to pray as a christian should, with fitting reverence to his God, without ostentatious piety, or fear of ridicule. At daylight, after their morning prayer, my companions prepared an excellent breakfast of bananas and fish. Neither of them would taste food without saying a short grace. Those travellers, who hint that a Tahitian prays only when the eyes of the missionaries are fixed on him, might have profited by similar evidence.

Second, and more important, Tahitian good qualities have been created, or substantially fostered, by missionary activity. They were a dubious lot, Darwin asserts, before Western civilization arrived.

On the whole, it is my opinion that the state of morality and religion in Tahiti is highly creditable. . . . Human sacrifices,—the bloodiest warfare,—parricide,—and infanticide,—the power of an idolatrous priesthood,—and a system of profligacy unparalleled in the annals of the world,—have been abolished,—and dishonesty, licentiousness, and intemperance have been greatly reduced, by the introduction of Christianity.

(On the subject of sexual freedom in women, so long an issue and legend for all Tahitian travelers from Captain Cook to Fletcher Christian, FitzRoy remarked: "I would scarcely venture to give a general opinion, after only so short an acquaintance; but I may say that I witnessed no improprieties." Nonetheless, FitzRoy did admit that "human nature in Tahiti cannot be supposed superior to erring human nature in other parts of the world." Darwin then added a keen observation on hypocrisy in Western male travelers who do not sufficiently credit missionaries as a result of their private frustration on this issue: "I do believe that, disappointed in not finding the field of licentiousness so open as formerly, and as was expected, they will not give credit to a morality which they do not wish to practise.")

Many arguments float back and forth through this interesting article, but the dominant theme can surely be summarized in a single word: paternalism. We know what is good for the primitives—and thank God they are responding and improving on Tahiti by becoming more European in their customs and actions. Praise the missionaries for this exemplary work. One comment, again by FitzRoy, captures this theme with special discomfort (to modern eyes) for its patronizing approach, even to royalty:

> The Queen, and a large party, passed some hours on board the *Beagle.* Their behavior was extremely correct, and their manners were inoffensive. Judging from former accounts, and what we witnessed, I should think that they are improving yearly.

Thus, we may return to my opening issue—the theme of juvenilia. Shall we rank this article on the "Moral State of Tahiti," Darwin's very first, in the category of severe later embarrassments? Did Darwin greatly revise his views on non-Western peoples and civilizations, and come to regard his early paternalism as a folly of youthful inexperience? Much traditional commentary in the hagiographical mode would say so—and isolated quotations can be cited from here and there to support such an interpretation (for Darwin was a complex man who wrestled with deep issues, sometimes in contradictory ways, throughout his life).

But I would advance the opposite claim as a generality. I don't think that Darwin ever substantially revised his anthropological views. His basic attitude remained: "They" are inferior but redeemable. His mode of argument changed in later life. He would no longer frame his attitude in terms of traditional Christianity and missionary work. He would temper his strongest paternalistic enthusiasm with a growing understanding (cynicism would be too strong a word) of the foibles of *human* nature in all cultures, including his own. (We see the first fruits of such wisdom in his comment, cited previously, on why sexually frustrated travelers fail to credit Tahitian missionaries.) But his basic belief in a hierarchy of cultural advance, with white Europeans on top and natives of different colors on the bottom, did not change.

Turning to the major work of Darwin's maturity, *The Descent of Man* (1871), Darwin writes in summary:

> The races differ also in constitution, in acclimatisation, and in liability to certain diseases. Their mental characteristics are likewise very distinct; chiefly as it would appear in their emotional, but partly in their intellectual faculties. Every one who has had the opportunity of comparison, must have been struck with the contrast between the taciturn, even morose, aborigines of S. America and the lighthearted, talkative negroes.

The most striking passage occurs in a different context. Darwin is arguing that discontinuities in nature do not speak against evolution, because most intermediate forms are now extinct. Just think, he tells us, how much greater the gap between apes and humans will become when both the highest apes and the lowest people are exterminated:

> At some future period, not very distant as measured by centuries, the civilized races of man will almost certainly exterminate and replace throughout the world the savage races.

At the same time the anthropomorphous apes . . . will no doubt be exterminated. The break will then be rendered wider, for it will intervene between man in a more civilized state, as we may hope, than the Caucasian, and some ape as low as a baboon, instead of as at present between the negro or Australian and the gorilla.

The common (and false) impression of Darwin's egalitarianism arises largely from selective quotation. Darwin was strongly attracted to certain peoples often despised by Europeans, and some later writers have falsely extrapolated to a presumed general attitude. On the *Beagle* voyage, for example, he spoke highly of African blacks enslaved in Brazil:

It is impossible to see a negro and not feel kindly towards him; such cheerful, open, honest expressions and such fine muscular bodies; I never saw any of the diminutive Portuguese with their murderous countenances, without almost wishing for Brazil to follow the example of Hayti.

But towards other peoples, particularly the Fuegians of southernmost South America, Darwin felt contempt: "I believe if the world was searched, no lower grade of man could be found." Elaborating later on the voyage, Darwin writes:

Their red skins filthy and greasy, their hair entangled, their voices discordant, their gesticulation violent and without any dignity. Viewing such men, one can hardly make oneself believe that they are fellow creatures placed in the same world. . . . It is a common subject of conjecture, what pleasure in life some of the less gifted animals can enjoy? How much more reasonably it may be asked with respect to these men.

On the subject of sexual differences, so often a surrogate for racial attitudes, Darwin writes in *The Descent of Man* (and with direct analogy to cultural variation):

It is generally admitted that with woman the powers of intuition, of rapid perception, and perhaps of imitation, are

268 | EIGHT LITTLE PIGGIES

more strongly marked than in man; but some, at least, of these faculties are characteristic of the lower races, and therefore of a past and lower state of civilization. The chief distinction in the intellectual powers of the two sexes is shown by man attaining to a higher eminence, in whatever he takes up, than woman can attain—whether requiring deep thought, reason, or imagination, or merely the use of the senses and hands.

Darwin attributes these differences to the evolutionary struggle that males must pursue for success in mating: "These various faculties will thus have been continually put to the test, and selected during manhood." In a remarkable passage, he then expresses thanks that evolutionary innovations of either sex tend to pass, by inheritance, to both sexes—lest the disparity between men and women become ever greater by virtue of exclusively male accomplishment:

It is, indeed, fortunate that the law of the equal transmission of characters to both sexes has commonly prevailed throughout the whole class of mammals; otherwise it is probable that man would have become as superior in mental endowment to woman, as the peacock is in ornamental plumage to the peahen.

Shall we then simply label Darwin as a constant racist and sexist all the way from youthful folly to mature reflection? Such a stiff-necked and uncharitable attitude will not help us if we wish to understand and seek enlightenment from our past. Instead I will plead for Darwin on two grounds, one general, the other personal.

The general argument is obvious and easy to make. How can we castigate someone for repeating a standard assumption of his age, however much we may legitimately deplore that attitude today? Belief in racial and sexual inequality was unquestioned and canonical among upper-class Victorian males—probably about as controversial as the Pythagorean theorem. Darwin did construct a different rationale for a shared certainty—and for this we may exact some judgment. But I see no purpose in strong criticism for a largely passive acceptance of common wisdom. Let us rather analyze why such potent and evil nonsense then passed for certain knowledge.

If I choose to impose individual blame for all past social ills, there will be no one left to like in some of the most fascinating periods of our history. For example, and speaking personally, if I place every Victorian anti-Semite beyond the pale of my attention, my compass of available music and literature will be pitifully small. Though I hold no shred of sympathy for active persecutors, I cannot excoriate individuals who acquiesced passively in a standard societal judgment. Rail instead against the judgment, and try to understand what motivates men of decent will.

The personal argument is more difficult and requires substantial biographical knowledge. Attitudes are one thing, actions another—and by their fruits ye shall know them. What did Darwin do with his racial attitudes, and how do his actions stack up against the mores of his contemporaries? By this proper criterion, Darwin merits our admiration.

Darwin was a meliorist in the paternalistic tradition, not a believer in biologically fixed and ineradicable inequality. Either attitude can lead to ugly statements about despised peoples, but practical consequences are so different. The meliorist may wish to eliminate cultural practices, and may be vicious and uncompromising in his lack of sympathy for differences, but he does view "savages" (Darwin's word) as "primitive" by social circumstance and biologically capable of "improvement" (read "Westernization"). But the determinist regards "primitive" culture as a reflection of unalterable biological inferiority, and what social policy must then follow in an era of colonial expansion: elimination, slavery, permanent domination?

Even for his most despised Fuegians, Darwin understood the small intrinsic difference between them in their nakedness and him in his regalia. He attributed their limits to a harsh surrounding climate and hoped, in his usual paternalistic way, for their eventual improvement. He wrote in his *Beagle* diary for February 24, 1834:

Their country is a broken mass of wild rocks, lofty hills and useless forests, and these are viewed through mists and endless storms. . . . How little can the higher powers of the mind come into play: what is there for imagination to paint, for reason to compare, for judgment to decide upon? To knock a limpet from the rock does not even require cunning, that lowest power of the mind. . . . Although essentially the same creature, how little must the mind of one of these beings resemble that of an educated man. What a scale of improvement is comprehended between the faculties of a Fuegian savage and a Sir Isaac Newton!

Darwin's final line on the Fuegians (in the *Voyage of the Beagle*) uses an interesting and revealing phrase in summary: "I believe, in this extreme part of South America, man exists in a lower state of improvement than in any other part of the world." You may cringe at the paternalism, but "lower state of improvement" does at least stake a claim for potential brotherhood. And Darwin did recognize the beam in his own shipmates' eyes in writing of their comparable irrationalisms:

Each [Fuegian] family or tribe has a wizard or conjuring doctor. . . . [Yet] I do not think that our Fuegians were much more superstitious than some of the sailors; for an old quartermaster firmly believed that the successive heavy gales, which we encountered off Cape Horn, were caused by our having the Fuegians on board.

I must note a precious irony and summarize (all too briefly) a bizarre and wonderful story. Were it not for paternalism, the *Beagle* might never have sailed, and Darwin would probably have lost his date with history. Regret paternalism, laugh at it, cringe mightily—but grant its most salutary, if indirect, benefit for Darwin. Captain FitzRoy had made a previous voyage to Tierra del Fuego. There he "acquired," through ransom and purchase, four Fuegian natives, whom he brought to England for a harebrained experiment in the "improvement" of "savages." They arrived at Plymouth in October 1830 and remained until the *Beagle* set sail again in December 1831.

One of the four soon died of smallpox, but the others lived at Walthamstow and received instruction in English manners, language, and religion. They attracted widespread attention, including an official summons for a visit with King William IV. FitzRoy, fiercely committed to his paternalistic experiment, planned the next *Beagle* voyage primarily to return the three Fuegians, along with an English missionary and a large cargo of totally incongruous and useless goods (including tea trays and sets of fine china) donated, with the world's best will and deepest naiveté, by women of the parish. There FitzRoy planned to establish a mission to begin the great task of improvement for the earth's most lowly creatures.

FitzRoy would have chartered a boat at his own expense to return York Minster, Jemmy Button, and Fuegia Basket to their homes. But the Admiralty, pressured by FitzRoy's powerful relatives, finally outfitted the *Beagle* and sent FitzRoy forth again, this time with Darwin's company. Darwin liked the three Fuegians, and his long contact in close quarters helped to convince him that all people share a common biology, whatever their cultural disparity. Late in life, he recalled in the *Descent of Man* (1871):

> The American aborigines, Negroes and Europeans differ as much from each other in mind as any three races that can be named; yet I was incessantly struck, whilst living with the Fuegians on board the 'Beagle,' with the many little traits of character, showing how similar their minds were to ours.

FitzRoy's noble experiment ended in predictable disaster. They docked near Jemmy Button's home, built huts for a mission station, planted European vegetables, and landed Mr. Matthews, avatar of Christ among the heathen, along with the three Fuegians. Matthews lasted about two weeks. His china smashed, his vegetables trampled, FitzRoy ordered him back to the *Beagle* and eventually left him in New Zealand with his missionary brother.

FitzRoy returned a year and a month later. He met Jemmy Button, who told him that York and Fuegia had robbed him of all his clothes and tools, and left by canoe for their own nearby region. Jemmy, meanwhile, had "reverted" completely to his former mode of life, though he remembered some English, expressed much gratitude to FitzRoy, and asked the captain to take some presents to his special friends—"a bow and quiver full of arrows to the schoolmaster of Walthamstow . . . and two spearheads made expressly for Mr. Darwin." In a remarkable example of stiff upper lip in the face of adversity, FitzRoy put the best possible spin upon a personal disaster. He wrote in conclusion:

> Perhaps a ship-wrecked seaman may hereafter receive help and kind treatment from Jemmy Button's children; prompted, as they can hardly fail to be, by the traditions they will have heard of men of other lands; and by an idea, however faint, of their duty to God as well as their neighbor.

But the strongest argument for admiring Darwin lies not in the relatively beneficent character of his belief, but in his chosen form of action upon these convictions. We cannot use a modern political classification—Bork vs. Marshall on affirmative action—as termini of an old spectrum. Thurgood Marshall's end did not exist for the policymakers of Darwin's day. All were racists by modern standards. On that spectrum, those we now judge most harshly urged that inferiority be used as an excuse for dispossession and slavery, while those we most admire in retrospect urged a moral principle of equal rights and nonexploitation, whatever the biological status of people.

Darwin held this second position with the two Americans best regarded by later history: Thomas Jefferson and Darwin's soulmate (for they shared the same birthdate) Abraham Lincoln. Jefferson, though expressing himself tentatively, wrote: "I advance it, therefore, as a suspicion only, that the blacks . . . are inferior to the whites in the endowment both of body and of mind." But he wished no policy of forced social inequality to flow from this suspicion: "Whatever be their degree of talents, it is no measure of their rights." As for Lincoln, many sources have collected his chilling (and frequent) statements about black inferiority. Yet he is national hero numero uno for his separation of biological assessment from judgments about moral issues and social policies.

Darwin, too, was a fervent and active abolitionist. Some of the most moving passages ever written against the slave trade occur in the last chapter of the *Voyage of the Beagle.* Darwin's ship, after calling at Tahiti, New Zealand, Australia, and South Africa (where FitzRoy and Darwin submitted their bit of juvenilia to a local paper) stopped for a last visit in Brazil, before setting a straight course to England. Darwin wrote:

> On the 19th of August we finally left the shores of Brazil. I thank God I shall never again visit a slave-country. . . . Near Rio de Janeiro I lived opposite to an old lady, who kept

screws to crush the fingers of her female slaves. I have stayed in a house where a young household mulatto, daily and hourly, was reviled, beaten, and persecuted enough to break the spirit of the lowest animal. I have seen a little boy, six or seven years old, struck thrice with a horse-whip (before I could interfere) on his naked head, for having handed me a glass of water not quite clean. . . . I was present when a kind-hearted man was on the point of separating forever the men, women, and little children of a large number of families who had long lived together.

In the next line, Darwin moves from description to refutation and plea for action:

I will not even allude to the many heart-sickening atrocities which I authentically heard of;—nor would I have mentioned the above revolting details, had I not met with several people so blinded by the constitutional gaiety of the negro as to speak of slavery as a tolerable evil.

Refuting the standard argument for benevolent treatment with a telling analogy from his own land, Darwin continues:

It is argued that self-interest will prevent excessive cruelty; as if self-interest protected our domestic animals, which are far less likely than degraded slaves to stir up the rage of their savage masters.

Though I have read them a hundred times, I still cannot encounter Darwin's closing lines without experiencing a spinal shiver for the power of his prose—and without feeling great pride in having an intellectual hero with such admirable human qualities as well (the two don't mesh very often):

Those who look tenderly at the slave owner and with a cold heart at the slave, never seem to put themselves into the position of the latter; what a cheerless prospect, with not even a hope of change! Picture to yourself the chance, ever hanging over you, of your wife and your little children— those objects which nature urges even the slave to call his own—being torn from you and sold like beasts to the first bidder! And these deeds are done and palliated by men, who profess to love their neighbors as themselves, who believe in God, and pray that his Will be done on earth! It makes one's blood boil, yet heart tremble, to think that we Englishmen and our American descendants, with their boastful cry of liberty, have been and are so guilty.

Thus, if we must convene a court more than 150 years after the event—a rather foolish notion in any case, though we seem driven to such anachronism—I think that Darwin will pass through the pearly gates, with perhaps a short stay in purgatory to think about paternalism. What then is the antidote to paternalism and its modern versions of insufficient appreciation for human differences (combined with too easy an equation of one's own particular and largely accidental way with universal righteousness)? What else but the direct and sympathetic study of cultural diversity—the world's most fascinating subject in any case, whatever its virtues in moral education. This is the genuine theme behind our valuable modern movement for pluralism in the study of literature and history—for knowing the works and cultures of minorities and despised groups rendered invisible by traditional scholarship.

I don't deny that occasional abuses have been perpetrated by people with strong emotional commitments to this good cause; what else is new? But the attempt by even more zealous conservatives to distort and caricature this movement as a leftist fascism of "political correctness" ranks as a cynical smokescreen spread to cover a power struggle for control of the curriculum. Yes, Shakespeare foremost and forever (Darwin too). But also teach about the excellence of pygmy bushcraft and Fuegian survival in the world's harshest climate. Dignity and inspiration come in many guises. Would anyone choose the tinhorn patriotism of George Armstrong Custer over the eloquence of Chief Joseph in defeat?

Finally, think about one more Darwinian line—perhaps the greatest—from the slavery chapter in the *Voyage of the Beagle.* We learn about diversity in order to understand, not simply to accept:

If the misery of our poor be caused not by the laws of nature, but by our institutions, great is our sin.

19 | Ten Thousand Acts of Kindness

THE VISITOR'S CENTER at Petrified Forest National Park, in Arizona, houses an exhibit both heartwarming and depressing. Signs throughout the park beg, exhort, order, and plead with visitors not to collect and keep any fossil wood, lest the park be denuded on less than a geological time scale. The exhibit contains pieces of wood stolen from the park, but returned in guilt—the heartwarming side.

The depressing side resides in the notes written to explain decisions to send the contraband back to its natural place. No note from an adult cites any moral principle or even a personal sense of guilt. All tell tales of bad luck, usually trivial rather than catastrophic, that occurred soon after the theft—Uncle Joe's broken hip or three hundred bucks worth of fender bending. The wood, as an evil talisman, must be returned. Apparently, these penitents understand neither principles of conservation nor laws of probability. A single exception—restoring one's faith in primal feeling—lies in the only letter from a child:

> Dear Mr. Ranger, I took this and felt bad later. I'm sending it back. I'm sorry.

I have often wondered why so many people feel compelled to take such a souvenir in the face of so many good reasons for abstaining. I know that the motives are varied and complex, but I believe that for many people a primary impetus arises from a common misunderstanding about fossils.

Many people think that fossils, almost by definition, are rare and precious. (Some are, of course—the six specimens of *Archae-opteryx* and our limited evidence of human ancestry, for example.) The urge to own something both uncommon and unusual must inspire many of the thefts. But most ordinary fossils, including petrified wood, are not single jewels on vast beaches of common sand, but intrinsic and abundant parts of their geological strata. Why purloin a piece from a national park, thereby committing both an illegal and an immoral act, when petrified wood can be found in abundance at so many places right outside the park boundaries? The fossils are beautiful, and they are tempting. But they are also plentiful.

Fossils are, for the most part, not comparable with single archeological sites—limited to one spot and easily exhausted without hope of replenishment. Destroy Troy by careless collecting, and that's that forever. (On this point, two tangential and contentious comments deserve essays in themselves, but shall have to pass by in epitome. First, most fossil localities should not be regulated like unique archeological sites. Fossils in the ground, wrapped in red tape, are worthless, and fossils exposed in an outcrop will quickly be weathered and destroyed if not collected. I abhor both careless collecting and commercial exploitation of fossils with scientific value, but misplaced regulation, based on a false taxonomy that equates paleontology with archeology, can be just as harmful. Second, paleontological expeditions are not called digs, because we so rarely go to a single spot and excavate. Since specimens are usually intrinsic to strata and spread throughout wide areas of outcrop, digging in one place would be a very foolish way to collect most fossils. But the archeologists' term *dig* has permeated pop culture, where it lurks as a snare for the unwary bluffer. If you wish to prove the opposite of your intention to impress, then ask a paleontologist, as I have been asked perhaps a thousand times, "Have you been on any interesting digs lately?" Sorry for those petty explosions, but grant me the catharsis of getting some parochial issues off my chest.)

Many fossils, then, are abundant components of their strata, exposed over miles of outcrop: Just consider the clam shells exposed by the millions on polished marble surfaces in the bathrooms of New York's finest art deco skyscrapers, or the thousands of *Turitella* shells weathering out of the limestone in older parts of Quasimodo's bailiwick at Notre Dame de Paris.

The Big Badlands of South Dakota replay the tale of the Petrified Forest. Fossil vertebrates can be outstandingly abundant, and these beds have been collected by professionals for more than a century. In much of the Brule Formation, source of the "worst" terrain, fossils are so common that every tiny pinnacle and elevation has a bone on top. (The fossils are harder than the enclosing sediments. Bones and teeth therefore weather out to form tops of tiny promontories, capping and protecting a column of sediment below, while surrounding rock crumbles away on all sides.) Yet visitors think they are seeing precious and tempting rarities on the official trails—and the stealing begins again. On the major nature trail, park officials covered the best specimens with plastic boxes. But people broke the boxes and took the bones underneath. So naturalists replaced the real fossils with casts and then put the plastic boxes back for good measure!

This extraordinary abundance of some fossils illustrates something important about the history of life. Evolution is a theory about change through time—"descent with modification," in Darwin's words. Yet when fossils are most abundant during substantial stretches of time, well-represented species are usually stable throughout their temporal range or alter so little and in such superficial ways (usually in size alone) that an extrapolation of observed change into longer periods of geological time could not possibly yield the extensive modifications that mark general pathways of evolution in larger groups. Most of the time, when the evidence is best, nothing much happens to most species.

Niles Eldredge and I have tried to resolve this paradox with our theory of punctuated equilibrium. We hold that most evolution is concentrated in events of speciation, the separation and splitting off of an isolated population from a persisting ancestral stock. These events of splitting are glacially slow when measured on the scale of a human life—usually thousands of years. But slow in our terms can be instantaneous in geological perspective. A thousand years is one-tenth of one percent of a million years, and a million years is a good deal less than average for the duration of most fossil species. Thus, if species tend to arise in a few thousand years and then persist unchanged for more than a million, we will rarely find evidence for their momentary origin, and our fossil record will only tap the long periods of prosperity and stability. Since fossil deposits of overwhelming abundance record such periods of success for widespread species living in stasis, we can resolve the apparent paradox that when fossils are most common, evolution is most rarely observed.

The abundant fossils of the classic Big Badlands strata provide an excellent illustration of this paradox. My colleague Donald Prothero has been studying all well-preserved mammalian species in these deposits. He finds that none change gradually during their residence in Big Badlands strata. New species enter with geological abruptness, either because they have evolved *in situ* as the theory of punctuated equilibrium predicts or because they have simply migrated into the area.

One of my graduate students, Tim Heaton, recently completed a thesis on the most common genus of rodents (themselves the most diverse group of mammals) from Oligocene sediments throughout western North America, prominently including the Big Badlands. Paleontologists divide the Oligocene into three "land mammal ages" called Chadronian, Orellan, and Whitneyan. Heaton's genus, *Ischyromys*, is relatively rare in the Chadronian, but fantastically abundant in the Orellan, where thousands of jaws have been collected (and nearly all—in an extended fit of admirable activity—photographed, measured, and statistically analyzed by Heaton).

The Orellan *Ischyromys* has a traditional interpretation consistent with conventional views of evolutionary gradualism. The Orellan sequence has been read as a tale of steady increase in size within a single species. But Heaton's statistical work on several thousand specimens has disproved this old idea in favor of an opposite interpretation. Heaton finds two separate species, one small and one large, in the lower Orellan; the small species then becomes extinct and only the large form persists into the upper Orellan. Neither species shows much, if any, change throughout its range (the large form may undergo a slight size increase in the upper beds). The old impression of gradual increase results from mixing the two species together and falsely treating the complex as a single form. As the small species decreases in abundance and finally dies off, average size of the whole complex increases because more and more (and finally all) specimens represent the stable large form—not because any gradual evolution is occurring.

On the other hand, in the older Chadronian beds, where *Is-*

chyromys is relatively rare, Heaton has discovered a previously unrecognized richness of taxonomic diversity: several species of *Ischyromys* and a related genus, *Titanotheriomys.* Although none of these species shows any change after its origin (most are too rare to provide much evidence for anything beyond simple existence), this diversity illustrates marked evolutionary activity for *Ischyromys* in the Chadronian, while Heaton has shown that nothing happens (beyond the extinction of the small species) in the overlying Orellan, where *Ischyromys* is so abundant. The small, isolated, and rapidly speciating populations that produced so much evolution among Chadronian *Ischyromys* did not often leave their calling cards in the fossil record.

Again, we note the paradox: Nothing much happens for most of the time when evidence abounds; everything happens in largely unrecorded geological moments. We could attribute this pattern to a devious or humorous God, out to confuse us or merely to chuckle at our frustration. But I choose to look upon this phenomenon in a positive light. There is a lesson, not merely frustration, in the message that change occurs in infrequent bursts and that stability is the usual nature of species and systems at any moment.

Being human, I love to toot my own horn in support of punctuated equilibrium. But I am writing this essay for another reason. What's past (in this essay), as the Bard says, is prologue—a prologue to make a point by analogy about the real subject of this essay: the vexatious issue of human nature.

Let us return to the irony of *Ischyromys* in the Chadronian and Orellan, and of punctuated equilibrium in general. Evolution has constructed the tree of life; yet, at almost any moment for any species, change is not occurring and stasis prevails. If we then ask, What is the normal nature of a species? the only possible reply is, stability. Yet exquisitely rare change has built the tree of life and made history on a broad scale. We now come to the nub of my argument: The defining property of a species, its normal state, its nature, its appearance at almost any time stands contrary to the process that makes history (and new species). If we tried to infer the nature of species from the process that constructs the history of life, we would get everything precisely backward!—for events of great rarity (but with extensive consequences) make history.

The same separation should be enforced between human nature and the events that construct our history. We have committed an enormous error in assuming that the behavioral traits involved in history-making events must define the ordinary properties of human nature. Must we not link the causes of our history, or so the false argument goes, to the nature of our being?

But if my analogy holds, precisely the opposite might be true. If rare behaviors make history, then our usual nature must be defined by our general actions in an everyday world that engulfs us nearly all the time, but does not set the fate of nations. The causes of history may be opposed to the ordinary forces that prevail at almost every moment—just as the processes that construct the tree of life are invisible and inactive nearly all the time within stable species.

History is made by warfare, greed, lust for power, hatred, and xenophobia (with some other, more admirable motives thrown in here and there). We therefore often assume that these obviously human traits define our essential nature. How often have we been told that "man" is, by nature, aggressive and selfishly acquisitive?

Such claims make no sense to me—in a purely empirical way, not as a statement about hope or preferred morality. What do we see on any ordinary day on the streets or in the homes of any American city—even in the subways of New York? Thousands of tiny and insignificant acts of kindness and consideration. We step aside to let someone pass, smile at a child, chat aimlessly with an acquaintance or even with a stranger. At most moments, on most days, in most places, what do you ever see of the dark side— perhaps a parent slapping a child or a teenager on a skateboard cutting off an old lady? Look, I'm no ivory-tower Pollyanna, and I did grow up on the streets of New York. I understand the unpleasantness and danger of crowded cities. I'm only trying to make a statistical point.

Nothing is more unfamiliar or uncongenial to the human mind than thinking correctly about probabilities. Many of us have the impression that daily life is an unending series of unpleasantnesses—that 50 percent or more of human encounters are stressful or aggressive. But think about it seriously for a moment. Such levels of nastiness cannot possibly be sustained. Society would devolve to anarchy in an instant if half our overtures to another human being were met with a punch in the nose.

No, nearly every encounter with another person is at least neutral and usually pleasant enough. *Homo sapiens* is a remarkably genial species. Ethologists consider other animals relatively peaceful if they see but one or two aggressive encounters while observing an organism for, say, tens of hours. But think of how many millions of hours we can log for most people on most days without noting anything more threatening than a raised third finger once a week or so.

Why, then, do most of us have the impression that people are so aggressive, and intrinsically so? The answer, I think, lies in the asymmetry of effects—the truly tragic side of human existence. Unfortunately, one incident of violence can undo ten thousand acts of kindness, and we easily forget the predominance of kindness over aggression by confusing effect with frequency. One racially motivated beating can wipe out years of patient education for respect and toleration in a school or community. One murder can convert a friendly town, replete with trust, into a nexus of fear with people behind barred doors, suspicious of everyone and afraid to go out at night. Kindness is so fragile, so easy to efface; violence is so powerful.

This crushing and tragic asymmetry of kindness and violence is infinitely magnified when we consider the causes of history in the large. One fire in the library of Alexandria can wipe out the accumulated wisdom of antiquity. One supposed insult, one crazed act of assassination, can undo decades of patient diplomacy, cultural exchanges, peace corps, pen pals—small acts of kindness involving millions of citizens—and bring two nations to a war that no one wants, but that kills millions and irrevocably changes the paths of history.

Yes, I fully admit that the dark side of human possibility makes most of our history. But this tragic fact does not imply that behavioral traits of the dark side define the essence of human nature. On the contrary, I would argue, by analogy to the ordinary versus the history-making in evolution, that the reality of human interactions at almost any moment of our daily lives runs contrary, and must in any stable society, to the rare and disruptive events that construct history. If you want to understand human nature, defined as our usual propensities in ordinary situations, then find out what traits make history and identify human nature with the opposite sources of stability—the predictable behaviors of nonaggression that prevail for 99.9 percent of our lives. The real tragedy of human existence is not that we are nasty by nature, but that a cruel structural asymmetry grants to rare events of meanness such power to shape our history.

An obvious argument against my thesis holds that I have confused a social possibility of basically democratic societies with a more general human propensity. This alternative view might grant my claims that stability must rule at nearly all moments and that much rare events make history. But perhaps this stability reflects the behaviors of geniality only in relatively free and democratic societies. Perhaps the stability of most cultures has been achieved by the same "dark" forces that make history when they break out of balance—fear, aggression, terror, and domination of rich over poor, men over women, adults over children, and armed over defenseless. I allow that dark forces have often kept balances, but still strongly assert that we fail to count the ten thousand ordinary acts of nonaggression that overwhelm each overt show of strength even in societies structured by domination and even if nonaggression prevails only because people know their places and do not usually challenge the sources of order. To base daily stability on anything other than our natural geniality requires a perverted social structure explicitly dedicated to breaking the human soul—the Auschwitz model, if you will. I am not, by the way, asserting that humans are either genial or aggressive by inborn biological necessity. Obviously, both kindness and violence lie within the bounds of our nature because we perpetuate both, in spades. I only advance a structural claim that social stability rules nearly all the time and must be based on an overwhelmingly predominant (but tragically ignored) frequency of genial acts, and that geniality is therefore our usual and preferred response nearly all the time.

Please don't read this essay as a bloated effort in the soft tradition of, dare I say it, liberal academic apologies for human harshness, or wishy-washy, far-fetched attempts to make humans look good in a world of woe. This is not an essay about optimism; it is an essay about tragedy. If I felt that humans were nasty by nature, I would just say, the hell with it. We get what we deserve, or what evolution left us as a legacy. But the center of human nature is rooted in ten thousand ordinary acts of kindness that define our days. What can be more tragic than the structural paradox that this Everest of geniality stands upside down on its pointed summit and can be toppled so easily by rare events contrary to our everyday nature—and that these rare events make our history. In some deep sense, we do not get what we deserve.

The solution to our woes lies not in overcoming our "nature" but in fracturing the "great asymmetry" and allowing our ordinary propensities to direct our lives. But how can we put the commonplace into the driver's seat of history?

20 | The Declining Empire of Apes

GOD MUST HAVE created mistakes for their wonderful value in illuminating proper pathways. In all of evolutionary biology, I find no error more starkly instructive, or more frequently repeated, than a line of stunning misreason about apes and humans. I have been confronted by this argument in a dozen guises, from the taunts of fundamentalists to the plaints of the honorably puzzled. Consider this excerpt from a letter of April 1981: "If evolution is true, and we did come from apes, then why are there still apes living. It seems if we evolved from them they should not be here."

If we evolved from apes, why are apes still around? I label this error instructive because its correction is so transforming: If you accept a false notion of evolution, the statement is a deep puzzle; once you reject this fallacy, the statement is evident nonsense (in the literal sense of unintelligible, not the pejorative sense of foolish).

The argument is nonsense because its unstated premise is false. If ancestors are groups of creatures that are bodily transformed, each and every one, into descendants, then human existence would preclude the survival of apes. But, plainly, we mean no such thing in designating groups as ancestors—lest no reptiles remain because birds and mammals evolved or no fishes survive because amphibians once crawled out upon the land.

Ladders and bushes, the wrong and right metaphors respectively for the topology of evolution, resolve the persistent nonpuzzle of why representatives of ancestral groups (apes, for example) can survive alongside their descendants (humans, for example). Since evolution is a copiously branching bush, the emergence of humans from apes only means that one branch within the bush of apes split off and eventually produced a twig called *Homo sapiens,* while other branches of the same bush evolved along their own dichotomizing pathways to yield the other descendants that share most recent common ancestry with us—gibbons, orangutans, chimps, and gorillas, collectively called apes. (These modern apes are, by genealogy, no closer than we are to the common ancestor that initiated the ape-monkey split more than 20 million years ago, but human hubris demands separation—so our vernacular saddles all modern twigs but us with the ancestral name ape. The figure and its caption should make this clear.)

The proper metaphor of the bush also helps us to understand why the search for a "missing link" between advanced ape and

A Few Branches on the Bush of Apes and Old World Monkeys

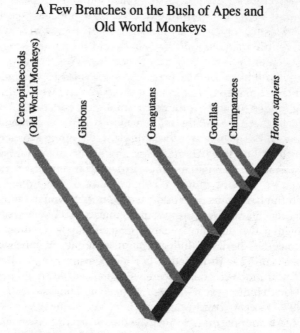

The genealogical sequence of branching in the evolution of apes and humans.

incipient human—that musty but persistent hope and chimera of popular writing—is so meaningless. A continuous chain may lack a crucial connection, but a branching bush bears no single link at a crucial threshold between no and yes. Rather, each branching point successively restricts the range of closest relatives—the ancestors of all apes separate from monkeys, then gibbon lineages from ancestors of other great apes and humans, then forebears of the orangutan from the chimp-gorilla-human complex, finally precursors of chimps from the ancestors of humans. No branch point can have special status as *the* missing link—and all represent lateral relationships of diversification, not vertical sequences of transformation.

An even more powerful argument on behalf of the bush arises from the reanalysis of classical ladders in our textbooks, particularly the evolution of modern horses from little eohippus and the "ascent of man" from "the apes." A precious irony—life's little joke—pervades these warhorses of the ladder: The "best" examples must be based upon highly *unsuccessful* lineages, bushes so pruned of diversity that they survive as single twigs. (See my essay "Life's Little Joke" in *Bully for Brontosaurus* for a fuller version of this argument.)

Successful bushes never enter our texts as classical trends, because they boast too many related survivors, and we can draw no rising ladder for the evolution of antelopes, rodents, or bats—although these are the three great success stories of mammalian evolution. But if only one twig survives, we apply a conceptual steamroller and linearize its labyrinthine path of lateral branching back to the main stem of its depleted bush. Horses, rhinos, and tapirs are not glorious culminations of ascending series within the Perissodactyla (odd-toed hoofed mammals) but three little twigs, barely hanging on, the remnants of a bush that once dominated the diversity of large mammalian herbivores. Similarly, we can specify a ladder of human ascent only because the bush of apes has dwindled to a few surviving twigs, all clearly distinct. If the bush of apes were vigorous and maintained a hundred branchlets evenly spaced at an expanding periphery, we would have many cousins and no chain of unique ancestors. Our vaunted ladder of progress is really the record of declining diversity in an unsuccessful lineage that then happened upon a quirky invention called consciousness.

This argument against human arrogance can be grasped well enough as an abstraction but becomes impressive only with its primary documentation—the record of vigorous diversity among apes in former times of greater success. The theme of previous vigor has recently received a boost from a new discovery—one that I had the great good fortune to witness last year.

The cercopithecoid, or Old World, monkeys are the closest relatives of the ape-human bush. Robert Jastrow, in his recent, popular book, *The Enchanted Loom: Mind in the Universe*, contrasts the evolutionary fate of these two sister groups:

> The monkey did not change very much from the time of his appearance, 30 million years ago, to the present day. His story was complete. But the evolution of the ape continued. He grew large and heavy, and descended from the trees.

This statement, so preciously wrong, so perfectly arse-backward, shows just how far astray the metaphor of the ladder can lead. There is no such creature, not even as a useful abstraction, as *the* monkey or *the* ape. Evolution's themes are diversity and branching. Most apes (gibbons and orangutans, and chimps and gorillas a good part of the time) are still living in trees. Old World monkeys have not stagnated; they represent the greatest success story among primates, a bush in vigorous radiation and including among its varied products baboons, colobins, rhesus and proboscis monkeys.

In fact, precisely opposite to Jastrow's claim, apes have been continuously losing and cercopithecoids gaining by the proper criteria of diversity and expansion of the bush. Let us go back to the early Miocene of Africa, some 20 million years ago, soon after the ape-monkey split, and trace the fate of these two sister groups. First of all, we would not find these Miocene ancestors as different from each other as their descendants are today—limbs of a bush usually diverge. Early Miocene apes were quite monkey-like in their modes of life. Compared with monkey forebears, early apes tended to be larger, more tree bound, more narrowly tied to fruit eating, and less likely to cope with a strongly seasonal or open environment.

Second—and the crucial point for this essay—apes were more common in two important senses during the early Miocene: more common than cercopithecoid monkeys at this early stage in their mutual evolution, and absolutely more diverse (just in Africa) than apes are today (all over the world). Taxonomic estimates vary, and this essay cannot treat such a highly technical and contentious literature, but early Miocene African apes have been placed in some three to five genera and perhaps twice as many species.

The next snapshot of time, the African middle Miocene, already records fewer species, although apes now appear for the first time in the fossil records of Europe and Asia. Old World monkeys meanwhile begin an acceleration extending right to our own time. Apes continue to decline and hang on in restricted habitats—yielding isolated groups of gibbons and orangutans in Asia, and chimps, gorillas, and the descendants of a small African group called australopithecines. If the resident zoologist of Galaxy X had visited the earth 5 million years ago while making his inventory of inhabited planets in the universe, he would surely have corrected his earlier report that apes showed more promise than Old World monkeys and noted that monkeys had overcome an original disadvantage to gain domination among primates. (He will confirm this statement after his visit next year—but also add a footnote that one species from the ape bush has enjoyed an unusual and unexpected flowering, thus demanding closer monitoring.)

We do not know why apes have declined and monkeys prevailed. We have no evidence for "superiority" of monkeys; that is, for direct struggles of Darwinian competition between apes and monkeys in the same habitat, with ape extinction and cercopithecoid prevalence as a result. Perhaps a greater flexibility in diet and environmental tolerance allowed monkeys to gain the edge, without any direct competition, in a world of changing climate and fewer stable habitats of trees and fruit. According to this interpretation, those few apes that could adapt to a more open, ground-living existence, had to develop some decidedly odd features, not in any way "prefigured" by their initial design—the knuckle walking of chimps and gorillas, and the upright gait of australopithecines and you know who.

This striking reversal of Jastrow's homily, and of all standard biases of the ladder, rests most forcefully upon the comparison of initial Miocene success with later restriction of the bush of apes.

But how great was this first flowering, and how severe, therefore, the later pruning? Unfortunately, this most crucial of all empirical questions encounters the cardinal problem of our woefully imperfect fossil record. We know the extent of later pruning; it is not likely that any living species of ape remains undiscovered on our well-explored earth. But what was the true diversity of early Miocene apes? Did they live only in Africa? What fraction of the African fauna has been preserved? What have we collected and identified of the material that has been preserved?

If our current collections contain most of what actually lived, then the pruning has been notable but modest. But suppose that we have only 10 percent or even only half the true diversity, then the story of decline and restriction among apes is far more pronounced. How can we know how much we have?

One rough indication—about the best we can do at this early stage of knowledge about Miocene primates in Africa—comes from the composition of new collections. Suppose that every time we find new early Miocene apes in Africa, they belong to species already in our collections. After several repetitions (particularly if our collections span a good range of geographies and environments), we might conclude that we have probably sampled a substantial amount of the true bush. But suppose that new sites yield new species most of the time—and that we can mark no real decline in the number of novelties. Then we might conclude that we have sampled only a small part of a much more copious bush—and that the story of decline and shortfall in the empire of apes has been more profound than we realized. Quite an effective antidote to the bias of the ladder and its attendant invitation to human arrogance!

In other words, we are seeking, as my colleague David Pilbeam, our leading student of fossil apes, said to me, "an asymptote" in the discovery of new apes. An asymptote is a limiting value approached by one variable of a curve as the other variable (often time or number of trials) increases towards infinity. When further collecting of fossils only yields more specimens of the same species, we have probably reached the asymptote in recoverable kinds of apes. We also reach asymptotes fairly quickly in training cats or cajoling children and should learn to recognize both the subtle point of diminishing returns and the actual asymptote not much further down the line.

An exciting discovery about the history of Miocene apes has recently furnished our best evidence that we have not yet come near the asymptote of the early bush of apes. This discovery provides the strongest possible evidence for an even greater intensity of life's little joke in our own evolution. The bush was bushier, the later decline in diversity more profound. We do not yet know the true extent of the initial success of apery.

In January 1986, I spent a week with Richard Leakey at his field camp on early Miocene sediments near the western shore of Lake Turkana in Africa's Great Rift Valley. Little vegetation obscures the geology of this arid region, and naked sediments stretch for miles, their eroding fossils littering the surface.

The data on genetic differences between chimps and humans suggest that our twig on the bush of apes last shared a common ancestor with chimps some 5 to 8 million years ago; in other words, the human lineage has been entirely on its own only for this short stretch of geological time. The oldest human fossils are less than 4 million years old, and we do not know which branch on the copious bush of apes budded off the twig that led to our lineage. (In fact, except for the link of Asian *Sivapithecus* to the modern orangutan, we cannot trace any fossil ape to any living species. Paleontologists have abandoned the once popular notion that *Ramapithecus* might be a source of human ancestry.) Thus, sediments between 4 and 10 million years in age are potential guardians of the Holy Grail of human evolution—the period when our lineage began its separate end run to later domination and a time for which no fossil evidence exists at all.

Richard Leakey almost surely has many square miles of good sediment from this crucial time in his field area at West Turkana. But he is not yet searching these beds. He is concentrating his efforts on older rocks of the early Miocene (15 to 20 million years ago) when the bush of apes had its great initial flowering in Africa. He is working before the time of maximal intrigue for several reasons. In part, he may be saving the best for later, perfecting his techniques and "feel" for the region before zeroing in on the potential prize. He also has the fine intuition and horse sense of any good historian—it may be best to begin at the beginning and work forward. But, most importantly, he has a professional's understanding that problems of maximal public acclaim are not always the issues of greatest scientific importance.

The public may yearn, above all, to know the status of our common ancestor with chimpanzees, but Richard Leakey recognizes that the early Miocene is also a time of mystery, promise, and conceptual importance: mystery because we know so little about the actual diversity of apes at this time of their greatest success; promise because he has sediments that can deliver many of the missing goods; conceptual importance because we have as much to learn from documenting the base of our ancestral bush as in searching for the little branchlet that led directly to us later on. The early Miocene is a good place to explore.

The ground of West Turkana glistens with crystals of quartz and calcite. The local Turkana children, passing time during long hours of tending goats under the relentless sun, collect geodes into piles and smash them to reveal the crystals inside. We are looking for duller fragments of bone.

There are no great secrets to success, no unusual basis for "Leakey's luck," beyond hard work and experience. In some areas, fossil-bearing strata are rare and must be traced through geological complexities of folding and faulting to assure that fieldworkers search only in profitable places. But here, the entire sequence is fair game (although some strata, as always, are richer than others), and all exposures of rock must be scrutinized. The key to success becomes patience and a trained workforce.

Leakey maintains a staff of trained Kenyan observers. He provides a long course in practical mammalian osteology (study of bones)—until they can distinguish the major groups of mammals from small scraps. The main ingredient of Leakey's luck is unleashing these people in the right place.

Kamoya Kimeu supervises this exploration. He has found more important fossils than any one else now alive. One night in camp, he told me his story. As a boy, he tended goats, sheep, and cattle for his father. He attended school for six years and then went to work for a farmer. His employer urged him to return to school and study to become a veterinary paramedic. Kamoya then walked for several days back to Nairobi, where his uncle told him that Louis Leakey, Richard's father, was recruiting people to "dig bones." His mother gave him only cautious approval, telling him to quit and come home if the task involved (as he then suspected) digging up human graves. But when he saw so many bones from so many kinds of creatures, he knew that nature had strewn these burial grounds. The sediments of West Turkana are, if anything, even more profuse.

When I arrived on January 16, Kamoya's team had just found a new and remarkably well-preserved ape skull (in a profession that usually works with fragments, mostly teeth, a skull more than half complete, and with a fully preserved dentition, is cause for rejoicing). The next day, we studied and mapped the geological context and then brought the specimen back to camp. I wrote in my field book: "Everyone is very excited because they have just found the finest Miocene ape skull known from Africa. It is quite new—with a long face, inflated nasal region, incisors worn flat with a diastema [gap] a finger wide to the massive canine—almost like a beaver among apes."

Research is a collectivity, and we all have our special skills.

The greatest of all fossil finders Kamoya Kimeu gathering material at West Turkana. *Photograph by Delta Willis.*

Kamoya's workers are the world's greatest spotters; Richard also has a hawk's eye, the intuition of a geologist who has lived with his land, and the organizational skills of a Washington kingpin; his wife, Meave, has an uncanny spatial sense and can beat any jigsaw champ in putting fossil fragments together; yours truly, I fear, is good for one thing only—seeing snails.

All field naturalists know and respect the phenomenon of "search image"—the best proof that observation is an interaction of mind and nature, not a fully objective and reproducible mapping of outside upon inside, done in the same way by all careful and competent people. In short, you see what you are trained to view—and observation of different sorts of objects often requires a conscious shift of focus, not a total and indiscriminate expansion in the hopes of seeing everything. The world is too crowded with wonders for simultaneous perception of all; we learn our fruitful selectivities.

I couldn't see bone fragments worth a damn—and Richard had to direct my gaze before I could even distinguish the skull from surrounding lumps of sediment. But could I ever see snails, the subject of my own field research—and no one else had ever found a single snail at that site. So I rest content with my minuscule contribution, made in character, to the collective effort. At the top right of page 143 in the November 13, 1986, issue of *Nature*—the article that describes the new skull—a few snails are included in the faunal list of the site, some added by my search image. (I also found, I believe, the first snails at the important South African hominid site of Makapansgat in 1984—where I also couldn't see a bone. I think I am destined to be known in the circle of hominid exploration as "he who only sees the twisted one.")

The traditions of nature writing demand that this personal narrative now continue at some length, with overwritten paeans to the wonder of this discovery, set in glowing clichés about the stark and fragile (two good adjectives) beauty of the countryside. But I desist. First of all, this isn't my style; it also doesn't match anything that actually happens in the field. People have varied reactions to such good fortune. Some may jump up and down, fall upon their knees to praise God, or wax eloquent about the new line wrested from nature's complex book. Most people I know, certainly including Richard, Kamoya, and myself as out-

sider, do not have personalities that match these romantic stereotypes. The conversation may flow more happily at dinner; some
kind of glow must form within. But you still have to make sure
that the trucks have gas, that the water jugs are full—and you do
have to get up at dawn the next day because it's too hot to work in
the afternoon. My favorite kind of excitement is quiet satisfaction.

Richard and Kamoya's team found a second, smaller ape skull
that field season at West Turkana. Both are new genera, not
merely variants on familiar themes of the ape's bush. Richard and
Meave Leakey published two papers in the November 13, 1986,
issue of *Nature* describing these new forms as *Afropithecus* (the one
I witnessed) and *Turkanapithecus*. In the most interesting line of
the *Afropithecus* paper, they write: "*Afropithecus* displays characters
typical of a variety of Miocene hominoids combined in a single
taxon." In other words, this new genus represents a unique combination of features known to vary among early apes—as if we
might shuffle the known variations into many more plausible
combinations as yet undiscovered. The bottom line after all this
exegesis is simplicity itself: We are not at, perhaps not even near,
the asymptote for true diversity of apes at their flourishing beginning. If one field season in uncharted lands could yield two new
genera, how many remain undiscovered in the hundreds of
square miles still open for exploration? Apes were bushier than
we had ever imagined during their early days; human evolution
seems even more twiggy, more contingent on the fortunes of
history (not enjoined like the successive rungs of a ladder), less
ordained, and more fragile. Our vaunted march to progress, the
standard iconography of our evolution, is just one more expression of life's little joke.

I have consciously permitted a professional's bias to permeate
this essay so far. I have been equating "success" with numbers of
branches on the bush—for paleontologists tend to view large-
scale evolution as the differential birth and death of species, and
we slip too easily into an equation of success with exuberance of
branching. But, of course, we must also consider the quality of
twigs, not merely their number. *Homo sapiens* is one small twig,
holding with just a few others all the heritage of a group once far
more diverse in branches. Yet our twig, for better or for worse,
has developed the most extraordinary new quality in all the his-

tory of multicellular life since the Cambrian explosion. We have invented consciousness with all its sequelae from Hamlet to Hiroshima. Life's little joke shows us our fragility, our smallness on the proper metaphor of the bush, but we have turned the joke upon itself with the power of one evolutionary invention.

The prophet Micah caught both sides of this tension with great understanding when he wrote that fragility and size of origin imply little about ultimate effect: "But thou, Bethlehem Ephrathah, though thou be little among the thousands of Judah, yet out of thee shall he come forth unto me that is to be ruler in Israel" (Micah 5:2). If we could merge the two themes, and if rulers could learn humility and respect from our common origins as fragile twigs on the bush of life, then we might break the equation between ability and right to dominate and might even fulfill that most famous of Biblical prophecies, which is, after all, about the proper nurturing of trees and bushes—"and they shall beat their swords into plowshares, and their spears into pruning hooks."

6 | Grand Patterns of Evolution

Two Steps towards a General Theory of Life's Complexity

21 | The Wheel of Fortune and the Wedge of Progress

CHARLES DARWIN was a master of metaphor, and much of his success may be attributed to his uncanny feel for timely comparisons that virtually compel understanding. We all know the two metaphors that Darwin invoked to define his theory: natural selection and the struggle for existence. We might also consider Darwin's three principal descriptions of nature, each wonderfully apt and poetic, and each a metaphor.

The tangled bank: To stress the intricacy of relationships among organisms as arising, somewhat paradoxically, by planless evolution. Darwin begins the last paragraph of the *Origin of Species:*

> It is interesting to contemplate an entangled bank, clothed with many plants of many kinds, with birds singing on the bushes, with various insects flitting about, and with worms crawling through the damp earth, and to reflect that these elaborately constructed forms, so different from each other, and dependent on each other in so complex a manner, have all been produced by laws acting around us.

The tree of life: Borrowed from other contexts to be sure (Proverbs 3:18, for example), but used brilliantly by Darwin to express the other form of interconnectedness—genealogical rather than ecological—and to illustrate both success and failure in the history of life. Darwin placed this famous passage at a crucial spot in his text—the very end of chapter four, marking the conclusion of his argument for natural selection (the rest of the book discusses problems and examples):

As buds give rise by growth to fresh buds, and these, if vigorous, branch out and overtop on all sides many a feebler branch, so by generation I believe it has been with the great Tree of Life, which fills with its dead and broken branches the crust of the earth, and covers the surface with its ever branching and beautiful ramifications.

The face of nature (and the darkness behind): To argue that apparent balance and harmony arise from the struggle and death of individuals:

We behold the face of nature bright with gladness, we often see superabundance of food; we do not see, or we forget, that the birds which are idly singing round us mostly live on insects or seeds, and are thus constantly destroying life; or we forget how largely these songsters, or their eggs, or their nestlings, are destroyed by birds and beasts of prey.

But if Darwin relied on metaphors to enlighten his readers, he also followed this good strategy in his private quest for understanding. Darwin's notebooks are, if anything, more awash in metaphor than his published works. I believe, along with many Darwin scholars (see, in particular, Ralph Colp's 1979 article in bibliography), that one metaphor stands out among all others in Darwin's own struggle to formulate the principles of natural selection—the metaphor of the wedge. At the very least, this comparison holds pride of place as the first image invoked by Darwin to explicate his initial statement of natural selection.

Great ideas, like species, do not have "eureka" moments of sudden formulation in all their subtle complexity; rather, they ooze into existence along tortuous paths lined with blind alleys (to invoke a metaphor). Still, not all moments are equal, and some may even be judged crucial. Darwin, at least, claimed that September 28, 1838, had been the key day for natural selection, and that his reading of Thomas Malthus's *Essay on the Principles of Population* had enabled him to put the disparate pieces of his puzzle together, as he recalled Malthus's argument that growth in population, if unchecked, must quickly outstrip food supply, leading to inevitable struggle for limited resources and death for losers. He wrote Malthus's principle into his notebook and appended, directly thereafter, his very first metaphor for his new theory of evolution:

> One may say there is a force like a hundred thousand wedges trying [to] force every kind of adapted structure into the gaps in the economy of nature, or rather forming gaps by thrusting out weaker ones.

Darwin honed and sharpened this metaphor throughout the next twenty years, as he prepared for the storm of publishing his ideas about evolution. He eventually settled upon the image of a surface absolutely chock-full with wedges, representing species in an economy of nature sporting a No Vacancy sign. Evolutionary change can only occur when one species manages to insinuate itself into this fullness by driving (wedging) another species out. Darwin developed this metaphor most fully in his manuscript for the long, unpublished version of the *Origin of Species* (a compressed account appears on page 67 of the shorter book that he eventually produced in 1859 under the pressure of A. R. Wallace's independent formulation of natural selection):

> Nature may be compared to a surface covered with ten thousand sharp wedges, many of the same shape and many of different shapes representing different species, all packed closely together and all driven in by incessant blows: the blows being far severer at one time than at another; sometimes a wedge of one form and sometimes another being struck; the one driven deeply in forcing out others; with the jar and shock often transmitted very far to other wedges in many lines of direction.

I have focused upon Darwin's commitment to the metaphor of the wedge because this unappreciated core belief, more than the notion of natural selection itself, shaped Darwin's conventional view about progress and predictability in the long-term history of life. Darwin, like any honest man in a world of such inordinate complexity, struggled hard but failed to resolve several crucial issues in the interpretation of nature. No question troubled him more than the common assumption, so crucial to Victorian Britain at the height of industrial and imperial success, that progress must mark the pathways of evolutionary change.

Darwin clearly understood that the basic mechanics of natural selection implied no statement about progress, for the theory only speaks of local adaptation to changing environments. But Darwin, as an eminent, if critical, Victorian himself, could not let go of progress so cleanly. He wished to validate predictable advance as a major theme of the fossil record, but knew that the bare bones of natural selection could not rationalize such a belief. How, then, could progress be affirmed as a fact of life's history if the fundamental theorem of organic change—natural selection—did not imply progress by itself?

To resolve this troubling discordance between the mechanics of his basic theory and his fundamental impression of pattern in life's history, Darwin called upon a second, basically ecological principle encompassed by the metaphor of the wedge. Nature, Darwin believed, is full of species ("a surface covered with ten thousand sharp wedges . . . all packed closely together"). All potential addresses are occupied, but new challengers continually arrive to compete for space. They can succeed in a full world only by driving other species out in overt competition for limited resources ("the one driven deeply in forcing out others"). It's a tough, crowded world out there; successful creatures claw their way to the top and remain there by constant vigilance and conquest. The *Origin of Species* contains several passages about progress in the history of life, and all are validated, not by the bare bones mechanism of natural selection, but by the second principle of the wedge, the vision of a full world ruled by overt competition among organisms:

> The more recent forms must, on my theory, be higher than the more ancient; for each new species is formed by having had some advantage in the struggle for life over other and preceding forms. . . . I do not doubt that this process of improvement has affected in a marked and sensible manner the organization of the more recent and victorious forms of life, in comparison with the ancient and beaten forms.

But what actual evidence do we have that long-term trends in the history of life arise by continuous and intense biological competition in a perennially crowded world? (Other models for the operation of natural selection are surely plausible. Perhaps most species do not fall victim to overt wedging by superior competitors but to changes in the physical environment that are too rapid or extensive to elicit an adaptive response.) I am persuaded that some cases in Darwin's preferred mode of organic competition have been documented. Biologist Geerat Vermcij (who lays out the brief for Darwin's view most elegantly in his book *Evolution and Escalation*), for example, has demonstrated a geological trend for thicker and stronger crab claws matched by ever more efficient defenses (spines, knobs, and thick shells) in the snails that crabs love to eat. I accept the interpretation of this lock-step escalation as an "arms race."

But just as we needed so great a scholar as Aristotle, in so weighty a place as the *Nicomachean Ethics,* to teach us that one swallow does not make a spring (yes, he said spring, not summer), a case or two in the fossil record does not establish a pattern. Directional trends produced by wedging do occur, but they scarcely cry for recognition from every quarry and hillslope. The overwhelming majority of paleontological trends tell no obvious story of conquest in competition. Why did the large and gorgeously complex ammonites crash to oblivion some 65 million years ago, while their rarer and apparently simpler closest cousins, the nautiloids, survived to our own day as the "ship of pearl" immortalized by Oliver Wendell Holmes (and Eugene O'Neill) as the builder of "more stately mansions." Why did all members of three great groups of crinoids die 225 million years ago, leaving all subsequent time to a fourth group that survives today but sports no feature ever identified as an improvement over the three losers? Why did a similar replacement of one group by another occur in reef-building corals at the same time, and why can no theme of improvement in competition be discerned here either? Why, for that matter, did dinosaurs die and mammals prevail after 100 million years of reptilian success and mammalian marginality? (If mammals were competitively superior, they certainly displayed no hurry in wedging out their "predecessors in the race for life," for 100 million years is almost two-thirds of mammalian history. Moreover, recent views on the sleekness and anatomical sufficiency of dinosaurs speak strongly against any notion of wedging by rat-sized mammals.)

I chose these four cases for two reasons. First, I think that they are far more characteristic of the fossil record—more numerous and more significant in import—than putative examples of progress by wedging. Second, they represent the kind of event that most directly challenges the metaphor of the wedge and therefore the favored rationale for progress by rigorous competition among organisms—"changing of the guard" across episodes of mass extinction. In fact, these four cases provide a small sample of events, a pair for each, in the two most celebrated mass extinctions: the Permian debacle that may have wiped out more than 95 percent of marine invertebrate species some 225 million years ago, and the Cretaceous event that removed remaining dinosaurs and gave mammals a chance some 65 million years ago.

Any reader with a good feel for the history of paleontology must now be intensely puzzled. How can I be suggesting that a study of mass extinction will alter our view about directionality in the history of life when widespread and coincident death of species is, perhaps, our oldest discovery about the stratigraphic record? The boundaries of the geological time scale, the alphabet of my profession, are defined by events of mass extinction. The two episodes cited in my cases separate the three great eras of life's multicellular history: the Permian extinction between the Paleozoic and the Mesozoic, and the Cretaceous extinction between the Mesozoic and the Cenozoic. How can something so canonical, and so long appreciated, now threaten to disrupt another cherished view about biologically based competition leading to progress in a crowded world?

A resolution to this historical puzzle calls upon a conventional interpretation of mass extinction that makes peace with, or even supports, the notion of biotic struggle as the driving vector of life. Mass extinction is a specter haunting the metaphor of the wedge, but the ghostbusters of denial and accommodation have held the fort—until recently. Darwin himself chose the more vigorous response of denial, trying his darndest to dissolve mass extinction (or at least to dilute it to insipidity) with his favorite argument about "the imperfection of the geological record." A hopelessly inadequate record can compress millions of years of missing data into an apparent "event." If we could recover the full flow of time, the individual items of a "mass extinction" would spread out on both sides of illusory suddenness, indicating a much longer period of successive disappearances at a rate little, if at all, accelerated beyond the usual tempo of wedging in ordinary times.

A more popular argument (the ghostbuster of accommodation) admits the unusual character of mass extinctions, acknowledging a substantial acceleration in the tempo of death, while arguing that the environmental stresses of these parlous times do not introduce a new regime in the causality of dying, but rather, only accentuate the power of the wedge. In these worst of times, the motor of biological competition runs faster as increased stress drives up the intensity of struggle. Mass extinction only "turns up the gain" on business as usual. If the "ancient and beaten" make their exit in ordinary times as superior wedges push themselves into the bustling economy of nature, then the rate of departure can only accelerate in the tough moments of mass extinction when dogs eat dogs and men must be men.

This honing of the wedge provides the traditional context, usually not well explained in general writing on mass extinction, that has made all the recent news about truly catastrophic causes based on extraterrestrial impact so controversial and so threatening. In 1980, the father and son, physicist and geologist team of Luis and Walter Alvarez, along with Frank Asaro and Helen Michel, first published their evidence that a large extraterrestrial object struck the earth some 65 million years ago and triggered the great Cretaceous extinction.

The explosion of fruitful scientific work that this hypothesis has engendered in just a decade must match the force of the impact itself. The Alvarezes' hypothesis has opened the small and arcane paleontological field of mass extinction into a grand arena of interdisciplinary cooperation. In October 1988, I spent three wonderful days at Snowbird, Utah, listening to several hundred excited scientists, from geochemists to planetary physicists, paleontologists to climatic modelers, ponder the causes of mass extinction. Debate on the Cretaceous event still swirls about two crucial issues: timing (one impact or many, one moment or an extended period of multiple bombardments) and cause (massive volcanic effusions have been proposed as an alternative, but impact now holds the upper hand according to most workers in this field). Moreover, the case for impact has not been established (or precluded) for other mass extinctions, so we do not know whether the Alvarezes' persuasive account of the Cretaceous episode ranks as the explanation for a single event or a general theory of mass extinction.

Nonetheless, I think we now know enough to summarize the new views in a statement that must suggest a radically revised perspective on the evolutionary meaning of mass extinction. These events, with their catastrophic causes, are *more frequent, more sudden, more profound in their extent,* and *more different* (from normal times) *in their results* than we had imagined. Mass extinction does not just turn up the gain on competition, so that wedging can proceed more ruthlessly and more efficiently; mass extinction entrains new causes that impart a distinctive stamp to evolutionary results. And if the history of life owes its shape more to the differential success of groups in surviving mass extinction than to accumulated victories by wedging in normal times, then a major component of Darwin's worldview—and the only sensible argument that he could supply for our deepest, culturally bound hope of progress—has been compromised or even overturned.

I can envision two models of causality in mass extinction that challenge wedging and consequent progress as a prominent vector of life. In the *random* model, species live or die by the roll of the dice and the luck of the draw; success reduces to little more than being in the right place at the right time when the comets hit, the fires roar, the earth darkens, and the oceans are poisoned. In the *different rules* model, species live or die for definite and specifiable reasons. But the causes of success are quirky and fortuitous with respect to initial reasons for evolving the features that secure survival. The wedge operates in normal times between mass extinctions. Organisms evolve features to enhance success in continuous ecological struggle. The cause of mass extinction then hits in all its sudden fury. Certain features are the passkeys to survival—tolerance of extreme climatic stress, for example. But these features must have evolved during normal times dominated by wedging. And they must, in principle, have arisen for reasons unrelated to their later (and lucky) use in guiding their possessors through the unanticipated debacle of mass extinction. (I say "in principle" because, unless our basic views on causality are seriously awry and the future can control the present, organisms cannot evolve a feature for its potential utility several million years later when the comet hits.)

I see a role for the random model especially in the most severe events. If we accept David Raup's estimate of 96 percent species extinction in the Permian debacle, then entire groups may have been lost by something akin to unalloyed bad luck. Yet I remain committed enough to a more conventional view of causality to think that, of my two proposals for radical reform, the different rules model must apply more often. We can specify causes for differential survival in mass extinction, but the features that secure success must have evolved for unrelated reasons, usually in the regime of wedging during normal times.

Kant told us that concepts without percepts are empty, and Harry Truman said, "Show me, I'm from Missouri." The different rules model, however interesting or elegant in the abstract, will have no power without empirical documentation.

We should begin with the most prominent group to die and the most widely cited cause of extinction. Most people think first of dinosaurs when they ponder the Cretaceous extinction. But our clearest and most extensive evidence comes from the opposite end of the discredited chain of being—the single-celled oceanic plankton. Extinctions are so prominent in these creatures that paleontologists speak of a "plankton line" marking the rapid and simultaneous termination of numerous lineages. As for "killing scenarios" in extraterrestrial impact, the most widely cited reason (though causes must have been complex, interacting, numerous, and varied) invokes a device of Moses against Pharaoh: "a thick darkness over all the land, even darkness which might be felt." An impacting comet or asteroid, the argument runs, would excavate a massive crater and send aloft a thick cloud of particles that would envelop the earth in sufficient darkness to shut down photosynthesis for several months.

A tie between the scenario of darkness and the death of plankton seems easy to formulate. A few months of darkness might not faze a tree (especially if the impact occurred in winter); a plant that lives for decades might shut down its factories for a few months. And even if plants die, seeds may survive to germinate when the dust cloud dissipates. But the single-celled photosynthetic plankton only live for a few days; months of darkness might easily destroy entire populations. These photosynthetic cells form the base of oceanic food chains. If they die, then the herbivorous plankton have nothing to eat and they perish; the carnivorous plankton find no herbivores, the tiny invertebrates no plankton, the fish no invertebrates—up and up to the fragile top of the ecological pyramid.

The rates of extinction for genera in several groups of planktonic organisms are staggering (and imply an even greater percentage of species deaths, for most genera contain several species, and all must be killed to efface a genus, while surviving genera may lose most of their species): 73 percent of coccolithophorids, 85 percent of radiolaria, and a whopping 92 percent of foraminifera. But an exploration of the different rules model must focus on winners and the reasons for their success. So let us consider the primary anomaly, one cited by many authors as an argument against the dust cloud hypothesis: The diatoms, perhaps the most prominent group of photosynthetic plankton, sailed through the Cretaceous debacle with a generic loss of only 23 percent.

We might first take an experimental approach and ask if a few months of darkness can cause differential death—with some species surviving for predictable reasons and others dying because they lack the tools of success. A recent study by Kathy Griffis and David J. Chapman supports this prerequisite for explanation under the different rules model (see bibliography).

Griffis and Chapman obtained cultures of several photosynthesizing planktonic species, including some closely related to forms that survived the extinction and others belonging to groups that first appeared after the extinction and might not have fared well in the preceding darkness. In a happy example of relatively "small" science (you don't need million-dollar grants from the National Science Foundation for all good work), they simulated the scenario of darkness by wrapping 500-milliliter Erlenmeyer flasks, each holding a population of one species, in aluminum foil. They controlled for other factors by providing constant temperatures and adequate supplies of nutrients. Three species belonging to groups that arose after the extinction died within a week. Two others, closely related to forms that pulled through the great dying, survived eight to ten weeks without light—a good estimate for the actual time of darkness in several dust cloud models.

We must then ask if a factor of success, consistent with the different rules model, can be identified. Jennifer A. Kitchell, David L. Clark, and Andrew M. Gombos, Jr., have recently made such a strong argument for the planktonic heroes of the Cretaceous debacle—the diatoms (see bibliography). Kitchell and colleagues studied a core of diatom-rich late Cretaceous sediments (before the extinction) from the High Arctic at 85.6° north latitude. They noted that a portion of the core contained finely laminated sediments. The alternating layers consisted almost entirely of diatom cells in different states: some layers made almost entirely of cells in their phase of growth (up to 96.4 percent); others of "resting spores" in a state of dormancy (up to 93.3 percent). These alternations record nothing about the extinction to come, but must represent an adaptive response to long seasonal fluctuations near the poles: With nearly six unbroken months of darkness every year, a cell that normally lives but a few weeks and subsists by photosynthesis must evolve some mechanism of dormancy—a kind of hibernation—in order to survive a long winter of worse than discontent for any organism dependent upon sunlight. Many species of diatoms have evolved such a complex life cycle; cells deprived of light or nutrients can shut down their metabolism, form a resting spore by encystment, increase in density, and sink to lower levels in the water column, awaiting the return of propitious times.

Removal of light is not the only factor that elicits transformation to a resting state. Diatoms build their cell walls of silica, which they must extract from seawater. They therefore thrive in areas of the ocean, called zones of upwelling, where deeper waters, rich in nutrients (including the vital silica), rise to the surface. But periods of upwelling are sporadic or seasonal. Diatoms must be flexible enough to take advantage of these infrequent and uncertain bounties—able both to enter a growth phase and produce a so-called diatom bloom when nutrients become available and to hunker down as resting spores when their own growth depletes the temporary building supply.

Kitchell and colleagues discerned a common theme behind this flexibility to exploit seasonal and sporadic sources of the two necessities in a diatom's world—silica for construction and light for growth and maintenance. Diatoms evolved the capacity to form resting spores in order to wait out predictably fluctuating seasons of inhospitable environments. They developed this key adaptation for ordinary life in normal times, not in anticipation of relative success should a comet strike the earth several million years in the future. Yet if months of darkness triggered death by extraterrestrial impact, then diatoms held a fortuitous leg up for survival (if you will pardon an inappropriate metaphor from the apex of the chain of being). Diatoms are not better than coccoliths or radiolaria; they are not fiercer competitors on an oceanic surface jammed full of wedges. They were just lucky enough to feature a physiological trick for survival, evolved for other reasons in different times. (Interestingly, Griffis and Chapman note only one feature held in common by all species that died within a week of darkness in their experiments: "None . . . appeared to produce resting cysts in response to the darkness.") Kitchell and colleagues end their paper with a strong defense of the different rules model for diatom success: "These data document an incidental, but causal, dependency between a biological character, selected for in normal background times of geologic history, and evolutionary survivorship during an exceptional time of crisis in earth history."

Life under the different rules model recalls the myth of Sisyphus, greedy king of Corinth, who is punished in Hades with an eternal task. He is compelled to roll a heavy stone up a steep hill; he groans and struggles, finally approaching the summit, but the stone always slips and rolls back down to the bottom, where Sisyphus must start all over again. Sisyphus, patiently and painfully rolling the stone up the mountain, works like life under Darwin's metaphor of the wedge—slow and steady progress by constant struggle, *ad astra per aspera.* But this work of normal times is undone by moments of catastrophe, and nothing ever happens in a larger sense.

This comparison of life with the Sisyphus myth works up to a point, but then fails in a crucial way. The undoing of the slow and patient work of the wedge does not imply demotion all the way down to square one. Catastrophes of mass extinction do not beat life back to an earlier starting point; rather, they deflect the stone of cumulative organic change into some unexpected and unfailingly interesting side channel. They create, by their imposition of different rules, a new regime of oddly mixed survivors imbued with opportunities that would never have come their way in a world of purposeful wedging.

We may be indifferent to most of these quirky shifts that eliminated highly successful groups of the moment and passed a potential torch to unheralded creatures in the wings—groups that fortuitously held a winning ticket purchased for a different reason long ago in other circumstances. Few people lament the loss of ammonites, so long as we still have nautiloids. (In fact, I have proof that few people have ever heard of nautiloids at all, and therefore don't give a damn in the fullest sense. Recently, the *World Weekly News*, king of the shopping-mall tabloids, published—with absolutely shameless faith in our ignorance—an unretouched photograph of a chambered nautilus labeled as a giant monster now on an earthbound path from Mars and scheduled to arrive well before the millennium.) Who cares (who even knows the names?) that all crinoids are now articulates and not inadunates, that reef corals are now scleractinians and not tabulates? Well, you may choose to disdain the details of marine invertebrate life, but you cannot be indifferent to the closest application of the different rules model—the death of dinosaurs and the resulting possibility of human evolution.

You may react to this essay by denying its claim to be conceptually troubling in the light of traditional hopes. You might say, after all, that the different rules model only validates a cliché so old and so widely appreciated that it became the motto of such straight arrow groups as the Boy Scouts and such jokers as Sancho Panza—"forewarned is forearmed; to be prepared is half the victory." Yes, "be prepared"; flexibility is a virtue. If you can keep a whole deck up your sleeve, you will surely have a useful card for any circumstance. But the cruel dilemma, the Catch-22, of evolution lies in recognizing that a species cannot consciously or actively prepare for future contingencies. A species can only evolve for current benefits and deliver its future fate to the wheel of fortune. Round and round she goes, and where she stops. . . . No, it's even worse than that. For the wheel never stops, but only speeds and slows, tacks and turns, bringing life along on the grandest and most sublime of all endless chases. What the hell! Two bucks on *Homo sap.* to win—at least for a little while.

22 | Tires to Sandals

DID YOU EVER WONDER what happens to old automobile tires? Since the United States boasts almost as many vehicles as people, our nation may feature nearly twice as many wheels as feet, and therefore (depending upon the Imelda Marcos factor of pairs per person) more tires than shoes. Well, I don't know what happens to our discarded wheel wear, but I do know the fate of many worn tires in third-world nations. They are cut apart and made into the soles and straps of sandals (I own pairs purchased in the open-air markets of three continents—Quito, Nairobi, and Delhi).

In our world of material wealth, where so many broken items are thrown away because the cost of repair now exceeds the price of replacement, we forget that most of the world fixes everything and discards nothing. (Streets in crowded Indian cities, contrary to our usual assumptions, are often spotlessly clean because any item has value in reuse or resale. Scraps of paper are immediately scavenged; even cow pies remain on the street for only a few seconds before they are collected for fuel and slapped against a wall to dry.) I have never visited a place more fascinating than the recycling market of Nairobi—a true testimony to human ingenuity. Here sandals are made from tires, bracelets from telephone wire, kerosene lamps from bisected tin cans, containers from scraps of metal, and cooking pots from the tops of oil drums.

This little prologue may strike you as interesting enough, but dubiously related to the intended subject of this essay—evolution and the history of life. Yet the theme of recycling for purposes almost comically different from original intent not only

313

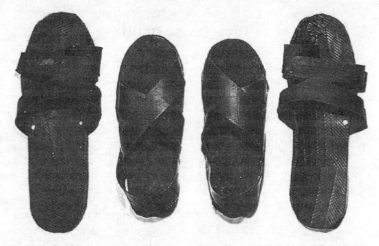

Two pairs of sandals made from recycled automobile tires. I bought the large pair at a street market in Delhi, India, and the smaller pair at the recycling market of Nairobi, Kenya.

Tire strips cut for sandals at the recycling market of Nairobi.

Sandal salesmen at a street market in Delhi.

occupies a central place in the history of life but also completes an argument that I left half-finished in the previous essay. The last piece examined Darwin's justification for predictable progress in the history of life and argued that recent discoveries about the frequency, rapidity, and extent of mass extinction suggest a much more quirky and uncertain path. I pointed out that Darwin did not locate the source of progress in the basic mechanics of natural selection itself—for he recognized natural selection as a theory of local adaptation only, not a statement about general advance. He justified progress with another argument about nature embodied in his favorite metaphor of the wedge. Nature is chock-full of species (like a surface covered with wedges) all struggling for a bit of limited space. New species usually win an address by driving out others in overt competition (a process that Darwin often described in his notebooks as "wedging"). This constant battle and conquest provides a rationale for progress, since victors, on average, may secure their success by general superiority in design.

I argued that mass extinction prevents wedging from establishing long-term patterns in the history of life. Progress by competition may occur in normal times, but episodes of mass extinction undo, disrupt, and redirect this process so frequently that wedging cannot put a dominant stamp upon the overall course of life. I do not believe that mass extinctions work with absolute randomness, treating each species as a coin to be flipped or a die to be rolled. Survivors probably prevail for reasons, but here's the rub (and the role for the wheel of fortune): The rules for survival change in these extraordinary episodes, and features that help species to prevail through catastrophes need not be the sources of success in normal times. Getting through a mass extinction may require a stroke of fortune in the following special sense: A feature evolved for one function in the wedging of normal times must, by good luck and for different reasons, provide a crucial benefit under the different rules of episodic catastrophe. (The diatoms of Essay 21, the only group of planktonic microorganisms to sail through the Cretaceous debacle, had evolved a life cycle with a dormant stage in order to weather predictable seasonal fluctuations of light and nutrients in normal times. If an impacting comet entrained a dust cloud that darkened the earth for several months, the survival of diatoms might be linked to a capacity for dormancy evolved for quite different reasons. Mammals may have prevailed, in large part, by virtue of small body size. But we can hardly label limited size as a preparation for long-term success or even as an active adaptation at all. Mammals may have remained small for primarily negative reasons—because dinosaurs dominated the ecospace of large creatures, and mammals could not displace them by wedging in normal times.) Thus, mass extinction sets a quirky and interesting course for life by opening opportunities to new groups and by basing success upon fortuitous side consequences of features evolved for other reasons. Who cares how well a tire works when the rules change and we run out of gas; continued existence now depends upon a fortuitous capacity for conversion to some other, utterly unanticipated role—retreading for sandals, for example.

But this argument—as far as I went in Essay 21—grants the wheel of fortune only half a loaf. For any champion of the wedge will swiftly reply: Mass extinction is a negative force. It makes nothing and can only pick and choose among creatures fashioned by natural selection. Sure, mass extinction can disrupt a trend, wipe out a complex group, or send life down an unpredicted channel—but evolution is about making, not about differential removal. The creative force in evolution, the motor of construction, must still reside in the processes of normal times, building creatures that will one day pass for review before the sieve of mass extinction. And the controlling process of normal times is wedging by competition.

I could rebut this argument in several ways. I might argue, for example, that sorting by differential death in mass extinction mimics, on a grander scale, the ordinary mode of operation for natural selection in normal times. Natural selection is a sieve, not a sculptor. Many are called; few are chosen. Natural selection is powered by differential birth and death; the few survivors accumulate and intensify their favorable features bit by bit on the relentless sieve of each generation. What is mass extinction but a grander sieve, sorting out species rather than organisms. If the sorting of organisms in normal times yields change that we label as positive, why not grant the same status to differential death in episodic catastrophes? Mass extinction need not be viewed as a negative force opposed to progress in normal times.

But I would choose an opposite tack to provide the other half-loaf to the wheel of fortune. Rather than promoting the wheel of mass extinction by its formal similarity to processes at smaller scale, I would proceed in the reverse direction. I will argue, in short, that the fundamental principle of quirky and unpredictable success by fortuitous side consequences pervades all scales. This vital principle of the wheel does not "click in" only at global catastrophe, leaving most of time to the wedge of ordered progress. The tires-to-sandals principle works at all scales and times, interacting with the wedge to permit odd and unpredictable initiatives at any moment—to make nature as inventive as the cleverest person who ever pondered the potential of a junkyard in Nairobi.

I will go further and make a statement that may seem paradoxical. The wedge of competition has been, ever since Darwin, the canonical argument for progress in normal times. I claim that the wheel of quirky and unpredictable functional shift (the tires-to-sandals principle) is the major source of what we call progress at all scales. Advance in complexity, improvement in design, may be mediated by the wedge up to a certain and limited point, but long-term success requires feinting and lateral motion, with each zig permitting another increment of advance, and progress crucially dependent upon the availability of new channels. Evolution is an obstacle course not a freeway; the correct analogue for long-term success is a distant punt receiver evading legions of would-be tacklers in an oddly zigzagged path toward a goal, not a horse thundering down the flat.

Let us take the broadest possible look at just three prerequisites for continuity in the sequence that, in our parochialism, we view as the crucial and representative case of evolutionary advance—the origin of human consciousness. At each true turning point, each leap in the capacity for substantial advance, we meet the wheel spinning its way toward a new use for old features (sandals from tires); the wedge can only promote, for a limited time and extent, what the wheel makes possible.

1. *Origin of the genetic flexibility for major advance in complexity.* For evolutionists, perhaps the most intriguing and unexpected discovery of molecular genetics emerged during the 1960s when study after study proved that, in multicellular organisms, only a small percentage of the total genetic material consists of functional genes in single copies. Most of the genetic material may be "junk" with respect to information needed to build and maintain a working body. Moreover, many genes exist in multiple copies for obscure reasons unrelated to the necessary functions of bodies.

Nonetheless, it soon became clear to evolutionists that redundancy of multiple copies might be *the* crucial prerequisite for evolution of complexity. Suppose that the original unicellular ancestor of all complex creatures had only one copy of each gene and that each gene coded for a vital function. (This claim is not mere conjecture, but represents the actual state of the simplest creatures alive today and the best models for ultimate multicellular ancestors.) These creatures work very well; they may be honed to an optimum by natural selection, shorn of all unneeded fat and flab. But now, a conundrum for evolution: These creatures may excite our admiration for efficiency, but how can they change in the crucial sense of adding new capacities in complexity? Each gene codes for something vital; it can only alter by improvement in its own channel. This kind of genetic system offers no flexibility, no play, no capacity for adding something truly new.

This conundrum led evolutionists to grasp the fundamental role of multiple copies in permitting the evolution of complexity. If genes exist in several copies, but only one supplies the body's functional need, then the other copies are free to experiment, vary, and add capacity through occasional good fortune.

Well and good so far, but we now face a logical puzzle: Unless we thoroughly misunderstand the fundamental nature of causality, multiple copies cannot arise "for" their potential use in permitting complexity millions of years in the future. Multiple copies are the key to complexity, but they must have evolved for other reasons—the tires-to-sandals principle of quirky functional shift.

This case is particularly interesting because the initial reason for duplication (harvesting of rubber for tires) may not, itself, have much to do with natural selection as traditionally conceived in terms of organisms struggling for reproductive success. Duplication may arise by selection at the lower level of genes, a process invisible in the larger world of the wedge. Genes also play the game of natural selection in their own realm, and those that develop the capacity to duplicate and move (transposons, or jumping genes) secure advantages at this lower level, just as organisms win in Darwin's world by leaving more surviving offspring. In fact, gene duplication may be abetted by producing no effect upon bodies, for invisibility at Darwin's level of the wedge provides hiding space and assures that no negative pressures of natural selection shall impede the accumulation of "unneeded" extra copies. Yet these "redundant" duplicate genes may house the latent source of later complexity.

2. *The evolution of complex cells.* Many biologists would place nature's fundamental distinction not between plants and animals, or even between unicellular and multicellular organisms, but at a division within unicellular creatures. The structurally simple prokaryotes, bacteria and cyanophytes, have no organelles within their cells—no nucleus or chromosomes, no mitochondria. The complex eukaryotes have evolved the array of internal structures that grace (or disfigure according to your view or status) nearly every high school biology final with its inevitable question: Label all the parts of the cell and state their functions.

This increment of complexity from prokaryote to eukaryote is deemed fundamental, in part because we view eukaryote organization as an absolute prerequisite to the later evolution of multicellular organisms. (To cite just one standard argument: Darwinian evolution of complexity requires copious variation to fuel natural selection; most variation arises from the mixture, via sexual reproduction, of two differing genetic systems in each offspring; sexual reproduction requires a mechanism for exact division of genetic material so that 50 percent of each parent reconstitutes the needed 100 percent in offspring; meiosis by reduction division of paired chromosomes is the biological invention that secured equal separation; prokaryotes, lacking chromosomes and other organelles, cannot produce an exact genetic halving.)

We now encounter the same conundrum faced in the last example: We can see why multicellular life required the evolution of organelles, but eukaryotic cells arose at least 800 million years before the origin of multicellular animals—so progress to multicellular complexity cannot be the reason why organelles evolved.

A favored theory for the origin of some organelles (the mitochondrion and chloroplast but not, alas, the nucleus, for which no good theory now exists) invokes the process of symbiosis. Mitochondria and chloroplasts look uncannily like entire prokaryotic organisms (they have their own DNA and are the same size as many bacteria). Almost surely, they began as symbionts within cells of other species and later became more highly integrated to form the eukaryotic cell (so that each cell in our body has the evolutionary status of a former colony). Now, one can argue that the wedge drove the ancestors of mitochondria to a life of symbiosis. These bacteria, gaining protection or whatever, did not enter the primordial eukaryote in order to provide an opportunity for multicellular complexity a billion years down the road. Symbiosis occurred for immediate Darwinian reasons; then the wheel turned and the rubber made for symbiosis put the first footprints on the path of multicellular complexity.

3. *The basic features of human consciousness.* The wheel and the wedge then interact for more than half a billion years to the separation of our lineage from the ancestry of chimpanzees some six to eight million years ago. The wedge produces some forward motion (and more blind alleys of overspecialization), but the wheel inaugurates each domain of change—the limbs on an odd group of fishes, by an unusual arrangement of fin bones, can bear

the body's weight on land (see Essay 4); mammals get a chance after 100 million years in the backwaters because dinosaurs succumb in a mass extinction.

Australopithecus now begins the process that textbooks used to call, before we reformed our language to include all people, the "ascent of man." Doesn't the wedge finally prevail? Isn't the unreversed trend of increasing brain size, from *Australopithecus* to *Homo habilis* to *Homo erectus* to us, driven by ordinary natural selection working on the advantages of superior cognition? Let me take the most conservative argument of the wedge (not my actual view) and reply: Yes, fine; I agree. The human brain got large because natural selection directly favored some traits of cognition that gave bigger-brained people advantages in competition.

Does such an admission imply that the foundations of human cognition, the universal traits that we define as "humanity" or "human nature," were built directly by the wedge? Of course not—and no argument is more important for our understanding of human nature, yet less widely appreciated, than this. Yes, the brain got big by natural selection. But as a result of larger size, and the neural density and connectivity thus imparted, human brains could perform an immense range of functions quite unrelated to the original reasons for increase in bulk. The brain did not get big so that we could read or write or do arithmetic or chart the seasons—yet human culture, as we know it, depends upon skills of this kind. If you label me as a hopelessly parochial academic for citing only the skills of an intellectual elite, I reply that the fortuitous side consequences of large brains include the defining activities of all people. What about language, the most widely cited common denominator and distinguishing factor of humanity? And I don't mean using sound or gesture for communication, as many complex animals do. I refer to the unique syntax and underlying universal grammar of all languages. I can't prove that language was not the selected basis of increasing brain size, but the universals of language are so different from anything else in nature, and so quirky in their structure, that origin as a side consequence of the brain's enhanced capacity, rather than as simple advance in continuity from ancestral grunts and gestures, seems indicated. (I lay no claim to originality for this argument about language. The reasoning follows directly as an evolutionary reading for Noam Chomsky's theory of universal grammar.)

To cite another example, consider Freud's argument on the origin of religion—or at least of a belief in some form of persistence after death as a common feature of this institution. Freud held that all religions maintain some belief in personal persistence after death—whether in heaven, by reincarnation, in a universal soul, or merely by continuity of tradition. This belief marks the common basis of religion because our large brains "forced" us to learn and acknowledge the fact of personal mortality (a concept not clearly grasped by any other animal). Now you cannot argue that our brains became large so that we would appreciate the fact of our death; knowledge of mortality is an inevitable (and largely unfortunate) side consequence of mental power evolved for other reasons. Yet this unwanted knowledge forms the basis of an institution often regarded as the most fundamental consequence of human nature.

If such features as language and the basis of religion are side consequences of the wheel, not direct gifts of the wedge, then is human nature a predictable product of organic improvement, honed in the fires of competition, or a set of oddly cobbled side consequences rooted in an unparalleled neural complexity built for other reasons? We are a bit of both—though more, I suspect, quirks of the wheel than boons of the wedge—and in this mixture lies our hope and our destiny.

If I have upset your equanimity by attributing the genuine complexity of human cognition to fortuity piled upon fortuity (with a little yardage for predictability after each spin of the wheel), then I must apologize for one further disturbance in conclusion. We talk about the "march from monad to man" (old-style language again) as though evolution followed continuous pathways of progress along unbroken lineages. Nothing could be further from reality. I do not deny that, through time, the most "advanced" organism has tended to increase in complexity. But the sequence from protozoan to jellyfish to trilobite to nautiloid to armored fish to dinosaur to monkey to human is no lineage at all, but a chronological set of termini on unrelated evolutionary trunks. Moreover, life shows no trend to complexity in the usual sense—only an asymmetrical expansion of diversity around a starting point constrained to be simple. Let me explain that last cryptic remark: For reasons of organic chemistry and the physics of self-organizing systems, life arose at or very near the lower limit of preservable size and complexity in the fossil record. Since diversity, measured as number of species, has increased through time, extreme values in the distribution of complexity can move in only one direction. No species can become simpler than the starting point, for life arose at the lower limit of preservable complexity. The only open direction is up, but very few species take this route. Increasing complexity is not a purposeful trend of an unbroken lineage but only the upper limit of an expanding distribution as overall diversity increases. We focus on this upper tail and call its expansion a trend because we crave some evolutionary rationale for our perception of ourselves as a predictable culmination.

But consider the system of variation as a whole, rather than focusing upon a few species at the right tail. What has ever changed besides overall diversity? The modal organism on earth is now, has always been, and probably will always be, a prokaryotic cell. There are more bacteria in the gut of each person reading this essay than there are humans on the face of the earth. And who has a better hope for long-term survival? We might do ourselves in by nuclear holocaust, but prokaryotes will probably hang tough until the sun explodes.

Progress as a predictable result of ordered causes therefore becomes a double delusion—first because we must seek its cause more in the quirkiness of the wheel, turning tires into sandals and big brains toward fear of death, than in the plodding predictability of the wedge, propelling monkeys into men; and secondly, because the supposed sweep of life toward progress only records our myopic focus on the right tail of a distribution whose mode has never moved from a prokaryotic cell.

Our reasons for profound unwillingness to abandon a view of life as predictable progress have little relation to truth, and all to do with solace. Ironically, while using the wedge to supply ultimate solace in his claim that "all corporeal and mental endowments will tend to progress towards perfection" (from the concluding section of the *Origin of Species*), Darwin also recognized a challenge in the bloodthirsty character of unrelenting battle. He therefore concluded chapter 3 of the *Origin* with one of the few soft statements of a very tough-minded thinker:

324 | <small>EIGHT LITTLE PIGGIES</small>

When we reflect on this struggle, we may console ourselves with the full belief, that the war of nature is not incessant, that no fear is felt, that death is generally prompt, and that the vigorous, the healthy, and the happy survive and multiply.

Our chances of understanding nature would improve so immensely if we would only shift our search for solace elsewhere. (Solace will always be a desperate need in this vale of tears, but why should the facts of our belated evolution be pressed into such inappropriate, if noble, service?) Perhaps I am just a hopeless rationalist, but isn't fascination as comforting as solace? Isn't nature immeasurably more interesting for its complexities and its lack of conformity to our hopes? Isn't curiosity as wondrously and fundamentally human as compassion?

*New Discoveries
in the Earliest
History of
Multicellular Life*

23 | Defending the Heretical and the Superfluous

SAMUEL TAYLOR COLERIDGE, in a reverie laced with laudanum, presented an image of striking incongruity in describing the pleasure palace of Kubla Khan:

> It was a miracle of rare device,
> A sunny pleasure-dome with caves of ice!

This vision of tropical languor mixed with arctic sternness recalls a juxtaposition of similar disparity from my own education—Marco Polo in Chinese summer palaces and Eric the Red conning settlers by describing inhospitable arctic real estate as "Greenland." This odd matching of China with Greenland records a key episode of "white man's history," taught as universal by New York City public schools in the late 1940s.

The history of civilization, we learned, is centrifugal—a process of outward expansion from European or near Near Eastern centers. Heroes of this process were called "explorers"—and they "discovered" land after land, despite the nagging admission that all these places featured indigenous cultures often more complex and refined than the European "source" (Kubla Khan vs. the Doge of Venice).

We worked through the panoply of explorers in strict chronology. Eric the Red, a tenth-century Norseman, came first, moving northwest into bleakness and chill. Marco Polo, Kubla Khan's most famous visitor (and Coleridge's source), followed, moving southeast into exotic splendor and warmth. (Eric's son Leif might have merited a chapter in between, especially since he reached

North America several hundred years before the official date for "discovery" of our well-populated continent. But, remember, I grew up in New York, not Lake Wobegon, and the Knights of Columbus had effectively put the kibosh on any Viking claims. Leif Ericson and Vineland ranked with Odin and Thor in the category of Scandinavian mythology.)

Thus, Greenland and China—lands of nearly maximal disparity in climate and geography—have always stood together in my mind as the one-two punch of initial discovery. And now, some forty years later, my own profession of more ultimate origins has juxtaposed these incongruous places again, this time in the legitimate service of discovery about true beginnings. During the last year, fossil finds in China and Greenland have penetrated the terra incognita of animal origins with an éclat to match the deeds of any old-time explorer.

I have written many essays and an entire book on the origin of multicellular animals. Yet, from a dominant perspective in evolutionary thought, such a subject should not exist at all, at least in the sense of "first" items that an explorer might discover. We inhabit a world of graded continuity, and transformation of single-celled microscopic ancestors to multicellular animals of modern design should occur by smooth transition over such a long time that no single organism or species should qualify as an unambiguous "first."

Life is continuous in the crucial sense that all creatures form a web of unbroken genealogical linkage. But connectivity does not imply insensible transition. Nothing breaks the continuity between caterpillar and butterfly, but stages of development are tolerably discrete. Similarly, the origin of animals reminds us, in outline, of an old quip about the life of a soldier—long periods of boredom punctuated by short moments of terror. In the evolution of multicellular animals, nothing much happens for very long periods of time, while everything cascades in brief geological moments. We can talk meaningfully about "firsts," and discoveries in Greenland and China qualify for this category of ultimate importance. A quick review of basic information will set a proper context:

Life on earth is as old as it could be—a striking fact that, in itself, points to chemical inevitability in origination (given proper conditions that may be improbable in the universe). Paleontological discoveries, starting in the mid-1950s, have shattered the previous consensus—never more than a sop to our hopes for uniqueness—that life is exceedingly improbable and only arose because so much geological time provided such ample scope for the linking of unlikely events (given enough trials, you will eventually flip thirty heads in a row). Under this discredited view, life arose relatively late in the earth's history, following a long geological era called "Azoic" (or lifeless, and representing the time needed for all those trials before the thirty fortunate successes).

But fossils of simple unicellular creatures have now been found in appropriate rocks of all ages, including the very oldest that could contain evidence of past life. The earth is 4.5 billion years old, but heat generated from two major sources—the decay of short-lived radioisotopes and bombardment by cosmic debris that pervaded the inner solar system during its early history—melted the earth's surface some 4 billion years ago. All rocks must therefore postdate this early liquefaction. The oldest known rocks on earth are a bit older than 3.8 billion years, but they have been so altered by heat and pressure that no fossils could have survived. The oldest rocks that could contain preserved organic remains are 3.5 to 3.6 billion years old from Australia and South Africa—and both deposits do feature fossils of single-celled creatures similar to modern bacteria. Hints and indications are not proofs, but I don't know what message to read in this timing but the proposition that life, arising as soon as it could, was chemically destined to be, and not the chancy result of accumulated improbabilities.

But if origination bears a signature of chemical inevitability, the pattern of later history tells a story of historical contingency dominated by portentous but unpredictable events. (I find nothing strange or unlikely in such a model of historical chanciness for subsequent pattern following a substrate of initial necessity. One might argue, for example, that the origin of speech and writing follows predictably from the evolved cognitive structure of the human mind. But the actual languages that developed, their timings and their interrelationships, would never unfold in the same way twice.)

Yet whatever attitude we adopt towards the total pattern, we must at least admit that one key event—the origin of multicellular animals—carries no prima facie signature of stately inevitability. If multicellular complexity is a predictable advance upon unicellular existence, then this salutary benefit surely took its time arising, and certainly burst upon the scene with unseemly abruptness by quirky and circuitous routes.

Nearly five-sixths of life's history is the story of single-celled creatures (with some amalgamation, towards the end to threads, sheets, and filaments of algal grade—an event entirely separate from the origin of animals in any case). Then, about 650 million years ago, the first multicellular assemblage appears in rocks throughout the world. This fauna, named Ediacara for an Australian locality, consists entirely of soft-bodied creatures with anatomical designs strikingly different from all modern animals (flattened disks, ribbons, and pancakes composed of strips quilted together). Some paleontologists have suggested that the Ediacara animals bear no relationship to modern creatures, and represent a separate, but failed, experiment in multicellular life.

Multicellular animals of modern design—and with hard parts readily preservable as fossils—first appear, also with geological alacrity, in an episode called the "Cambrian Explosion" some 550 million years ago. Trilobites, a group of fossil arthropods beloved of all collectors, provide the principal signature for this first fauna of modern design. The full flowering of this initial fauna reaches its finest expression in the exquisite, soft-bodied fossils of the Burgess Shale, subject of my recent book, *Wonderful Life*.

This basic pattern has been well publicized and is now known to most nonprofessionals with strong interests in the history of life: a long period of unicellular creatures only; followed by a rapid appearance of the Ediacara fauna, perhaps with no relationship to living animals; and the final, equally quick, origin of modern anatomical designs in the Cambrian Explosion, with maximum expression soon thereafter in the Burgess Shale.

Less well known is the fine-scale geological anatomy of the Cambrian Explosion itself. Trilobites do not appear in the earliest Cambrian strata with hard-bodied fossils; they enter the geological record in the second phase of the Cambrian, called Atdabanian. The initial phase, called Tommotian after a Russian locality, contains a fauna with an interesting balance of the familiar and the decidedly strange. The new discoveries in China and Greenland give us our first decent insight into the anatomical character of the strange component—hence the great importance of these new finds, for we cannot grasp the ordinary (so designated only because they survived to yield modern descendants) without the surrounding context of creatures that left no progeny and therefore appear to us like products of a science fiction novel.

The earth's first hard-bodied fauna of the Tommotian does include several fossils of modern design—sponges, echinoderms, brachiopods, and mollusks, for example. It also features an outstanding group of large, reef-building creatures that died out well before the end of the Cambrian. These enigmatic animals, called archaeocyathids, resemble a two-layered cone. Put one ice-cream cone within another, leave a small space between, and you have a reasonable anatomical model for an archaeocyathid. The affinities of archaeocyathids have been debated for more than a century, with uncertain results. Most paleontologists would probably vote for a position near sponges, but scientific issues are not settled at the ballot box, and other opinions enjoy strong minority support.

But by far the most enigmatic, and most mind-boggling, component of the Tommotian faunas includes a set of bits and pieces with a catch-all name that spells frustration. These tiny spines, plates, caps, and cups tell us so little about their origin and affinity that paleontologists dub them the "small, shelly fauna," or SSF for short. "Small shellies" may be a charming phrase, when issued from the mouth of a professional who usually spouts incomprehensible Latin jargon, but please remember that this name conveys ignorance and frustration rather than delight.

We may envision two obvious potential interpretations for the SSF. Perhaps they are the coverings of tiny, entire organisms, a diminutive fauna for a first try at modernity. But perhaps—and this second alternative has always seemed more likely to paleontologists—they are bits and pieces representing the disarticulated coverings of larger multicellular organisms studded with hundreds or thousands of these SSF elements. This second position certainly makes sense. We can easily imagine that the ability to secrete hard skeletons had not fully developed in these earliest days, and that many of the first skeletonized organisms did not bear a discrete, fully protective shell, but rather a set of disconnected, or poorly coordinated fragments that only later coalesced to complete skeletons. These fragments, disarticulating after death, would form the elements of the SSF.

If this second interpretation prevails, then paleontologists are in deep trouble, and well up the proverbial creek named for the droppings of these and all later creatures. For how can we possibly reconstruct a complete animal from partial fragments that didn't even form a coherent skeleton, and that clothed a creature of entirely unknown shape and form? Yet we can obtain no real insight into the full nature of this crucial, first Tommotian fauna until we can reanimate these most important components of the SSF. Jigsaw puzzles are hard enough when we have all the pieces and their ensemble forms a picture that can guide us as we assemble the parts. But the SSF fragments set a daunting and almost hopeless task, for they probably represent pieces from one hundred different jigsaw puzzles all mixed together. The pieces contain no pictures, and we probably have less than one piece in ten of the total covering for each frame. Moreover, to make matters even worse, we don't know the sizes or shapes of the frames.

In this light, the reanimation of a complete SSF animal from preserved skeletal fragments seems truly hopeless—and so it has been, as two decades of work have produced no plausible reconstructions. We must adopt another strategy—unfortunately passive in one sense, though active in another. We must hope to discover a different kind of fossil—not the common disarticulated bits that cannot be reassembled, but a rare preservation of an entire SSF organism with all its elements in place. I call such a change in focus passive because we must wait for the discovery of a basically soft-bodied creature with its covering bits of shell still in place—and soft-bodied preservation is rare in the fossil record. But this strategy is also active because we now have good guidelines for exploration; we now know where and how to look for soft-bodied fossils.

The discoveries in Greenland and China can now be placed into proper context and excitement in a single sentence: They represent the first remains of entire SSF organisms, preserved with full coverings of their separated skeletal elements. The second interpretation of the SSF has prevailed. These cups, caps, cones, and spines are bits and pieces of incomplete skeletons upon larger organisms—and we finally have some insight into the nature of these important creatures; the dominant component of the earth's first skeletonized fauna. (The SSF elements arise in the earliest Tommotian beds, but persist into subsequent Cambrian strata. The SSF animals of China and Greenland were found in later rocks containing trilobites as well, but their SSF elements are identical with those found in earliest Tommotian sediments, so the two organisms are true representatives of this heretofore mysterious first fauna.)

Microdictyon is a classic element of the SSF. The hard parts, and previously only-known components, are round to oval, gently convex, phosphatic caps, no more than 3 mm in diameter. Each cap is a meshwork of hexagonal cells with round holes in the center of each cell (see figure). How could a paleontologist possibly move from this limited morphology to a reconstruction of the animal that secreted these partial coverings?

Since scientists, having no access to divine inspiration or the magical arts, cannot make such a move, *Microdictyon* has simply stood as a stratigraphic marker of its time and a complete mystery in anatomical terms. *Microdictyon* has been found worldwide in rocks of Tommotian to middle Cambrian age in Asia, Europe, North and Central America, and Australia.

In 1989, three Chinese colleagues from the Nanjing Institute

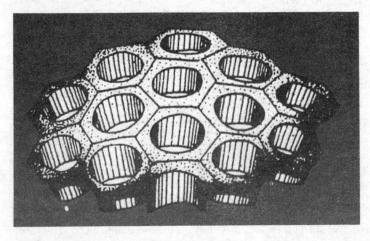

A plate of *Microdictyon* showing the characteristic meshwork. Acta Palaeontologica Sinica, *Vol. 28, No. 1, p.5.*

of Geology and Paleontology—Chen Jun-yuan, Hou Xian-guang, and Lu Hao-zhi—published a remarkable article in volume 28 of the *Acta Paleontologia Sinica.* (I remain profoundly grateful for the international character and cooperative traditions of paleontological work. Our science is global, and we would be stymied if we lost access to information from selected parts of the world. I thank both Drs. Chen and Hou for sending me reprints of their work along with letters providing further valuable data about their discoveries.)

Drs. Chen, Hou, and Lu have been working with the remarkable Chengjiang fauna of south-central China, an equivalent in age and soft-bodied preservation of the famous Burgess Shale in western Canada. Among other stunning creatures of unknown affinity, they discovered several specimens of a worm-like animal, some 8 cm in length.

This creature (see figure) bore ten thin pairs of leglike appendages, generally decreasing in strength from front to back. Traces of a simple, tubular gut can be seen on most specimens. But, most remarkable of all, this animal carried pairs of rounded phosphatic caps, inserted in pairs on the body sides, just above the joining points of the legs with the trunk. Each pair of legs, in other words, sports a corresponding pair of caps on the trunk above. These caps, *mirabile dictu,* are the elements previously

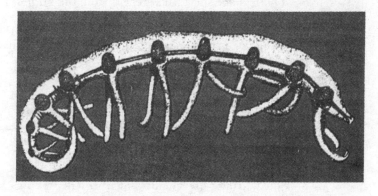

Restoration of *Microdictyon sinicum* showing the paired side plates. Acta Palaeontologica Sinica, *Vol. 28, No. 1, p. 5.*

named *Microdictyon,* but known only from the uninterpretable
hard-part dabs. Their discovery on the Chinese animal not only
adds a fascinating and mysterious creature to the roster of earliest animals, but also confirms our long-held suspicions about the
SSF. *Microdictyon,* at least, is just an element covering a much
larger body. Moreover, the hard parts enclose only a small portion of the body and do not articulate with each other (the space
between pairs of caps is about double the diameter of the caps
themselves). How could we possibly have inferred the character
of the animal from the caps alone?

Halkieria forms an even better known and more frustrating element of the SSF. Collected from lower Cambrian rocks throughout the world, halkieriids are preserved as sclerites (flattened
blades and spines of calcification, just a millimeter or two in
length). The sclerites assume several characteristic shapes,
named siculate (narrow, crescentric, and asymmetrical), cultrate
(elongate and more symmetrical), and palmate (wider and flattened like the palm of a hand). Although some paleontologists
have tried to reconstruct halkieriids as tiny creatures, each living
within or around a single sclerite, most agree that the halkieriid
animal must have been substantially bigger and covered with
large numbers of sclerites.

In July 1989, in a most inhospitable spot on Peary Land in
northern Greenland, some twenty-one specimens of a halkieriid
animal were finally unearthed from another deposit capable of
preserving soft parts. (The spectacular results from the Burgess
Shale have inspired paleontologists to devote attention to the
discovery and exploitation of these rare and precious soft-bodied
fossil faunas. Science, at its best, not only answers questions, but
provokes new problems and guides fruitful research by posing
issues previously unconsidered.)

In July 1990, S. Conway Morris and J. S. Peel published the first
report on the halkieriid animal. Again, paleontological suspicions
are confirmed, but with an amazing twist and surprise. The halkieriid is, as anticipated, a large animal (up to 7 cm), bearing many
sclerites—up to 2,000 or more. The body is elongate, flattened,
and wormlike, with the sclerites arranged in zones corresponding
to forms previously named (see figure). Siculate sclerites surround the base of the animal (the underlying bottom surface

probably carried no hard parts, as the animal crawled on a naked sole). A groove separates the zone of siculate sclerites from a lateral region of cultrates. The top surface of the creature bears more flattened, palmate sclerites.

So far, so good—and quite in line with predictions. But nature always throws us a surprise or two. Each end carries a prominent and entirely unanticipated shell. These are found in the same position on every specimen, and therefore represent no fluke of juxtaposition or odd preservation. The anterior shell is roughly rectangular, the larger posterior shell (up to 1 cm in length) more oval and flattened. With apparent growth lines and an apex near the margin, this posterior shell, if found separately, would surely have been called a brachiopod or mollusk valve. (I suspect that several named mollusks and brachiopods of the Tommotian will

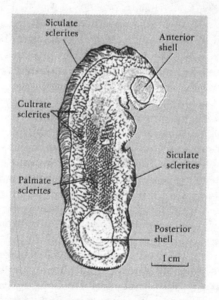

Figure of a halkieriid showing the fields of sclerites and the two end plates. *Reprinted by permission from* Nature, *June, p. 808; Copyright © 1990 Macmillian Magazines Limited.*

turn out to be halkieriid end plates). Conway Morris and Peel wisely offer no interpretation of these elements, though others have suggested that the terminal shells might have plugged the ends of a U-shaped tube, if halkieriids burrowed as do many modern worm-shaped organisms.

We are, of course, enormously gratified to know, for the first time, some prominent animals of the SSF fauna, the earth's initial complement of modern creatures with hard parts. But should we be surprised? (I realize that the phrase "SSF fauna" is as redundant as pizza pie and AC current, but abbreviations achieve a life of their own and may then be modified, even by one of their own elements).

One misguided reply might proffer little surprise (and relative indifference to my efforts in this essay). After all, we knew that the scattered SSF elements had to represent some kind of animal or other, and now that we have found two of the creatures, they turn out to be something rather familiar after all. Just a couple of worms—and as Mr. Reagan once said about redwoods, when you've seen one, you've seen 'em all. But such an attitude would be more than just deplorably Philistine; it would be dead wrong as well. *Wormlike* is a functional term used to describe flexible, soft-bodied organisms that are basically bilaterally symmetrical, with sensory organs in a head at front, and excretory organs at the rear end. Wormlike is not a genealogical concept uniting a group of organisms related in any evolutionary sense of common parentage. *Microdictyon* and *Halkieria* are wormlike only in this functional meaning, and no anatomical plan is more common and more often evolved by radically different creatures. Wormlike bodies are good designs for any mobile creature that must move with efficiency towards food and away from enemies—and no mode of life is more common in nature. Modern wormlike creatures include animals of such genealogical disparity as truly segmented earthworms, slugs of the snail lineage, sea cucumbers of the echinoderm phylum, *Amphioxus* of our own parentage (or at least cousinship), and a host of phyla that we all once learned in high school—Platyhelminthes (including laboratory planaria and tapeworms in vertebrate intestines), Nematoda, Kinorhynchia, Pogonophora, Chaetognatha, and so forth.

The proper evolutionary perspective is genealogical. Bats may be functionally similar to birds, but they are mammals by descent. Ichthyosaurs may look and work like fishes, but they are reptiles by ancestry. In this more fundamental context of genealogy, both *Microdictyon* and *Halkieria* are puzzling. The *Microdictyon* animal looks like an onychophoran, a small modern group considered by some as transitional between the Annelida (segmented worms) and the Arthropoda (insects, spiders and crustaceans—see next essay). *Halkieria* has been compared with the later *Wiwaxia* from the Burgess Shale, but *Wiwaxia* itself is an enigma, and the two shells at the end of *Halkieria* are just plain odd. Perhaps better evidence will establish some homologies with known groups, but for now, *Halkieria* must be viewed as a unique creature of unknown affinity with any other animal.

Thus, we may dismiss the "seen one worm, seen 'em all" argument as simply wrong, but a more sophisticated version of "should we be surprised" does have potential merit. Consider any genealogical system that ends up with a few well-differentiated survivors, all rather distant one from the other. Modern life surely displays this cardinal feature. Our modern phyla represent designs of great distinctness, and our diverse world contains nothing in between sponges, corals, insects, snails, sea urchins, and fishes (to choose standard representatives of the most prominent phyla). A distant past must have included many linking forms, now extinct. These links would not resemble fanciful hybrids between living organisms (a cat-dog or a cow-horse), because modern lineages have been separate for so long. They would, instead, be odd animals with veiled hints of several lineages to come and many unique features of their own (as we actually find in mammals like *Hyracotherium*, the 50-million-year-old ancestor of both horses and rhinoceroses).

Consider a figure and a nonbiological analogy (with thanks to R. T. Simmonds of Nordland, Washington, who wrote to me about this example in another context). The modern Romance languages—French, Spanish, Portuguese, Italian, and Romanian—all derive from Latin and represent clearly separate entities, despite evident similarities. But if we could—as we cannot—trace all the lineages leading from Latin, we would find a forest of village dialects linking all these end-points together. Many would be odd and unique, others smoothly transitional. We would learn that our modern descendants are just a small sample of the total richness, most now lost. We do get some hints of the full tree in survival of a few "minor phyla" (Catalan and Romansh, for example), and in historical records of a few extinct lineages (Provençal and Burgundian). But if we could go back to the beginnings of the spread, Dr. Simmonds conjectures, we would probably encounter a veritable Cambrian Explosion of lost variants.

In this sense, a phenomenon like the Cambrian Explosion must generate a majority of lineages that will seem peculiar in comparison with modern survivors—for these form the web of intermediary links that must die out if we are to emerge (as we have) with a limited set of widely separated designs (see figure). But I would raise two strong arguments against any boredom about *Microdictyon* and *Halkieria* on these grounds.

First of all, the forest of extinct lineages includes two categories of differing degrees of strangeness with respect to modern survivors. Unless modern survivors include forms at all the ancient peripheries—and this seems most unlikely, since peripheries are tenuous places—then many extinct lines will lie outside the range of all modern designs, and will feature more than an amalgam of primitive, but intermediary, characters. The point may sound abstract, but can be easily grasped in the diagram. Only lineages 1, 2, and 3 have emerged from the forest of this "Cambrian Explosion" to yield modern descendants. Now consider the lettered representatives from an early time of maximal diversity. Some of these lineages (e–l of the diagram) do lie within the bounds of modern groups; in our retrospective view, we will regard them as unique, but not fetchingly odd. But other lineages (a–d and m–p of the diagram) lie outside the limits of modern groups, often well beyond (p, for example, lies further from lineage 3 than 3 does from any modern survivor). These creatures will be read in our parochial light (recognizing only 1, 2, and 3 for "standard" animals) as bewitchingly peculiar—and all but the most benighted dolt will take a keen interest. (The Burgess Shale excites our imagination largely because several of its "weird wonders" probably lie in this exterior domain, well outside the boundaries of modern groups).

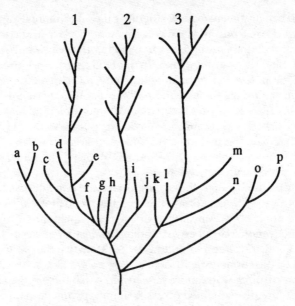

A hypothetical genealogy of the early history of multicellular life, illustrating the likelihood that many of these early forms had anatomies that would be judged outside the range of modern survivors. *Iromie Weeramontry. Courtesy* Natural History.

Thus, on my first argument, we cannot exclude *Microdictyon* and *Halkieria* from fascination just because we recognize that any genealogical system—like our diagram, like modern life, like the Romance languages—must include a great majority of early lineages deemed unique by modern standards. For *Microdictyon* and *Halkieria* may belong to the special group of *outsiders* (a–d and m–p), truly resident in the world of science fiction, and not to the more comfortable *insiders* (e–l) that only mix and match the cardinal features of later groups.

But suppose that *Microdictyon* and *Halkieria* do turn out to rank among the insiders? Do we then lose interest, shrug our shoulders, put down this essay, and move on to the horoscopes and gossip columns? We now come to the second, and I think more important, argument—an aesthetic or moral claim really, not an empirical proposition. What is fascination? Do we invest our interest only in unknown things beyond the boundaries of current categories? Do we not yearn for more beauty, more diversity, more examples, more wrinkles of novelty, more cases for inspiration, in the things we love and partially know? Do we not grieve for one hundred lost cantatas of Bach even though we may listen to more than two hundred? Would we not give our eye tooth (what's a canine more or less) for the unknown works of Aristotle? Would we not trade half our GNP for tapes of Socrates in conversation with his students?

Why do intellectuals feel such special pain in the destruction of the library of Alexandria—the greatest repository of ancient texts, begun by Alexander the Great, maintained by the Ptolemaic monarchs of Egypt, and finally destroyed, according to the legend you choose to follow, by the Romans, the early Christians, or the conquering Moslems? In part, we lament the loss of the utterly unknown. But we miss just as much the opportunity to relish a greatly expanded diversity among people and ideas that we already know and love. We miss the joy of making concrete, the pleasure of holding what has disappeared forever. What is history all about if not the exquisite delight of knowing the details, and not only the abstract patterns. Even if *Microdictyon* and *Halkieria* are only "inside" animals between surviving phyla, they are still prominent creatures of our earliest multicellular world. They have unique forms and peculiar features—shells on both ends, or lateral dabs in pairs. We want to know as many of these creatures as we can, for they are papyrus rolls in the great and largely lost library of our own past.

One legend of Alexandria, almost surely false, states that the library was still intact when Moslems captured the city in the seventh century. The emir Amrou Ibn el-Ass, having conquered Alexandria in 640, wrote to the caliph Omar asking (in part) what should be done with the library (and hoping against hope that the caliph would spare this great treasure). But Omar replied with the most stunning statement of "heads I win, tails you lose" in all human history. The books, he proclaimed, are either contrary to the Koran, in which case they are heretical and must be destroyed—or they are consonant with the Koran, in which case they are superfluous and must also be destroyed. The contents of the library were therefore burned to heat water in the public baths of Alexandria. The books and scrolls kept the fires going for six months.

The Omar of this legend will never win any praise from intellectuals, but I do grasp his point in an entirely reversed way. *Microdictyon* and *Halkieria* are, in a sense, either heretical (if lying outside the range of modern forms) or superfluous (if lying inside). But they are equally wonderful, and worthy of our most cherished interest and protection, in either case—and in this judgment lies the difference between most of us and the enemies of the light. In this lies the turf that we must defend at all costs.

24 | The Reversal of *Hallucigenia*

YOU CAN GENERATE a lot of mischief just by strolling. When God asked Satan what he'd been doing, the foremost of the fallen angels responded: ". . . going to and fro in the earth and . . . walking up and down in it" (Job 1:7). But you can also do a lot of good. Aristotle preferred to teach while ambling along the covered walk, or *peripatos*, of his Lyceum in Athens. His followers were therefore called peripatetics. In Greek, a *patos* is a path, and *peri* means "about." The name for Aristotle's philosophical school therefore reflects the master's favorite activity.

The same etymology lies behind my all-time favorite technical name for an animal—the genus *Peripatus*. I just love the sound, especially when pronounced by my Scottish friends who really know how to roll their *r*'s. I can hardly ever bring myself to write about this animal without expressing delight in its name. The only reference in my book *Wonderful Life* speaks of the "genus with the lovely name *Peripatus*."

Peripatus is an elongated invertebrate with many pairs of stout, fleshy legs—hence the chosen name for this obligate walker. The Reverend Lansdown Guilding—quite a name itself, especially given the old stereotype of English clergymen as amateur natural historians—discovered and designated *Peripatus* in 1826. He falsely placed his new creature into the mollusk phylum (with clams, snails, and squids) because he mistook the antennae of *Peripatus* for the tentacles of a slug. Since true mollusks don't have legs, Guilding named his new beast for a supposed peculiarity.

Peripatus is the most prominent member of a small group

known as Onychophora. Modern onychophorans are terrestrial invertebrates of the Southern Hemisphere (with limited extension into a few regions of the Northern Hemisphere tropics)—hence little known and never observed in natural settings by residents of northern temperate zones.

About eighty species of living onychophorans have been described. They live exclusively in moist habitats, usually amid wet leaves or rotting wood. Most species are one to three inches in length, although the size champion from Trinidad, appropriately named *Macroperipatus,* reaches half a foot. They resemble caterpillars in outward appearance (although not in close evolutionary relationship). They are elongated, soft bodied, and unsegmented (the ringlike "annulations" on antennae, legs, and sometimes on the trunk are superficial and do not indicate the presence of segments, or true divisions of the body). The onychophoran head bears three paired appendages: antennae, jaws, and just adjacent to the jaws, the so-called slime papillae. Onychophorans are carnivores and can shoot a sticky substance from these papillae, thus ensnaring their prey or their enemies. Behind the head, and all along the body, onychophorans carry fourteen to forty-three pairs (depending on the species) of simple walking legs, called lobopods. The legs terminate in a claw with several spines—the source of their name, for Onychophora means "talon bearer."

The Onychophora present the primary case for a classical dilemma in taxonomy: How do we classify small groups of odd anatomy? (Oddness, remember, is largely a function of rarity. If the world contained a million species of onychophorans and only fifty of beetles, we would consider the insects as bizarre.) The chief fault and foible of classical taxonomy lies in its passion for clean order—an imposition bound to distort a messy world of continuity and complexity. A small group of distinctive anatomy sticks out like the proverbial sore thumb, and taxonomists yearn to heal the conceptual challenge by enforcing an alliance with something more familiar. Two related traditions have generally been followed in this attempt, both misleading and restrictive: the shoehorn ("cram 'em in") and the straightening rod ("push 'em between").

The shoehorn works by cramming odd groups into large and well-established categories, usually by forced and fanciful comparison of one or two features with characteristic forms of the larger group. For example, the Onychophora have sometimes been allied with the Uniramia, the dominant arthropod group that includes insects and myriapods (millipedes and centipedes), because both have single-branched legs (never mind that arthropod legs are truly segmented and that onychophoran lobopods are constructed on an entirely different pattern).

The straightening rod tries to push a jutting thumb of oddness back into a linear array by designating the small and peculiar group as intermediary between two large and conventional categories. The Onychophora owe whatever small recognition they possess to this strategy—for they have most commonly been interpreted as living relics of the evolutionary transition between two great phyla: the Annelida (segmented worms, including leeches and the common garden earthworm) and the Arthropoda (about 80 percent of animal species, including insects, spiders, and crustaceans). In this argument, *Peripatus* is a superworm for its legs and a diddly fly for building these legs without true segments.

A third possibility obviously exists and clearly bears interesting implications. This third way has been supported, often by well-respected taxonomists, but our general preference for shoehorns and straightening rods has given it short shrift. The Onychophora, under this view, might represent a separate group, endowed with sufficient anatomical uniqueness to constitute its own major division of the animal kingdom, despite the low diversity of living representatives. After all, the criterion for separate status should be degree of genealogical distinctness, not current success as measured by number of species. A lineage may need a certain minimal membership just to provide enough raw material so that evolution can craft sufficient difference for high taxonomic rank. But current diversity is no measure of available raw material through geological history. Evolution is ebb and flow, waxing and waning; once-great groups can be reduced to a fraction of their past glory. A great man once told us that the last shall be first, but just by the geometry of evolution, and not by moral law, the first can also become last. Perhaps the Onychophora were once a much more diverse group, standing wide and tall in their distinctness, while *Peripatus* and its allies now form a pitifully reduced remnant.

(By speaking of potential distinctness, I am not making an untenable claim for total separation without any relationship to other phyla. Very few taxonomists doubt that onychophorans, along with other potentially distinct groups known as tardigrades and pentastomes, have their evolutionary linkages close to annelids and arthropods. But this third view places onychophorans as a separate branch of life's tree—splitting off near the limbs of annelids and arthropods and eventually joining them to form a major trunk—whereas the shoehorn would stuff onychophorans into the Arthropoda, and the straightening rod would change life's geometry from a tree to a line and place onychophorans between primitive worms and more advanced insects.)

We can only test this third possibility by searching for onychophorans in the fossil record—a daunting task because they have no preservable hard parts and therefore do not usually fossilize. I write this essay because several striking new discoveries and interpretations, all made in the past year or two, now point to a markedly greater diversity for onychophorans right at the beginning of modern multicellular life, following the Cambrian explosion some 550 million years ago. These discoveries arise from two fortunate circumstances: First, onychophorans have been found in the rare soft-bodied faunas occasionally preserved by happy geological accidents in the fossil record; second, some ancient onychophorans possessed hard parts and can therefore appear in ordinary fossil deposits.

I fully realize that this expansion in onychophoran diversity at the beginning of multicellular animal life can scarcely rank as the hottest news item of the year. Most readers of these essays, after all, have probably never heard the word *onychophoran* and, lamentably, have no acquaintance with poor, lovely *Peripatus.* So why get excited about old onychophorans if you never knew that modern ones existed in the first place? Do hear me out if you harbor these doubts. Much more than *Peripatus* lies at stake, for validation of the third position—that onychophorans represent a separate branch of life's tree—has broad and interesting implications for our entire concept of evolution and organic order. I also think that you will marvel at the details of these early onychophorans for their own sake—and their weirdness.

We have actually known a bit about ancient onychophorans for most of this century, thanks once again to that greatest of treasure troves for soft-bodied fossils, the Burgess Shale. In 1911, two years after discovering the Burgess Shale, C. D. Walcott gave the unpronounceable name *Aysheaia* (we generally call it "a-shy-a" in the trade) to an animal that he described as an annelid worm. Many taxonomists, just viewing Walcott's illustrations, immediately saw that the creature looked much more like an onychophoran. In 1931, G. Evelyn Hutchinson, who became the world's greatest ecologist and was, perhaps, the finest person I have ever had the privilege of knowing, published a definitive account on the onychophoran affinities of *Aysheaia*. Hutchinson had studied *Peripatus* in South Africa and he knew onychophoran anatomy intimately. As an ecologist, he was powerfully intrigued by the issue of how an ordinary marine invertebrate like *Aysheaia* could evolve into a terrestrial creature like *Peripatus* with such minimal change in outward anatomy. (*Aysheaia* had fewer pairs of legs and fewer claws per leg than do modern onychophorans. It also bore a terminal mouth at the body's end, while living onychophorans have a ventral mouth on the underside. In addition, *Aysheaia* had no slime papillae and could not use such a device to shoot sticky stuff through ocean waters in any case. But, all told, the differences are astonishingly slight for more than 500 million years of time and a maximal ecological shift from ocean to land.)

One other ancient onychophoran was recognized before last year—a European form named *Xenusion*, found during the 1920s. But *Aysheaia* and *Xenusion* did not shake the shoehorn or the straightening rod. Only two fossils, both so similar to modern forms, do not make an impressive show of diversity. Onychophorans remained a tiny and uniform group, ripe for stuffing in or between larger phyla and not meriting a status of its own.

In the last essay, I described the beginning of the onychophoran coming of age (I was going to say "renaissance," but a renaissance is a rebirth, and onychophorans never had an earlier period of flowering in our consciousness). I described the discovery in China of the animal that bore the small, circular, meshwork plates known for many years from lowermost Cambrian rocks as *Microdictyon*. This fossil comes from the remarkable Chengjiang fauna of south-central China, a Burgess Shale equivalent (although slightly older), with beautiful soft-bodied preservation of many animals already known from the more famous Canadian site (and several novelties as well, including the *Microdictyon* animal). The plates called *Microdictyon* are attached in pairs to the side of the animal just above the junction of paired lobopods with the trunk of the body. The animal itself looks like an onychophoran. If this interpretation holds, then some ancient onychophorans had hard parts. The Chengjiang fauna also contains a second probable onychophoran with plates, named *Luolishania.*

Thus, the early fossil record of onychophorans had begun to expand in numbers and anatomical variety, including fully soft-bodies forms like *Aysheaia* and creatures with small pairs of plates like *Microdictyon* and *Luolishania.* But the big boost, the event that might finally push onychophorans over the border of distinct respectability, finally occurred on May 16, 1991, when the Swedish paleontologist L. Ramsköld and his Chinese colleague Hou Xianguang published an article in the British journal *Nature* (science, at its best, is truly international—see bibliography).

Ramsköld and Hou dropped a bombshell that makes elegant sense of a major paleontological puzzle of recent years. In 1977, Simon Conway Morris described the weirdest of all Burgess Shale organisms with the oddest of all monikers: *Hallucigenia,* named, as Simon wrote, for "the bizarre and dream-like appearance of the animal." He described *Hallucigenia* (see figure) as a tubular body supported by seven pairs of long, pointed spines—not jointed arthropod appendages or fleshy lobopods, but rigid spikes. In Conway Morris's reconstruction, a single row of seven fleshy tubes, each ending in a pair of little pincers, runs along the back, with a tuft of six smaller tubes, perhaps in three pairs, behind the larger seven. The head, not well preserved on any specimen, was depicted as a bulbous extension and the tail as a straight, upward-curving tube.

Hallucigenia was bizarre enough in appearance, but even more puzzlement attended the issue of how such a creature could function. In particular, how could a tiny animal, no more than an inch in length, be stable on seven pairs of rigid spikes for "legs"? And if stable, how could it possibly move? Some of our best functional morphologists, including Mike Labarbera of the University of Chicago, struggled with this issue and found no resolution.

The unlikely morphology, and the even more troublesome question of function, led many paleontologists to dispute Conway Morris's reconstruction (and Simon himself also began to doubt his original conclusions). In my book *Wonderful Life,* I presented Conway Morris's original version and then opted for a

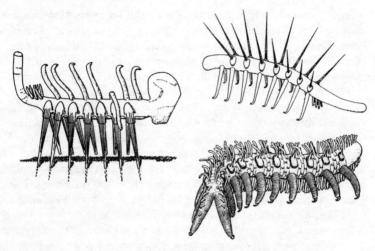

Left: Conway Morris's original reconstruction of *Hallucigenia*. *Simon Conway Morris (1977): Reprinted by permission of* Palaeontology. Right top: Ramsköld and Hou's inversion of *Hallucigenia* as an onychophoran. *Reprinted by permission from* Nature; Copyright © 1991 Macmillan Magazines Limited. Right bottom: Ramsköld and Hou's reconstruction of the new Chengjiang onychophoran with side plates and spines. *Reprinted by permission from* Nature; *Copyright © 1991, Macmillan Magazines Limited.*

different interpretation proposed by several colleagues before me—that *Hallucigenia* is a part broken off from a larger (and still unknown) animal. I wrote:

> *Hallucigenia* is so peculiar, so hard to imagine as an efficiently working beast that we must entertain the possibility of a very different solution. Perhaps *Hallucigenia* is not a complete animal, but a complex appendage of a larger creature, still undiscovered. The "head" end of *Hallucigenia* is no more than an incoherent blob in all known fossils. Perhaps it is no head at all, but a point of fracture, where an appendage (called *Hallucigenia*) broke off from a larger main body.

I received several dozen much appreciated letters from readers of my book, suggesting different reconstructions for some of the oddball creatures of the Burgess Shale. *Hallucigenia* received the lion's share of attention—and one suggestion cropped up again and again, in at least twenty separate letters. These correspondents, nearly all amateurs in natural history, pointed out that *Hallucigenia* would make much more sense turned upside down—for the spines, which never made any sense as organs of locomotion, could then function far more reasonably for protection!

I responded to these letters with, for me, the decisive rejoinder that a single row of tentacles (Simon's version of the upper surface) would work even more poorly than paired spines as devices of locomotion. How could an animal balance, not to mention hop around, on a single row of tentacles? Yet I couldn't deny that everything else made more sense upside down.

It doesn't happen often, but if Ramsköld and Hou are correct—and I think they are—then the gut feeling of amateurs has triumphed over the weight of professional opinion. For Ramsköld and Hou have, unbeknownst to them of course, followed the advice of my correspondents. They have turned *Hallucigenia* upside down, but with an added twist (intellectual, not geometric) as well—they have inverted it into an onychophoran!

Ramsköld and Hou present two major arguments for their inversion of *Hallucigenia*. First, they must tackle the issue that hung me up: How can a single row of tentacles function as legs? They acknowledge the problem, of course, but suggest that Conway Morris was wrong and that two rows of paired tentacles are actually present along the surface that he called dorsal, or topmost. If Ramsköld and Hou are correct, then the major objection to reversing *Hallucigenia* disappears—for two rows of flexible tentacles look like the ordinary legs of a bilaterally symmetrical creature. Moreover, when you turn *Hallucigenia* upside down on the assumption that two rows of tentacles adorn the topside of Conway Morris's version, then the inverted beast immediately says "onychophoran" to any expert, for the little paired pincers at the end of each tentacle become dead ringers for onychophoran claws. Ramsköld and Hou have not yet developed enough evidence to prove the double row of tentacles conclusively, but our museum at Harvard contains the sample best suited for resolving this issue—a slab of rock with more than a dozen *Hallucigenia* specimens. I have lent this slab to Ramsköld and Hou, and I suspect that an answer will soon be forthcoming.

Second, they must explain how an onychophoran could possess the several pairs of long, pointed, upward-protruding spines that an inversion of *Hallucigenia* places along the top edge of the animal—for some fossil onychophorans bear plates (*Microdictyon* and *Luolishania* as previously discussed), but none yet described carry spines. Here, Ramsköld and Hou present compelling evidence in a form much favored by natural historians—a sequence or continuum linking a strange and unexpected form to something familiar through a series of intermediates.

Start the series with *Microdictyon*. This animal, probably an onychophoran, carried pairs of flat plates along the side of its body just above the insertion of lobopods. Go next to a new and as yet unnamed "armored lobopod," again from the prolific Chengjiang fauna. This clear onychophoran also bore paired plates in the same position as in *Microdictyon*. But each plate now carries a small spine (see figure)—nothing like the elongation in *Hallucigenia*, but evidence that onychophoran plates can support spines. For a third step, go to isolated plates collected in lower Cambrian rocks of North Greenland by J. S. Peel and illustrated by Swedish paleontologist Stefan Bengtson in a commentary in *Nature* written to accompany Ramsköld and Hou's paper. These Greenland plates have the same meshwork structure as those of *Microdictyon*—and onychophoran affinity seems a reasonable conjecture (although in this case, we have only found preserved plates, not the entire body). But the Greenland plates carry spines verging on the length of *Hallucigenia* spikes. We now require only a small step to a fourth term in the series—to an onychophoran bearing plates with highly elongated spines: in other words, to *Hallucigenia* turned upside down!

We are witnessing a veritable explosion of Cambrian onychophorans—*Aysheaia* and *Xenusion* with their soft bodies, *Microdictyon* and *Luolishania* with plates, the unnamed Chengjiang creature with plates and short spines, the Greenland form with longer spines, and finally, inverted *Hallucigenia* with greatly extended spines.

The reversal of *Hallucigenia* has capped and sealed the tale. The larger conclusion seems inescapable: In the great period of maximal anatomical variety and experimentation that followed right after the Cambrian explosion first populated the earth with multicellular animals of modern design, the Onychophora repre-

sented a substantial and independent group of diverse and successful marine organisms. The modern terrestrial species are a
tiny and peripheral remnant, a bare clinging to life for a lineage
that once ranked among the major players. The shoehorn and
straightening rod have served us poorly as strategies of interpretation. Groups with few species may be highly distinct in genealogy.

Onychophorans, moreover, are not the only small cluster of
straggling survivors within groups that were once major branches
of life's tree. The distinct phylum of priapulid worms, for example, contains fewer than 20 species worldwide, compared with
some 8,000 for marine polychaete worms, members of the dominant phylum Annelida. Yet, in the Cambrian period at the beginning of multicellular history, priapulids and polychaetes were
equally common and similarly endowed (or so it seemed) with
prospects for long-term success. Moreover, just as onychophorans have held on by surviving in the peripheral habitat of terrestrial life (for a formerly marine group), modern priapulids all live
in harsh and marginal environments—mostly in cold or deep
waters and often with low levels of oxygen.

In recognizing the Onychophora as a distinct group with an
ancient legacy of much greater breadth, we may regret the loss of
tidiness provided by the shoehorn and straightening rod, but we
should rejoice in the interesting conceptual gains. For by our
latest reckoning of life's early history, "uncomfortable" groups
like the Onychophora should exist today. We once thought that
the history of life moved upward and outward from simple beginnings in a few primitive, ancestral lines to ever more and ever
better—the conventional notion that I have called the cone of
increasing diversity. On this model, an ancient and distinctive
genealogical status for several small groups (like the Onychophora) makes no sense—for life's early history, at the point of the
cone, shouldn't have featured many distinct anatomies at all. The
large living groups of mollusks, arthropods, annelids, vertebrates, etc.—all of which have fossil records extending back to
this beginning—provide quite enough material for legacies from
these early times of limited simplicity. But the reinterpretation of
the Burgess Shale, and our burgeoning interest in the early history of multicellular life in general, have indicated that the cone
model is not only wrong but also backward. Life may have

352 | EIGHT LITTLE PIGGIES

reached a maximal spread of anatomical experimentation in these early days—and later history may be epitomized as a diminution of initial possibilities by decimation, rather than a continual expansion.

In this reversed model of a grass field, with most blades clipped off and just a few proliferating wildly thereafter, we should expect to find some blades that survived the mower but never flowered extensively again—whereas, in the cone model, the forest of blades never existed, and the early history of life provides insufficient raw material for many distinct modern groups like the Onychophora.

However much I may regret the loss of a wonderful weirdo in the reversal of *Hallucigenia,* and in its consequent change in status from oddball to onychophoran, I am more than compensated by fascinating insight into the history of ancestry for my favorite name bearer, *Peripatus.* I revel in the knowledge that these marginal and neglected animals belong to a once-mighty group that included armored members with plates and long spines. And I rejoice in the further knowledge thus provided about the strange and potent times of life's early multicellular history. (My regret, in any case, could not possibly be more irrelevant to nature's constitution, either now or 500 million years ago. *Hallucigenia* was what it was. My hopes, and those of any scientist, are only worth considering as potential biases that can block our understanding of nature's factuality.)

Peripatus may walk prouder in the pleasures of pedigree. We humans, as intellectual descendants of Aristotle, the original peripatetic, might consider a favorite motto from "the master of them that know"—well begun is half done (from the *Politics,* book 5, chapter 4). Apply it first to the onychophorans themselves—for in a tough world dominated by contingent good fortune in surviving extinction, a strong beginning of high diversity affords maximal prospect for some legacy long down the hard road. But apply it also to us, the paleontologists who strive to understand this complex history of life. By turning *Hallucigenia* upside down, we have probably taken a large step toward getting the history of life right side up.

7 | Revising and Extending Darwin

25 | What the Immaculate Pigeon Teaches the Burdened Mind

TWO SUCCESSIVE symbols of Saint Louis typify the passages of our century. Saarinen's magnificent arch, gleaming and immaculate, seems to soar from the Mississippi River into heaven (an optical illusion, in large part, cleverly produced by a gradual decrease in edge length from 54 feet at the base of the structure to 17 feet at the summit. Our minds expect a constant size, and the marked decrease therefore makes the summit seem ever so much higher than its actual 630 feet—the size of an ordinary skyscraper in a modern city). By contrast, St. Louis's older symbol, an equestrian statue of the eponymous Louis IX, the only canonized king of France, still stands in front of the art museum in Forest Park. It is anything but immaculate, thanks to that primary spotting agent of all cities—pigeons.

As a team, pigeons and the statue of Saint Louis go way back. The current statue is a 1906 recasting in bronze of the impermanent original made for the main entrance to one of the world's greatest expositions—the 1904 World's Fair held to celebrate (if just a bit late) the centenary of the Louisiana Purchase. The fair must have been spectacular; my wife's family, raised in Saint Louis, still mentions it with awe in stories passed down through three generations. The fair gave us iced tea, ice cream cones, and a great song, "Meet Me in Saint Louis, Louis" (many folks don't even know the next line—"meet me at the fair").

A ferris wheel stood twenty-five stories high; Scott Joplin played his rags. The Pike, main street of the amusement area, featured daily reenactments of the Boer War and the Galveston Flood. The world's greatest athletes came to participate in the third Olympic Games. The fairgrounds, bathed at night in the newfangled invention of electric lighting, inspired Henry Adams to write: "The world has never witnessed so marvelous a phantasm; by night Arabia's crimson sands had never returned a glow half so astonishing [a statement that will need revision after the night bombing of Desert Storm], as one wandered among long lines of white palaces, exquisitely lighted by thousands and thousands of electric candles, soft, rich, shadowy, palpable in their sensuous depths" (from *The Education of Henry Adams*). This statement also makes sense of the next two lines of the famous song: "Don't tell me the lights are shining/Any place but there."

Intellectuals must be constantly clever and industrious. We know that we are peripheral to society's main thrust, and we must be constantly vigilant in seeking opportunities to piggyback on larger enterprises—to find something so big and so expensive that prevailing powers will grant us a bit of space and attention at the edges. The hoopla and funding of major exhibitions often gives us a little room for a smaller celebration in our own style. I was invited to give a speech at something called the "Academic Olympiad" in Seoul during the last Olympic games. (I wasn't able to go and never heard boo about the outcome—though television deluged us with details about javelins and the hop, step, and jump.) Similarly, since the 1904 World's Fair set up shop right next to Washington University, academicians rallied to hold a "Congress of Arts and Science" at the Universal Exposition (as the fair was officially called). At this collocation, the great American biologist Charles Otis Whitman gave a leading address with the general title: "The Problem of the Origin of Species." He spoke primarily about pigeons.

Whitman's work, while treating so humble a subject, had a certain panache and boldness. He wrote at a time when biologists, though fully confident about the fact of evolution, had become very confused and polemical about the causes of evolutionary change. Darwin's own theory of natural selection had never commanded majority support (and would not emerge as a general consensus until the 1930s). As visitors ate their ice-cream cones on the Pike, at least three other theories of evolutionary change enjoyed strong support among biologists—(1) the inheritance of acquired characters, or *Lamarckism*; (2) the origin of species in sudden jumps of genetic change, or *mutationism*; (3) the unfolding of evolution along limited pathways set by the genetic and developmental programs of organisms, or *orthogenesis* (literally, "straight line generation"). Whitman, who had been raising and breeding pigeons for decades, wrote his article to defend the last alternative of orthogenesis, thereby relegating Darwinian natural selection to a small and subsidiary role in evolution.

Whitman's boldness did not lie in his choice of the orthogenetic theory—for this argument was a strong contender in his day, though probably the least popular of the three major alternatives to Darwinism. We judge a man intrepid when he uses his adversary's tools or data to support a rival system. A famous story about Ty Cobb tells of his disgust with Babe Ruth's new style of power hitting (Ruth swatted more home runs per year all by himself than most entire teams had formerly managed in a season). Cobb, the greatest and most artful practitioner of the earlier style of slap, hit, and scramble for a run at a time, held his hands apart, slipped them together high on the bat as the pitch came in (thus achieving maximal control while sacrificing power), and then slapped the ball to his chosen spot; Ruth, by contrast (and following the strategy of all sluggers), held the bat at the end and swung away, missing far more often than he connected. Cobb regarded this style as easy and vulgar, however effective. One day, near the end of his career, and to show his contempt in the most public way, Cobb ostentatiously held the bat in Ruth's manner, hit three home runs in a single game and then went right back to his older, favored style forever after.

Whitman's assault on Darwin's theory from within was far bolder and more sustained, if not quite so showy. For Whitman had chosen, for study over decades, the very organisms that Darwin had selected as the primary empirical support for his own theory—pigeons.

Darwin stated in chapter 1 of the *Origin of Species*:

> Believing that it is always best to study some special group, I have, after deliberation, taken up domestic pigeons. I have kept every breed which I could purchase, or obtain, and have been most kindly favored with skins from several quarters of the world. . . . I have associated with several eminent fanciers, and have been permitted to join two of the London Pigeon Clubs.

Darwin states an excellent reason for his choice in the next sentence:

> The diversity of the breeds is something astonishing. Compare the English carrier and the short-faced tumbler, and see the wonderful difference in their beaks. . . . The common tumbler has the singular and strictly inherited habit of flying at a great height in a compact flock, and tumbling in the air head over heels. . . . The Jacobin has the feathers so much reversed along the back of the neck that they form a hood. . . . The fantail has thirty or even forty tail-feathers, instead of twelve or fourteen, the normal number in all members of the great pigeon family; and these feathers are kept expanded, and are carried so erect that in good birds the head and tail touch.

These breeds are so different that any specialist, if "he were told that they were wild birds," would assume major taxonomic distinction based upon substantial differences. "I do not believe," Darwin writes, "that any ornithologist would place the English carrier, the short-faced tumbler, the runt, the barb, pouter, and fantail in the same genus."

And yet, demonstrably by their interbreeding and their known history, all these pigeons belong to the same species, and therefore have a common evolutionary parent—the rock-pigeon, *Columba livia.* Darwin writes: "Great as the differences are between the breeds of pigeons, I am fully convinced that the common opinion of naturalists is correct, namely, that all have descended from the rock-pigeon (*Columba livia*)." (Darwin might have chosen the even more familiar example of dogs to make the same point, but he was not convinced that all dog breeds came from a common wolf source, whereas the evidence for a single progenitor of all pigeons seemed incontrovertible.) Only one step—the key analogy that powers the entire *Origin of Species*—remained to secure the most important argument in the history of biology, and to make pigeons the heroes of reform: If human breeding, in a few thousand years at most, could produce differences apparently as great as those separating genera, then why deny to a vastly more potent nature, working over millions of years, the power to construct the entire history of life by evolution? Why acknowledge the plain fact of evolution among pigeons, and then insist that all natural species, many no more different one from the other than pigeon breeds, were created by God in their permanent form?

Whitman, of course, did not disagree with Darwin's focal contention *that* pigeons had evolved, but he strongly questioned Darwin's opinion on *how* they and other species had arisen. Charles Otis Whitman (1842–1910), though scarcely a household name today, was the leading American biologist of his generation. He was, perhaps, the last great thinker to span the transition from the pre-Darwinian world to the rise of twentieth-century experimental traditions—for he had studied with Louis Agassiz, the last true creationist of real stature, and he lived to found and direct the symbol of rigorous modernism in American biology, the Marine Biological Laboratory at Woods Hole, still very much in vigorous operation. In his research, Whitman gained fame for meticulous work on *cell-lineage* studies in embryology—tracing the eventual fate of the first few embryonic cells in forming various parts of the body. In promoting this form of research as canonical in Woods Hole, and in establishing at his new laboratory the finest young American biologists, Whitman succeeded in bringing to this country the experimental and mechanistic traditions championed as the soul of modernism in Europe.

In this light, I have always had trouble remembering that Whitman's main love in research lay in the opposite camp of "old-fashioned" and largely descriptive natural history—decades of work on the raising, breeding, and observation of pigeons. This passion even led to his death. In Chicago (where he served as professor of Biology at the University), on the first chill day of December, 1910, Whitman worked all afternoon in his backyard, hastily preparing winter quarters for his pigeons to save them from the cold. As a result, he developed pneumonia and died five days later. F. R. Lillie, once his assistant and then his successor at Woods Hole, eulogized his old boss: "In his zeal for his pigeons, he forgot himself."

Unfortunately, Whitman died before completing and integrating his lifelong work on pigeons. He had published a few preliminary addresses (most notably, his offering in Saint Louis), but never the promised major statement. I can't help thinking that the history of evolutionary thought might have been different had

Whitman lived to promulgate and proselytize his non-Darwinian evolutionary ideas. His colleagues did gather his notes and data into a three-volume posthumous work on pigeons, finally published in 1919. But this work (the basis for my essay) was too disjointed, too incomplete, and, above all, too late to win its potential influence.

Darwin's pigeon agenda extended beyond the simple demonstration of evolution. He also wished to promote his own theory about how evolution had occurred—natural selection. Again, he relied on argument by analogy: Pigeon breeds had been made by artificial selection based on human preferences for gaudiness (pouters, fantails) or utility (carriers, racers)—see figure. "When a bird presenting some conspicuous variation has been preserved, and its offspring have been selected, carefully matched, and again propagated, and so onwards during successive generations, the principle is so obvious that nothing more need be said about it." (This quotation comes from Darwin's most extensive discussion of pigeons—two long chapters in his 1868 book on *The Variation of Animals and Plants under Domestication.* Other statements in this essay are cited from the *Origin of Species,* 1859). But if selection is so undeniably the cause of small-scale evolution over millennia, why deny to nature the power for similar, but far greater, transformation over eons: "May not those naturalists who . . . admit that many of our domestic races have descended from the same parents—may they not learn a lesson of caution, when they deride the idea of species in a state of nature being lineal descendants of other species?"

Orthogeneticists like Whitman did not deny natural selection, but viewed Darwin's force as too weak to accomplish anything beyond a bit of superficial tinkering. Natural selection, they argued, can make nothing and can only accept or reject the variation that arises naturally among differing organisms in an interbreeding population. If the genetic and embryological systems of organisms prescribe a definite direction to this variability, then natural selection cannot deflect the course of evolutionary change. Suppose, for example, that size in offspring only varied in a single direction from parental dimensions—that is, all kids ended up either the same size or taller than their folks. What could natural selection do? Darwin's force could hasten an inherent trend by favoring the taller offspring. At most, selection could

Darwin's illustration of the showiest of all pigeon breeds—the English fantail. This figure comes from his 1868 book: *The Variation of Animals and Plants Under Domestication. Courtesy of Department of Library Services, American Museum of Natural History.*

prevent a trend and keep the population stable, by eliminating the taller offspring and preserving only those of parental dimensions. But selection could not work counter to the inherent direction of variation because no raw material would be available for trends in any direction other than increasing size. Thus, selection would be a force subsidiary to an internal tendency for directional variation—or orthogenesis.

Darwin, of course, was well aware of the logic of this destructive argument. He countered by claiming that, in fact, variation has no inherent direction and occurs "at random" relative to the favored path of natural selection. (This debate set the context for Darwin's confusing claim that variation is random, a statement that has led many people into the worst vernacular misconception about Darwinism: the false belief that Darwin viewed evolutionary change itself as random, and that the manifest order of life therefore disproves his theory. In Darwin's scheme, variation is random, but natural selection is a deterministic force, adapting organisms to changes in their local environments. In fact, Darwin upheld the randomness of variation in order to empower natural selection as a directional agent.) If variation is only random raw material, occurring in no favored direction relative to environmental advantages, then some other force must shape this formless potential into adaptive change. Random raw material requires another mechanism to supply direction by carving out and preserving the advantageous portion—and natural selection plays this role in Darwin's system. But orthogenetically directed variation requires no other shaping force and can set an evolutionary trend all by itself.

Whitman therefore set out to prove that an inherent trend in variation pervades the pigeon lineage—a trend too powerful for natural selection to alter in direction or even to slow substantially. Whitman based his argument on patterns of coloration and began by reversing Darwin's assumption about the plumage of parental forms.

The feral pigeons that speckle our public statuary show two basic color patterns in their extensive repertoire of variation. Some have two black bars on the front edges of their wings and a uniform gray color elsewhere; others develop black splotches, called checkers, on some or all wing feathers, but also retain the two bars (though usually in more indistinct form). The bars, in any case, are composed of checkers (on several adjacent feathers) that line up to produce the impression of a broad continuous stripe (see figures taken from Whitman's posthumous monograph).

Darwin had assumed that ancestral pigeons were two-barred and that checkers represent an evolved modification by intensification of coloration. Whitman cleanly reversed Darwin's perspective:

> The latter [two-barred] was regarded by Darwin as the typical wing pattern for *Columba livia*; the former [checkered] was supposed to be a variation arising therefrom, a frequent occurrence but of no importance. Just the contrary is true; the checkered pigeon represents the more ancient type, from which the two-barred type has been derived.

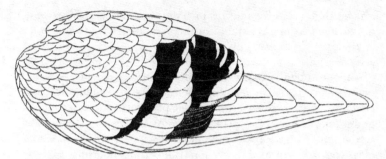

Whitman's figure of the two-barred wing pattern, which he regarded as an advanced state in his orthogenetic sequence. *Courtesy of Department of Library Services, American Museum of Natural History.*

Whitman's illustration of the checkered pattern—the primitive state in the evolution of pigeon wing colors. *Courtesy of Department of Library Services, American Museum of Natural History.*

This reversal is of no great significance in itself, unless you happen to be a pigeon fancier devoted to the peculiarities of these generally unloved creatures. But Whitman properly made much of his inversion because his favored sequence of checkers-to-bars formed the major part of his proposed (and inexorable) orthogenetic sequence of internally prescribed variation—an evolutionary pattern inherent in the biological organization of pigeons and quite beyond the power of natural selection to deflect. The orthogenetic trend, Whitman argued, moved from original diffusion to later concentration of color. Checkers plus indistinct bars must precede clear bars and no checkers. In the lines following the quotation just cited, Whitman writes:

The direction of evolution in pattern in the rock pigeons has been from a condition of relative uniformity to one of regional differentiation.

But Whitman had an even bolder vision, based on the same orthogenetic pattern. He did not interpret the pathway from checkers to bars as a circumscribed peculiarity of domestic pigeons, but rather as part of a much more extensive orthogenetic trend pervading the entire pigeon family (including all other species from mourning dove to passenger pigeon, which became extinct in the decade between Whitman's death and the posthumous publication of his monograph), and perhaps even all coloration in birds—an inherent and ineluctable progression from an original homogeneous checkering on all feathers; to the concentration of color in certain areas (checkers plus bars, and then to bars alone); to the reduction and weakening of these concentrations; to the final elimination of all color. Whitman located the ancestral state in the uniform checkering of some species—the "turtle dove pattern" in his terminology (see figure)—and the final goal in some idealized, albinized version of the Holy Ghost, depicted as a pure white dove in so many medieval paintings: in short, and in his words, from uniform spotting to "immaculate monochrome"—a most unpigeonlike state (in both appearance and deed), at least in our metaphors. In a remarkable vision of inexorable movement through the entire great family of pigeons, Whitman writes:

When we look around among allied species and see the same bars reduced to about half dimensions in the rock pigeon of Manchuria, reduced to mere remnants of two to six spots in the stock dove, carried to complete obsoletion [sic] or to a few shadowy reminiscences in . . . *Columba rufina* of Brasil, gone past return in some of our domestic breeds and in many of the wild [doves and pigeons]—when we see all these stages multiplied and varied through some 400 to 500 wild species and 100 to 200 domestic breeds, and in general tending to the same goal, we begin to realize that they are . . . slowly passing phases in the progress of an orthogenetic process of evolution, which seems to have no

The fully checkered turtledove pattern, which Whitman regarded as the original state in his orthogenetic sequence. *Courtesy of Department of Library Services, American Museum of Natural History.*

fixed goal this side of an immaculate monochrome—possibly none short of complete albinism.

Moreover, while the progress of the trend may be "lengthened or shortened, strengthened or weakened" by such subsidiary forces as natural selection, the orthogenetic sequence is invariable: "The steps are seriated in a causal, genetic order—an order that admits no transpositions, no reversals, no mutational skips, no unpredictable chance intrusions." When we discern the proper sequence of orthogenetic stages, evolution may become a predictive science: "Not only is the direction of the change hitherto discoverable, but its future course is predictable."

I have not resurrected Whitman's largely forgotten work in

order to defend orthogenesis as a replacement for natural selection, for this is truly a dead issue decided long ago in Darwin's favor. Rather, from my deep admiration of Whitman's keen intelligence and my abiding respect for his decades of careful work with pigeons, I wish to point out that his conclusions were not foolish, and that several aspects of his work would repay our close study, even today. Consider three points that might prompt a proper sympathy and interest.

1. *The false charge of teleology.* The standard one-liners of evolution texts dismiss orthogenesis as a theistically inspired last-gasp effort to salvage some form of inherent goal and purpose in Darwin's new world. In this canard, supporters of orthogenesis abandoned rationality itself to embrace a woolly mysticism of "vital forces" and "inherent directions"—the very concepts that science had struggled to discard in field after field, from cosmology to physics to chemistry. Whitman has been viewed as a particularly sad example of this slippage and surrender, for he had been such a committed mechanist in his earlier embryological work.

This hurtful charge is not only wrong, but entirely backwards. Whitman and nearly all other prominent supporters of orthogenesis maintained as firm a commitment to mechanical causation, and as strong an aversion to mystical or spiritual direction, as any contemporary in this late-nineteenth-century age of industrial order. In the opening paragraph of his 1919 monograph on pigeons, Whitman wrote:

> His [Darwin's] triumph has won for us a common height from which we see the whole world of living beings as well as all inorganic nature; phenomena of every order we now regard as expressions of natural causes. The supernatural has no longer a standing in science; it has vanished like a dream, and the halls consecrated to its thraldom of the intellect are becoming radiant with a more cheerful faith.

In fact, Whitman's orthogenesis arose from his mechanical perspective, not in opposition to his former life's work. The orthogenetic trend did not follow a mystical impulse from outside, but a mechanistic drive from within, based upon admittedly unknown laws of genetics and embryology. His work on cell lineages had mapped the fate of earliest cells in the embryo, and had indicated that the source of eventual organs could be specified even in the tiny and formless clump of initial cells. If embryos grew so predictably, why should evolutionary change be devoid of similar order. "I venture to assert," Whitman wrote, "that variation is sometimes orderly and at other times rather disorderly, and that the one is just as free from teleology as the other. . . . If a designer sets limits to variation in order to reach a definite end, the direction of events is teleological; but if organization and the laws of development exclude some lines of variation and favor others, there is certainly nothing supernatural in this."

2. *Whitman's evidence.* Modern detractors who misconstrue orthogenesis as old-fashioned teleology often assume that its supporters could only have been working on hope and the flimsiest of supposed evidence. But Whitman spent decades gathering reams of fascinating data (not all properly interpreted in our view, but still full of interest). He marshaled three major sources of evidence to support his orthogenetic theory: breeding, comparative anatomy, and ontogeny (the growth of individual birds).

In breeding, he found that he could develop a two-barred race from checkered parents by selecting birds with the weakest checkers in each generation. But he could never produce checkered birds from two-barred progenitors. In comparative anatomy, he argued that species judged more ancestral on other criteria grew plumages of early stages in the orthogenetic series. In ontogeny, he found that juvenile plumage exhibited earlier stages than adult feathers (a juvenile bird might moult its checkered feathers and then grow a two-barred adult plumage). Most nineteenth-century biologists believed that "ontogeny recapitulates phylogeny"—a mouthful meaning that individuals, in their embryology and growth, tend to pass through stages representing adult forms of their ancestry. Juvenile plumage would therefore represent an ancestral condition. The law of recapitulation is wrong (see my book, *Ontogeny and Phylogeny*), but you can't blame Whitman for accepting the consensus of his time.

3. *Channels versus one-way streets.* Whitman conceived his series of orthogenetic stages as a forced pathway—a one-way street with pigeons as the cars. He was clearly wrong in this vision, and two major errors invalidate his form of orthogenesis. First, the cars can go in either direction; Whitman's series may carve a road into a complex landscape, but the traffic flows both ways. Pigeons

can either gain or lose color. Second, I doubt that either the checkered or two-barred condition represents a primitive state for domestic pigeons. The ancestor of domestic races was a population, not an individual—and populations are variable. I suspect that the parental population included both checkered and two-barred birds within a spectrum of variation—and that the spectrum represents the ancestral condition.

But think about Whitman's vision in a slightly modified form, and we encounter an idea that is not only valid, but also full of potential insight for correcting a major misconception and teaching a fundamental truth about evolution. Think of his one-way street as a channel instead—a favored pathway of evolutionary variation set by the inherited genetic and developmental programs of organisms.

If natural selection controlled evolution entirely, then no such limits and pathways would exist, and organisms would be like billiard balls, capable of movement in any direction and subject to any change in position induced by the pool cue of natural selection. But, to borrow an old metaphor from Francis Galton (see Essay 27 for a full explication), suppose that organisms are polyhedrons rather than billiard balls—and that they can only move by flipping from one side (on which they now rest) to an adjacent facet. They may need a push from natural selection to move at all, but internal limitations and possibilities set the direction of possible change. If inherited genetic and developmental programs build the facets of Galton's polyhedron, then strong internal constraints upon evolutionary change must exist and Whitman's insight is correct, so long as we convert his one-way streets into channels—that is, strong biases in the direction of variation available for evolutionary change. Moreover, Whitman probably identified the most important internal channel of all—the pathway of ontogeny, or the growth of individuals. Evolutionary change proceeds most readily in directions already established in growth—lengthening a bit here, cutting out a stage or two there, changing the relative timing of development among organs and parts.

The most serious of vernacular misconceptions views evolution as an inexorable machine, working to produce optimal adaptations as best solutions to problems posed by local environments and unconstrained by the whims and past histories of organisms. For example, I have monitored the "Ask the *Boston Globe*" science query column for years and have never seen anything but adaptationist answers to evolutionary questions. One correspondent asked, "Why do we have two breasts?" and the paper responded that the "right" number of nipples (for optimal adaptation) is one more than the usual complement of offspring, thus providing a margin of safety, but not so large a surplus to become a burden. Since human females almost always have but a single child, two become the right number of breasts by this argument rooted in natural selection. I couldn't help but laugh when I read this conclusion. I do grasp the logic, but surely the primary channel of our anatomical design—bilateral symmetry—has some relevance to the solution. Most externalities come in twos on our bodies—eyes, nostrils, ears, arms, legs, etc.—and the reason cannot be singleton births. Isn't this prior channel of architecture more likely to supply the primary reason for two breasts?

If the purely adaptationist vision were valid, we might gain the comfort of seeing ourselves, and all other creatures, as quintessentially "right," at least for our local environments of natural selection. But evolution is the science of history and its influence. We come to our local environments with the baggage of eons; we are not machines newly constructed for our current realities. These historical baggages are expressed as the genetic and developmental channels that led Whitman too far. But these same channels, properly interpreted as strong biases in variation rather than one-way streets of change, would give us a much richer view of evolution as a subtle balance between constraints of history and reshapings by natural selection.

The power of these channels provides a key to understanding our bodies and our minds; we will never grasp the evolutionary contribution to our nature if we persist in the naive view that natural selection builds best solutions. We can accept the idea more readily for our bodies; hernias and lower back pain are the price we pay for walking upright with bodies evolved for quadrupedal life and not optimally redesigned. But how much of the quirkiness and limitations of our modes of thinking might also record a structure evolved during eons for other uses, and only recently adding the varied phenomena of higher consciousness and its primary tool of expression in language? Why are we so bad at so many mental operations? Why do we seem so singularly unable to grasp probability? Why do we classify by the painfully inadequate technique of dichotomy? Why can we not even conceive of infinity and eternity? Is the limit of current cosmological thinking a defect of data, or a property of mind that gives us no access to more fruitful kinds of answers?

I do not intend this list as a statement of despair about our mental midgetry. To recognize a potential limit is to think about tools of possible transcendence. Freedom, as Spinoza said, is the recognition of necessity. Let us return once again to the proper metaphor of channels and remember the finest statement in literature about emerging from ruts: "There is a tide in the affairs of men, which, taken at the flood, leads on to fortune."

26 | The Great Seal Principle

TENNYSON'S *In Memoriam,* published in 1850, was surely the most popular of Victorian poems. The good queen herself remarked to her poet laureate, following the death of her beloved husband, Prince Albert: "Next to the Bible, *In Memoriam* is my comfort." As a paradoxical and ultimate testimony of success, many lines became so popular, so much a part of everyday speech, that their relatively recent source was forgotten and a false Shakespearian or biblical origin often assumed. Be honest now; didn't you think that Shakespeare wrote:

> 'Tis better to have loved and lost
> Than never to have loved at all.

(*In Memoriam* also gave us, nearly a decade before Darwin's book, the classic metaphor for natural selection: "nature red in tooth and claw.")

After loving and losing, the most famous misattributed line from *In Memoriam* must be:

> He seems so near, and yet so far.

Tennyson's image provides an excellent epitome for that constant and unwelcome companion of intellectual life: frustration. I may be fascinated by big questions—the ultimate origin of the universe, for example—but I am not frustrated because I expect no near or immediate solution. Frustration lies just beyond the finger tip—the solution that is almost palpable, but for one little, stubborn obstacle.

Scientific frustration takes two primary forms. In the usual, empirical variety, deeply desired data lie just beyond our reach. Remember that we looked at the moon for millennia, but never knew the form of her back face (and couldn't really develop a decent theory of origin and subsequent history without this information). So near (if we could only grab hold of the damned thing and turn it around)—and yet so far (a good quarter of a million miles). One space probe and a camera resolved this frustration of the ages.

A second species of frustration arises from logical problems, and these sometimes seem more intractable because solutions must come from inside our heads. Consider the classics, Zeno's paradoxes, or the puzzles of our primers:

> Brothers and sisters have I none
> But that man's father is my father's son.

Again, the answers seem so close (after all, the arrow does move and Achilles does pass the tortoise), yet the structure of resolution eludes us.

Empirical frustrations are resolved by evidence; I don't know that they present much of a general message beyond the obvious value of data over casuistry. Logical frustrations have more to teach us because solutions require a reorientation of mental habits (if only the minor realization that problems need not be viewed as external to their posers, and therefore "objective": The man in the puzzling couplet is pointing to his own son).

The study of evolution is beset with frustration, most of the empirical variety (inadequacy of the fossil record, our inability to track and document enough members of a population). But the profession also features some persistent logical puzzles, most treated (and some resolved) by Darwin himself. Several take a similar form, roughly: "I can figure out why a particular feature is useful to an organism once it develops, but how could it arise in the first place?" I have treated one standard form of this puzzle in several essays—the "10 percent of a wing" problem, or how can wings evolve if tiny initial stages could confer no aerodynamic benefit? Darwin's solution, now experimentally confirmed (see Essay 9 in *Bully for Brontosaurus*), argues that initial stages functioned in a different manner (perhaps for thermoregulation in the case of incipient wings), and were later coopted, when large enough, for their current use in flight.

A related and equally thorny problem asks why a useful evolutionary trend can begin in the first place and why one pathway is taken in a large potential field. The knee-jerk adaptationist answer—"because the evolved feature works so well (and must therefore, in some sense, have been prefavored as a solution)"— simply will not do, for current utility and historical origin are entirely separate issues. (What, in nature, works better than a wing?—and yet we all agree that benefit in flight did not initiate the trend).

Darwin also thought about this issue and proposed a solution. His argument features a trio of important properties: it is interesting, probably correct, and largely unappreciated. Darwin considered the classic case of mimicry in butterflies—the convergence of a tasty species on the pattern of a noxious form, all the better to fool predators (viceroy on monarch, for example). A potential mimic may share an environment with one hundred other butterfly species. Why converge upon one, rather than any of the ninety-nine others? And why initiate such a trend at all among so many other evolutionary possibilities? Darwin argued that the inception must reside in accident, whatever the predictable character of the trend once it starts, and whatever the resulting benefit. The mimic's ancestor must begin with a slight and fortuitous resemblance to the species eventually copied. Such predispositions can only be chancy, for a species cannot know its complex future. A beginning "leg up" can nudge the trend into a particular path. The path itself will be carved by the deterministic force of natural selection, but the push into the path requires a bit of luck. Without an initiation in nonadaptive good fortune, the final and stunning adaptation could never evolve.

I am tempted to call this logical solution the "great seal principle," to honor the motto of our national emblem (engraved on the flip side of a dollar bill)—*annuit coeptis.* The agent is usually construed as God and the line, following our gender-biased tradition, therefore translated as "he smiles on our beginnings." But the Latin third-person singular is androgynous, and I prefer to think of the agent as Lady Luck—therefore, "she smiles on the initiations."

Darwin's argument is theoretically sound as an abstract resolution of a conceptual puzzle—one of the logical frustrations of my introduction. But have we any evidence that nature actually bows to the reason of our arguments? This issue is particularly important in evolutionary biology because we so often make the mistake of assuming that we understand the origin of a feature just because it now works so well. Consider, for example, the large set of "showy" male organs—from peacocks' tails to deer antlers to elaborate behavioral displays in birds-of-paradise—that presumably evolved in the process identified by Darwin as "sexual selection." In one category of sexual selection, called "female choice" by Darwin, these elaborate structures (encumbrances in any other context) develop and enlarge because females prefer the bigger or more decorated males. Female choice may explain the extensive and gaudy patterning, but why tail feathers in the first place? Why not one of a hundred alternatives—head plumes, elaborate calls, or the more common mammalian analog of old-fashioned male bullying?

Several evolutionists, in the past few years, have thought more deeply about the difference between origins and later pathways and have taken Darwin's problem and solution more seriously. They have realized that the pathways set by female choice must often involve an important initiating component of preexisting bias in sensory and cognitive systems. Females, after all, perceive and process information in a limited number of ways based on broad features of brain and sense organs that obviously did not evolve in order to prefer showy tail feathers in some unspecified future. Relative to tail feathers, or antlers, or complex behavioral displays, these biases are components of good fortune that permit the initiation of a particular trend. Two recent studies have provided excellent evidence for the great seal principle by combining experimental data on female choice, with a documentation of preexisting sensory biases in a genealogical context that validates an evolutionary argument.

Fish tails. When we think of the conjunction of weaponry and fishes, we usually picture a large and graceful marine species with a sword for a snout and a lovely name of classical redundancy— *Xiphias gladius,* the swordfish (*xiphias* is a Greek sword; *gladius,* a Latin counterpart, as in the gladiator who wields it in combat). But a much smaller, Central American, fresh-water relative of the guppy, also bears a sword—this time at the rear end, formed after sexual maturity, and only in males, by an elongation of rays at the base of the caudal (tail) fin.

Alexandra Basolo of the University of California at Santa Barbara performed behavioral experiments on the swordtail, *Xiphophorus helleri,* and proved that females do prefer males with longer swords, thus establishing the efficacy of female choice in maintaining and enlarging the male sword. But such information tells us very little about the origin of swords. These projections do males a world of Darwinian good, but why swords, rather than big eyeballs, funny fins, or elaborate swimming displays? Fortunately, we have enough information about the genealogy of swordtails to reconstruct a historical sequence, and to recognize an important component of preexisting female sensory bias in the evolution of swords.

The close relatives of *X. helleri* are all swordless, and we may conclude that ordinary swordless tails represent the original state of this lineage. In particular, *Xiphophorus maculatus,* the closest living relative of the swordtail, lacks a rear projection (despite its taxonomic residence in the same genus, with its etymology of "sword bearer"). Basolo therefore performed a series of ingenious and elegant experiments on the swordless *X. maculatus.*

In her basic procedure, she placed a female in the center section of an aquarium constructed with two side compartments of equal volume. She then put a single male into each of the side compartments and noted female preference by time spent in the vicinity of males, and by performance of courtship behaviors. In these particular experiments, she surgically implanted, into the tail of swordless *X. maculatus* males, swords of the same relative length and form as in *X. helleri.* In some males, the swords had the same distinctive yellow color and bold black border as in *X. helleri*; in others, the swords were transparent (and shown in behavioral experiments to be invisible to females).

Basolo placed a male with a colored tail into one side compartment and a male with an invisible tail into the other chamber. (She followed this elaborate procedure of implanting invisible tails, rather than simply using ordinary tailless males, because she needed to control for the results of surgery and the effect of a tail upon swimming and other male behaviors. If females just reacted differently to an ordinary, unoperated male, than to a male with an experimentally implanted tail, we would not know whether this disparity recorded the tail's presence or the results of surgical intervention.)

Basolo used six pairs of males, each containing one fish with a visible and the other with a transparent sword. She tested each pair with nine to sixteen female fishes. Invariably, females preferred males with visible swords—even though males of this species have no swords at all in nature. As in all good experiments, Basolo then performed a variety of additional tests to eliminate other interpretations. She changed sides for males with visible and invisible swords—just in case females were choosing left or right sides of the aquarium, rather than visible or invisible swords. The females preferred the visible sword, regardless of position in the aquarium. She even performed a second operation and switched swords—placing the transparent sword into the fish that previously carried the colored version, and implanting the colored sword into the fish that had borne the invisible addition. The fish that had previously been shunned (presumably for its invisible sword) was now favored when bearing the sword with the prominent black border. Basolo concludes: "These data suggest that the females were basing their choice on sword preference and not other traits."

The title of Basolo's article says it all—"Female preference predates the evolution of the sword in swordtail fish" (see bibliography). Something in the sensory system of ancestral fishes evidently predisposed females of the *X. helleri* line to prefer males with swords. Since no previous member of this large and successful group of fishes possessed swords (so far as we know), this sensory and cognitive bias exists for other reasons, and must be regarded as fortuitous with respect to the evolution of swords. Male *Xiphophorus helleri* must, in this sense, thank Lady Luck for their graceful extensions.

Frog calls. So much of what we view as most aesthetic and charming in nature—the singing and plumage of birds, as a prime example—actually functions as part of the great Darwinian struggle for reproductive success. Chorusing of males in crickets, frogs, or birds, is no paean of praise to the night, no hosanna to the joys of life, but a complex tapestry of challenge (to other males) or advertisement (to females).

In many frogs, the female choice model of sexual selection seems to apply, and males call to win the sexual attention of females—all in the service of the great Darwinian attempt to avoid croaking (vernacular sense) of family lines. Michael J. Ryan and colleagues at the University of Texas in Austin have applied the preexisting bias model ("sexual selection for sensory exploitation" in their terms) to the complex call of the Tungara frog, *Physalaemus pustulosus* (see bibliography).

This Panamanian frog has an unusually complex call, consisting of two sequential components with expressive names: a whine and a chuck. The call begins with the longer whine, about 350 milliseconds in duration, that gradually decreases in fundamental frequency from 900 to 400 Hertz. Although the whine contains up to three harmonics, most energy resides in the fundamental. (Harmonics are overtones generated from the fundamental and having higher frequencies at integral multiples of the fundamental. If that sounds like a mouthful, the first harmonic of a 220 Hz fundamental frequency is 440 Hz, the second 660 Hz, the third 880 Hz, etc.). The whine is followed by a series of one to six chucks. These chucks are much shorter in duration (about 40 milliseconds) and have a lower fundamental frequency of about 220 Hz. But, unlike the whine, chucks have much higher energy in the fifteen harmonics above the fundamental. In fact, some 90 percent of the energy resides in harmonics above 1500 Hz, with the peak frequency above 2000 Hz.

This complexity becomes important in Ryan's argument for an interesting reason based on the physiology of amphibian hearing. Uniquely among terrestrial vertebrates, amphibians possess two inner-ear organs that pick up airborne vibrations—the amphibian papilla and the basilar papilla. The amphibian papilla is most sensitive to frequencies below 1200 Hz, while the basilar papilla responds best to higher frequencies above 1500 Hz.

Direct study of the inner ear in *Physalaemus pustulosus* shows that the most sensitive fibers of the amphibian papilla are tuned to about 500 Hz, while all fibers in the basilar papilla are most sensitive to about 2100 Hz. These facts suggest an obvious hypothesis for evolution of the complex call in *P. pustulosus,* particularly for the addition of chucks to the presumed ancestral call of whine alone—namely, that the whine only stimulates the amphibian papilla, while addition of the chuck takes advantage of a latent capacity already present but unutilized in ancestral calls: the acoustical properties of a basilar papilla tuned to higher frequencies concentrated in the chuck. The basilar papilla provides the preexisting sensory bias (sensitivity to higher frequencies), and the chuck finally contacts this everpresent, but initially unexploited, capacity.

Since the calls elicit female attention (with approach and eventual mating), Ryan and colleagues performed an interesting and successful experiment. They synthesized a variety of calls and broadcast two different versions from opposite ends of an indoor arena measuring 3 square meters. They put a female in the center of the arena and covered her with an opaque cone. They gave her five minutes to acclimate as they played the calls. A remote device then lifted the cone, and the female was free to approach a speaker. If she consistently preferred one to another, then the relative evolutionary value of whines and chucks might be assessed.

Females consistently favored the complex call of whine plus chuck over the whine alone. This preference is not a simple result of adding more total energy by including the chucks, but seems to be set by distinctive characters of the chuck. If females are given a choice between whine plus chuck and an enhanced whine alone (with 50 percent more total energy than whine plus chuck), they still prefer the complex call of whine plus chuck, despite its lower energy. Finally, Ryan and colleagues determined that females respond equally well to both the low and the high harmonics of the chuck. In other words, they are equally positive towards components of the chuck that stimulate either the amphibian or the basilar papilla.

So far so good. The added chuck does elicit female preference, and the component of the chuck that stimulates the previously unexploited basilar papilla also appeals to females. But to argue for the preexisting sensory bias model, we need to know that ancestral species (or relatives maintaining the ancestral state) are also inclined to react favorably to the chuck, even though their distinctive call contains no such component—just as Basolo showed that females in unarmored species prefer males with surgically implanted swords. Ryan was able to supply this final piece of evidence by measuring the tuning of basilar papillas in seven individuals of the closely related species, *Physalaemus coloradorum*. Ryan writes: "This species does not produce chucks, and the ability to produce chucks was derived in *P. pustulosus* after these species diverged." Ryan found no statistically significant difference between the most sensitive frequencies for *P. pustulosus* (2130 Hz) and for the chuckless *P. coloradorum* (2230 Hz). The basilar papilla of *P. coloradorum* is therefore tuned to high frequencies not found in their actual call. A preexisting sensory bias for potentially advantageous evolutionary change can therefore be specified—a pathway actually followed by *P. pustulosus.*

These conclusions about preexisting sensory bias, while satisfying, also present another paradox. How can something so specific as the preference for an extension to a tail or a funny sound be encoded by accident into ancestral behaviors? Can we seriously believe that an animal might be adapted to favor swords never seen or chucks never heard?

The probable resolution of this paradox (another logical frustration in the terms of my introduction) may be illustrated by a famous experiment on quail, done ten years ago by the British ethologist Patrick Bateson (see bibliography). Avoidance of incest is very common in vertebrates with complex behaviors and high cognitive capacity. The evolutionary rationale is easy to express: Mating with closest kin produces a high frequency of genetically compromised offspring with disadvantageous traits in double recessive doses (a phenomenon called *inbreeding depression*). But quail don't know Mendelian genetics (and neither did people before this century). So what can be leading animals to this evolutionarily advantageous behavior?

Bateson built an ingenious device that exposed individual quail to five birds of the opposite sex, but of different degrees of relationship: a sibling nestmate, a sibling never seen before, a first cousin, a third cousin, and an unrelated bird. Both males and females generally preferred first cousins over all alternatives.

One popular hypothesis (applicable to humans in some interpretations) holds that we avoid closest kin by a simple learning rule that derails later sexual feelings towards individuals reared with us from our earliest days (as one wag said, if we share potties early, we don't party later—and remember, where I live in Boston, the two words are pronounced nearly alike). On this argument, since rearingmates are usually sibs, the simple learning rule turns the proper evolutionary trick. But this explanation will not suffice for Bateson's data, for quail prefer first cousins over true siblings never seen before.

Bateson therefore concludes, from this and other arguments, that quail may be following a highly abstract aesthetic rule—prefer intermediary degrees of familiarity: not so close as to be cloying, not so distant as to be overly strange. If he is right, an elegant solution to the problem of avoiding incest suggests itself. Quail are not Mendelian calculators. They are, rather, following a deeper, and more abstract, rule of aesthetic preference that may be common to a wide range of animals and neurologies. Maximal attraction to intermediate familiarity will automatically exclude disadvantageous closest kin as potential mates. Natural selection need not work for the specific goal of avoiding incest. By good fortune, a deeper cognitive principle engenders this result as a consequence. (Of course, one might turn the argument around and claim that the aesthetic principle arose because incest avoidance is so important, and animals could only achieve this result by such an indirect route. But I prefer to view the specific as a manifestation of the general, rather than the rule as a surrogate for the example.)

This same style of argument makes the preexisting bias hypothesis more sensible. We need not postulate a preexisting bias for seeing and favoring swords on tails. We only require a general behavioral rule (like intermediate familiarity), that might render the specific result (avoiding incest) as a reasonable manifestation. In fact, Basolo suggests that swords may be preferred by females in swordless species because the implanted weapon makes the male look larger in general, and bigger size is a potent spur to female choice in many regimes of sexual selection. Thus, the general cognitive rule would proclaim: Prefer larger males. The specific solution in this case would be: Extended swords give an impression of larger size with little addition of actual bulk. A thousand other pathways might have satisfied the same broad rule, but *Xiphophorus* evolved a sword. Similarly, the preexisting bias in frogs is a basilar papilla tuned to high frequencies, not an irresistible urge to hear a chuck. Again, this bias might have been exploited in many other ways, but *P. pustulosus* evolved a chuck.

The solution is elegant (and probably even true in these cases—what a rare and lovely combination). Evolution is always a wondrous mixture of the quirky and unpredictable (usually expressed as historical legacies brought to different modern contexts) with the sensible adaptive improvements wrought by natu-

ral selection. The quirky component of historical legacy con-
strains the predictable force of immediate selection, so we usually
think of history as restrictive and selection as flexible. But the
stories of this essay reverse the usual perspective. Here, historical
legacy is a broad cognitive rule bursting with potential along a
thousand possible pathways—prefer larger males, or prefer individuals of intermediate familiarity. And adaptation then clamps
the limit by choosing one manifestation—a sword, a chuck, or a
first cousin. If Lady Luck smiles on the beginning of such an
evolutionary trend from one side of our great seal, I am tempted
to quote the more familiar motto from the other side to describe
the final choice of a singular solution: *E pluribus unum,* one from
many.

27 | A Dog's Life in Galton's Polyhedron

IN THE OPENING sentence of *Hereditary Genius* (1869), the founding document of eugenics, Francis Galton (Charles Darwin's brilliant and eccentric cousin—see Essay 31 for another tale of this remarkable man) proclaimed that "a man's natural abilities are derived by inheritance." He then added, making an appeal by analogy to changes induced by domestication:

> Consequently, as it is easy . . . to obtain by careful selection a permanent breed of dogs or horses gifted with peculiar powers of running, or of doing anything else, so it would be quite practicable to produce a highly-gifted race of men by judicious marriages during several consecutive generations.

Darwin had also invoked domestication as his first argument in the *Origin of Species.* Darwin began his great treatise, not with fanfare or general proclamation, but with a discussion of breeding in domestic pigeons (see Essay 25).

Darwin attributed the wondrous variety among pigeons, dogs, and other domesticated animals to the nearly limitless power of selection: "Breeders habitually speak of an animal's organization as something plastic, which they can model almost as they please." He quotes one authority on the "great power of this principle of selection": "It is the magician's wand, by means of which he may summon into life whatever form and mold he pleases."

This optimistic notion—that diligence in selection can produce almost any desired trait by artificial selection of domesticated animals or cultivated plants—has inspired the customary extrapolation into nature's larger scales, leading to a conclusion that natural selection must work even more inexorably to hone wild creatures to a state of optimal design. As Darwin wrote:

> How fleeting are the wishes and efforts of man! how short his time! and consequently how poor will his products be compared with those accumulated by nature during whole geological periods. Can we wonder, then, that nature's productions . . . should be infinitely better adapted to the most complex conditions of life, and should plainly bear the stamp of far higher workmanship. It may be said that natural selection is daily and hourly scrutinizing, throughout the world, every variation, even the slightest; rejecting that which is bad, preserving and adding up all that is good; silently and insensibly working.

This common claim for organic optimality cannot be reconciled with a theme that I regard as the primary message of history—the lesson of the panda's thumb and the flamingo's smile: The quirky hold of history lies recorded in oddities and imperfections that reveal pathways of descent. The allure of perfection speaks more to our cultural habits and instructional needs than to nature's ways (good design inspires wonder and provides excellent material for boxed illustrations in textbooks). Optimality in complex structure would probably bring evolution to a grinding halt, as flexibility disappeared on the altar of intricate adaptation (how might we change a peacock for different environments of its unknown future?).

In any case, leaving aside the abstractions of how nature ought to work, we have abundant empirical evidence that enormous effort in husbandry does not always bring its desired reward. Poultrymen have never broken the "egg-a-day barrier" (no breed of hen consistently lays more than one egg each day), and we are now trying to produce frost-resistant plants by introducing foreign DNA with techniques of genetic engineering because we have not been able to develop such traits by selection upon the natural variation of these plants. We do not know whether such failures represent our own stupidity or lack of sufficient diligence (or time) or whether they record intrinsic structural and genetic limits upon the power of selection. In any case, selection, either natural or artificial, is not the agent of organic optimality that our newspapers and textbooks so often portray. Limits are as powerful and interesting a theme as engineering triumph.

Francis Galton himself, in the same book that promised so much for human futures by controlled breeding, presented our most incisive metaphor for the other side of the coin—the limits to improvement imposed by inherited form and function. (Darwin was also intrigued by the subject of limits and devoted as much attention to this aspect of growth and development as to natural selection itself—see his longest book, the two-volume *Variation in Animals and Plants Under Domestication*, 1868.) Following the optimistic notion of unrestricted molding, we might view an organism as a billiard ball lying on a smooth table. The pool cue of natural selection pushes the ball wherever environmental pressure or human intent dictates. The speed and direction of motion (evolutionary changes) are controlled by the external force of selection. The organism, in short, does not push back. Evolution is a one-way street; outside pushes inside.

But suppose, Galton argues, organisms are not passive spheres but polyhedrons resting upon stable facets.

> The changes are not by insensible gradations; there are many, but not an infinite number of intermediate links. . . . The mechanical conception would be that of a rough stone, having, in consequence of its roughness, a vast number of natural facets, on any one of which it might rest in "stable" equilibrium. . . . If by a powerful effort the stone is compelled to overpass the limits of the facet on which it has hitherto found rest, it will tumble over into a new position of stability. . . . The stone . . . can only repose in certain widely separated positions.

Galton proposes no new force. The polyhedral stone will not move at all unless natural selection pushes hard. But the polyhedron's response to selection is restricted by its own internal structure; it can only move to a limited number of definite places. Thus, following the metaphor of Galton's polyhedron to its conclusion, the actual directions of evolutionary change record a dynamic interaction of external push and internal constraint. The constraints are not merely negative limits to Panglossian possibilities, but active participants in the pathways of evolutionary change. St. George Mivart, whom Darwin acknowledged as his most worthy critic, adopted Galton's polyhedron as the basis of his argument and wrote (1871):

> The existence of internal conditions in animals corresponding with such facets is denied by pure Darwinians. . . . The internal tendency of an organism to certain considerable and definite changes would correspond to the facets on the surface of the spheroid.

If Galton's polyhedron ranks as more than mere verbiage, then we must be able to map the facets of genetic and developmental possibility. We must recast our picture of evolution as an interaction of outside (selection) and inside (constraint), not as an untrammeled trajectory toward greater adaptation. We can find no better subject for investigating facets than Darwin's own prototype for evolutionary arguments—changes induced in historical time through conscious selection by breeders upon domesticated animals. I can imagine no better object than our proverbial best friend—the dog—Galton's own choice for comparison in the very first sentence of his manifesto for human improvement.

We should begin by asking why dogs, cows, and pigs, rather than zebras, seals, and hippos are among our domesticated animals? Are all creatures malleable to our tastes and needs, and do our selections therefore reflect the best possible beef and service? Or do some of the strongest and tastiest not enter our orbit because selection cannot overcome inherited features of form or behavior that evolved in other contexts and now resist any recruitment to human purposes?

From the first—or at least since Western traditions abandoned the idea that God had designed creatures explicitly for human use—biologists have recognized that only certain forms of behavior predispose animals to domestication and that our successes represent a subset of available species, not by any means an optimally chosen few amidst unlimited potential. In particular, we have recruited our domestic animals from social species with hierarchies of rank and domination. In the basic "trick" of domestication—what we call "taming" in our vernacular—we learn the animal's own cues and signals, thus assuming the status of a dominant creature within the animal's own species. We tame creatures by subverting their own natural behavior. If animals do not manifest a basic sociability and propensity to submit under proper cues, then we have not been able to domesticate them, whatever their potential as food or beast of burden. As Charles Lyell wrote in 1832:

> Unless some animals had manifested in a wild state an aptitude to second the efforts of man, their domestication would never have been attempted. . . . It conforms itself to the will of man, because it had a chief to which in a wild state it would have yielded obedience . . . it makes no sacrifice of its natural inclinations. . . . No solitary species . . . has yet afforded true domestic races. We merely develop to our own advantage propensities which propel the individuals of certain species to draw near to their fellows.

The dog is our primary pet because its ancestor, the wolf *Canis lupus,* had evolved behaviors that, by a fortunate accident of history, included a predisposition for human companionship. Thus, our story begins with a push onto a facet of Galton's polyhedron. Domestication required a preexisting structure of behavior.

We might readily admit this prerequisite, yet marvel at the stunning diversity of domestic breeds and conclude that any shape or habit might be modeled from the basic wolf prototype. We would be wrong again.

We can usually formulate "big" questions easily enough; the key to good science lies in our ability to translate such ideas into palpable data that can help us to decide one way or the other. We can readily state the issue of limits versus optimality, but how shall we test it? In most cases, we approach such generalities best by isolating a small corner that can be defined and assessed with precision. This tactic often disappoints nonscientists, for they feel that we are being paltry or meanspirited in focusing so narrowly on one particular; yet I would rather tackle a well-defined iota, so long as I might then add further bits on the path to omega, than meet a great issue head-on in such ill-formed complexity that I could only waffle or pontificate about the grand and intangible.

A standard strategy for the study of limits lies in the field of *allometry,* or changes in shape associated with variation in size. Two sequences of size differences might be important for studying variation in form among breeds of domestic dogs: *ontogeny,* or changes in shape that occur during growth of individual dogs from fetus to adult; and *interspecific scaling,* or differences in shape among adults of varying sizes within the family Canidae, from small foxes to large wolves. We might search for regularities in the relationship between size and shape in these two sequences and then ask whether variation among dog breeds follows or transcends these patterns. If, for all their stunning diversity, dogs of different breeds end up with shapes predicted for their size by the ontogenetic or interspecific series, then inherited patterns of growth and history constrain current selection along channels of preferred form. Growth and previous evolution will act as facets of Galton's polyhedron, favored positions imposed from within upon the efforts of breeders.

The biological literature includes a large but obscure series of articles (mostly *auf Deutsch*), dating to the early years of this century, on allometry in domestic breeds. These themes have been neglected by English and American evolutionists during the past thirty years, primarily because an overconfident, strict Darwinism had so strongly emphasized the power of adaptation that the older subject of limits lost its appeal. But exciting progress in our understanding of genetic architecture and embryological development has begun to strike a proper balance between the external strength of selection and the internal channels of inherited structure. I sense a welcome reappearance of Galton's polyhedron in the primary technical literature of evolutionary biology. As one example, consider a recent study of ontogenetic and interspecific allometry in dogs by Robert K. Wayne.

Wayne asks how inherited patterns of allometry might constrain the variety of domestic breeds. He finds, for example, that all measures of skull length (face, jaws, and cranium) show little variation in three senses: First, the ontogenetic and interspecific patterns are similar (baby dogs look like small foxes in the proportions of length elements); second, we note little change of shape as size increases (baby dogs are like old dogs, and small foxes are like large wolves); third, we find little variation among breeds or species at any common size (all young dogs of the same size have roughly equal length elements).

These three observations suggest that natural variation among canids offers little raw material for fanciers to create breeds with exotic skull lengths. Wayne has confirmed this suspicion by noting that few adults of different breeds, from toy poodle to Great Dane, depart far from the tight relationship predicted by ontogeny or interspecies scaling: The length elements of a small breed may be predicted from the proportions of puppies in larger breeds or from small adult foxes.

Wayne points out that the criteria of artificial selection in domesticated races (the quirky human preferences imposed upon toy or fancy breeds, for example) must differ dramatically from the basis of natural selection in wild species—"the dog's ability to catch, dismember, or masticate live prey." If length elements are so constant in such radically different contexts, then their invariance probably reflects an intrinsic limit on variability rather than a fortuitous concurrence in different circumstances. Wayne concludes:

> Despite considerable variability in the time, place and conditions of origination of dog breeds, the scaling of skull-length measurement components is relatively invariant. All [small] dog breeds are exact allometric dwarfs with respect to measures of skull length. It is unlikely that such a specific morphological relationship has been the direct result of selection by breeders. Rather, a lack of developmental variation seems a better explanation.

When we turn to skull widths, however, we note variation where lengths showed constancy: Puppies differ from adult foxes of the same size; puppies also turn into dogs of greatly altered shape, and small foxes are easily distinguished from large wolves by proportions of width elements. The material available to dog breeders should be extensive.

Wayne finds that dog breeds do vary greatly in width elements. (We might be tempted to say, "So what; doesn't everyone know this from a lifetime of casual inspection?" Yet our intuitions are often faulty. Wayne shows that small, snubnosed breeds have wide faces, not short skulls or jaws. Readers might be chuckling and saying, what's the difference—doesn't overwide amount to the same thing as undershort, as in the fat man's riposte that he is only too short for his weight? But the statements are not equivalent, for we are comparing lengths and widths with a common standard of body size. Small breeds are the right length, but unusually wide, for their size.) The great variation among dog breeds is not uniform among all parts, but concentrated in those features that supply raw material in growth and evolutionary history. Dog breeds form along permitted paths of available variation.

This study of internal potential helps us to predict which features will form the basis for variety among breeds (not simply what the selector wishes, but what the selectee can provide). But we can extend this insight much further to encompass the great differences in variety among domesticated species. If some features of dogs differ more among breeds because internal factors must supply the requisite variation, then perhaps, by extension, some entire species develop more diversity, not because human selectors have been more assiduous or because human needs require such variety, but because the internal facets of Galton's polyhedron are many, closely spaced, and varied.

Why do breeds of dogs differ so greatly, and those of cats relatively little? (Cats vary widely in color and character of coat but not much in shape. Nothing in the world of felines can approach the disparity in skull form between a stubby Pekingese and an elongated borzoi.) Before we speculate about diminished human effort or desire to explain Garfield versus Lassie, we should consider the more fundamental fact that available variation in the ontogeny and interspecific scaling among cats offers very little for breeders to select. Lions differ from tabbies far less than large from small dogs while, more importantly, kittens grow to adult cats with only a fraction of the change in shape that accompanies the transformation of puppy to grown dog. Consider the accompanying figure taken from Wayne's article and comparing, at the same size, neonates and adults of cats and dogs. Dogs have contributed to their own flexibility; cats, as ever, are recalcitrant.

Wayne makes a persuasive case that comparative diversity among domesticated species depends more upon available variation in the growth of wild ancestors than in the extent of human

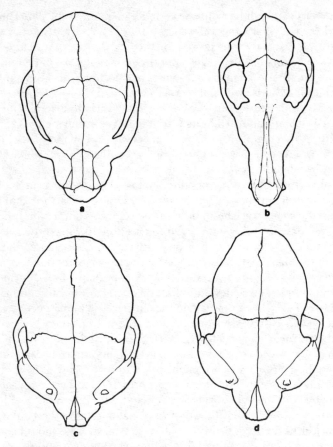

Skulls of dogs (upper row) and cats (lower row) show the
differences between neonates (left) and adult animals (right).
Note the much greater range of change in dogs vs. cats.
Evolution *40 (2), 1986.*

efforts. Horses change relatively little in shape during growth,
and the heads of Shetland ponies do not differ much from those
of the largest workhorses. Pigs, on the other hand, are second
only to dogs in diversity of breeds. They are also unmatched
among farm animals for marked change of shape during growth.

Amounts of intrinsic variation therefore set limits and supply possibilities to breeders. But even the most variable of wild species do not become putty in the hands of breeders. Pigs and dogs vary in definite ways during their growth, and only certain shapes are available for selection at definite sizes. Wayne's most persuasive case for internal limits lies in his demonstration that sets of traits in a standard "ontogenetic trajectory" (a sequence of stages from puppy to adult) tend to hang together. Dog breeds are not a hodgepodge of isolated traits, each taken at will from any stage of ontogeny. Traits of juvenility remain associated, and many breeds, particularly among small dogs, continue to look like puppies when adult—an evolutionary process called *paedomorphosis* (child-shaped) or *neoteny* (literally, "holding on to youth").

Wayne has shown that—without exception—adults of small breeds resemble the juvenile stages of large dogs more than the adults of other wild canid species (small foxes)—a convincing demonstration that inherited patterns of growth set possibilities of change. Dogs resolutely stick to their own trajectories of growth. "To some extent," Wayne concludes, "many dog breeds represent morphological snapshots between these developmental endpoints. . . . This suggests that small domestic dogs differ from foxes because puppies of small dogs cannot grow out of their distinctive neonate morphology."

We know, of course, that breeders can do many wonderful and peculiar things, from making a dachshund into a frankfurter to turning a chihuahua into a hairless rat or a sheepdog into a woolly mimic of its charges. But these peculiarities are imposed upon a basic and unaltered pattern set by constraints of inherited growth. The trajectory of ontogeny provides, as Raymond Coppinger states, a "rough first draft" for all breeds.

If ordinary variation in growth provides the main source for breeders, then a wild species' own juvenile stages are the primary storehouse of available change. Under this basic theme of limits, we may understand an old and otherwise puzzling observation about domesticated versus wild species. Over and over again, we note that domestic species develop more juvenile proportions than their wild ancestors. We cannot explain this difference by smaller size (since domestic breeds are often larger than their ancestors) or by conscious selection on the old theme of planned optimality, for what possible common adaptive advantage could have inspired breeders to produce such a similarly shortened face in Middle white pigs, the Niatu oxen of South America, and the Pekingese of the Chinese imperial court (see figure). The only sensible coordinating theme behind these similarities is retention by neoteny of juvenile traits common to most vertebrates.

If these shared neotenous traits of domesticated species are not products of direct selection by breeders, then what is the common basis of their origin? Most experts argue that these juvenile traits are spinoffs from the true object of selection—tame and playful behavior itself. Organisms may vary as much in rates of development as in form. By breeding only those animals that retain the favored juvenile traits of pliant and flexible behavior past the point of sexual maturity, humans hasten the process of domestication itself. Since traits are locked together in development, not infinitely dissociable as hopes for optimality require, selection for desired juvenile behavior brings features of juvenile morphology along for the ride. (We may, in some cases, also select juvenile morphology for its aesthetic appeal—large eyes and rounded craniums, for example—and obtain valuable spinoffs in behavior). In short, these common juvenile features are facets of Galton's polyhedron—correlated consequences of selection for something else—not direct objects of human desire.

This theme of correlated consequences brings me to a final point. We have explored the role of developmental constraint in shaping many features of dogs—their difference in diversity from other domesticated species; the disparate contributions made by various parts of their bodies to the unparalleled variety among breeds; the restricted sources of variation for construction of breeds, particularly the limits imposed by tendencies for traits to "hang together" during growth.

We might view these themes in a pessimistic light—as a brake upon the power of selection to build with all the freedom of a human sculptor. But I suggest a more positive reading. Constraints of development embody the twin and not-so-contradictory themes of limits and opportunities. Constraints do preclude certain fancies, but they also supply an enormous pool of available potential for future change. So what if the pool has borders; the water inside is deep and inviting. The pessimist might view correlated characters in growth as a sad foreclosure of certain combinations. But an optimist might emphasize the power for rapid change provided by the possibility of recruiting so many

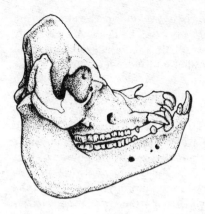

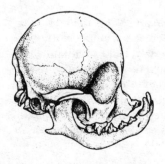

Neotenically shortened faces
of domesticated animals.
From top to bottom: pig, ox,
and Pekingese dog. *Courtesy of
The Natural History Museum, London.*

features at once. The great variation that dogs develop during their growth builds an enormous pool all by itself. A wide range of juvenile stages becomes available for recruitment by neoteny—a potential precluded if youngsters look and act like adults.

This great pool of potential has been used over and over again by breeders of domestic dogs. We should view this restricted variation as the main source of their success, not the tragic limit to their hopes. My colleague Raymond Coppinger, of Hampshire College, has spent ten years promoting the use of guarding dogs (a great European tradition that never caught on in America) as an alternative to shooting, poisoning, and other mayhem in the protection of sheep from coyotes and other predators. He has placed more than five hundred dogs with farmers in thirty-five states. Coppinger notes the great, and usually unappreciated, difference between guarding and herding breeds. Herders control the movement of sheep by using predatory behaviors of adult dogs—stalking, chasing, biting, and barking—but inhibiting the final outcome. These breeds feature adult traits of form and behavior; they display no neotenous characteristics.

Guard dogs, on the other hand, simply move with and among the flock. They work alone and do not control the flock's motion. They afford protection primarily by their size, for few coyotes will attack a flock accompanied by a one-hundred-pound dog. They behave toward sheep as puppies do towards other dogs—licking the sheep's face as a puppy might in asking for food, chasing and biting with the playfulness of young dogs, even mounting sheep as young dogs mount each other in sexual play and rehearsal. This neotenous behavior accompanies a persistently juvenile morphology, as these dogs grow short faces, big eyes, and floppy ears.

Coppinger has raised guarding and herding dogs together from babyhood. They show little difference in behavior until puberty. Herders then develop the standard traits of adulthood—border collies begin to stalk, while retrievers and pointers live up to their names. But the guarders develop no new patterns and simply retain their youthful traits. Thus, a valuable set of features can be recruited together because they already exist as the normal form and behavior of juvenile dogs. Patterns of growth are rich reservoirs, not sterile strictures.

One tradition of argument identifies neoteny with all that is good and kind—"Except ye be converted, and become as little children, ye shall not enter into the kingdom of heaven." Yet I resist any facile transference between natural realities and human hopes if only because the dark side of social utility should teach us caution in proposing analogies.

Neoteny certainly has its dark side in social misconstruction. Konrad Lorenz, who, to put it as kindly as possible, made his life in Nazi Germany more comfortable by tailoring his views on animal behavior to the prevailing orthodoxy, often argued during the early 1940s that civilization is the analogue of domestication. Domestic animals are often neotenous; neotenous animals retain the flexibility of youth and do not develop the instinctive and healthy aversion that mature creatures feel toward deformed and unworthy members of their race. Since humans have therefore lost this instinctive power to reject the genetically harmful, Lorenz defended Nazi racial and marriage laws as a mirror of nature's mature ways.

Still, I cannot help noting, since dogs are descended from wolves, and humans really are neotenous in both form and behavior (without justifying Lorenz's fatuous and hateful reveries), that the neoteny of sheep-guarding dogs does fulfill, in a limited sense, one of the oldest and most beautiful of all prophecies: "The wolf also shall dwell with the lamb . . . and a little child shall lead them."

28 | Betting on Chance— and No Fair Peeking

DOUBLE ENTENDRE can be delicious. Who does not delight in learning that Earnest, in Oscar Wilde's play, is a good chap, not a worthy attitude. And who has ever begrudged that tragic figure his little joke. But double meanings also have their dangers—particularly when two communities use the same term in different ways, and annoying confusion, rather than pleasant amusement or enlightenment, results.

Differences in scientific and vernacular definitions of the same word provide many examples of this frustrating phenomenon. "Significance" in statistics, for example, bears little relation to the ordinary meaning of something that matters deeply. Mouse tails may be "significantly" longer in Mississippi than in Michigan—meaning only that average lengths are not the same at some level of confidence—but the difference may be so small that no one would argue for significance in the ordinary sense. But the most serious of all misunderstandings between technical and vernacular haunts the concepts of probability and particularly the words *random* and *chance*.

In ordinary English, a random event is one without order, predictability, or pattern. The word connotes disaggregation, falling apart, formless anarchy, and fear. Yet, ironically, the scientific sense of *random* conveys a precisely opposite set of associations. A phenomenon governed by chance yields maximal simplicity, order, and predictability—at least in the long run. Suppose that we are interested in resolving the forces behind a large-scale pattern of historical change. Randomness becomes our best hope for a maximally simple and tractable model. If we flip a penny or throw a pair of dice, once a second for days on end, we achieve a rigidly predictable distribution of outcomes. We can predict the margins of departure from 50-50 in the coins or the percentage of sevens for our dice, based on the total number of throws. When the number of tosses becomes quite large, we can give precise estimates and ranges of error for the frequencies and lengths of runs—heads or sevens in a row—all based on the simplest of mathematical formulas from the theory of probability. (Of course, we cannot predict the outcome of any particular trial or know when a run will occur, as any casual gambler should—but so few do—know.)

Thus, if you wish to understand patterns of long historical sequences, pray for randomness. Ironically, nothing works so powerfully against resolution as conventional forms of determinism. If each event in a sequence has a definite cause, then, in a world of such complexity, we are lost. If A happened because the estuary thawed on a particular day, leading to B because Billy the Seal swam by and gobbled up all those fishes, followed by C when Sue the Polar Bear sauntered through—not to mention ice age fluctuations, impacting asteroids, and drifting continents of the truly long run—then how can we hope to predict and resolve the outcome?

The beauty (and simplicity) of randomness lies in the absence of these maximally confusing properties. Coin flipping permits no distinctive personality to any time or moment; each toss can be treated in exactly the same way, whenever it occurs. We can date geological time with precision by radioactive decay because each atom has an equal probability of decaying in each instant. If causal individuality intervened—if 10:00 A.M. on Sunday differed from 5:00 P.M. on Wednesday, or if Joe the uranium atom, by dint of moral fiber, resisted decay better than his brother Tom, then randomness would fail and the method would not work.

One of the best illustrations for this vitally important, but counterintuitive, principle of maximal long-term order in randomness comes from my own field of evolutionary biology—and from a debate that has greatly contributed to making professional life more interesting during the past twenty years. Traditional Darwinism includes an important role for randomness—but only as a source of variation, or raw material, for evolutionary change, not as an agent for the direction of change itself. For Darwin, the predominant source of evolutionary change resides in the deterministic force of natural selection. Selection works for cause and adapts organisms to changing local environments. Random variation supplies the indispensable "fuel" for natural selection but does not set the rate, timing, or pattern of change. Darwinism is a two-part theory of randomness for raw material and conventional causality for change—*Chance and Necessity*, as so well epitomized by Jacques Monod in the title of his famous book about the nature of Darwinism.

In the domain of organisms and their good designs, we have little reason to doubt the strong, probably dominant influence of deterministic forces like natural selection. The intricate, highly adapted forms of organisms—the wing of a bird or the mimicry of a dead twig by an insect—are too complex to arise as long sequences of sheer good fortune under the simplest random models. But this stricture of complexity need not apply to the nucleotide-by-nucleotide substitutions that build the smallest increments of evolutionary change at the molecular level. In this domain of basic changes in DNA, a "neutralist" theory, based on simple random models, has been challenging conventional Darwinism with marked success during the past generation.

When the great Japanese geneticist Motoo Kimura formulated his first version of neutral theory in 1968 (see bibliography), he was impressed by two discoveries that seemed difficult to interpret under the conventional view that natural selection overwhelms all other causes of evolutionary change. First, at the molecular level of substitutions in amino acids, measured rates indicated a constancy of change across molecules and organisms—the so-called molecular clock of evolution. Such a result makes no sense in Darwin's world, where molecules subject to strong selection should evolve faster than others, and where organisms exposed to different changes and challenges from the environment should vary their evolutionary rates accordingly. At most, one might claim that these deterministic differences in rate might tend to "even out" over very long stretches of geological time, yielding roughly regular rates of change. But a molecular clock surely gains an easier interpretation from random models. If deterministic selection does not regulate most molecular changes—if, on the contrary, most molecular variations are neutral, and therefore rise and fall in frequency by the luck of the draw—then mutation rate and population size will govern the tempo of change. If most populations are large, and if mutation rates are roughly the same for most genes, then simple random models predict a molecular clock.

Second, Kimura noted the recent discovery of surprisingly high levels of variation maintained by many genes among members of populations. Too much variation poses a problem for conventional Darwinism because a cost must accompany the replacement of an ancestral gene by a new and more advantageous state of the same gene—namely, the differential death, by natural selection, of the now disfavored parental forms. This cost poses no problem if only a few old genes are being pushed out of a population at any time. But if hundreds of genes are being eliminated, then any organism must carry many of the disfavored states and should be ripe for death. Thus, selection should not be able to replace many genes at once. But the data on copious variability seemed to indicate a caldron of evolutionary activity at far too many genetic sites—too many, that is, if selection governs the changes in each varying gene. Kimura, however, recognized a simple and elegant way out of this paradox. If most of the varying forms of a gene are neutral with respect to selection, then they are drifting in frequency by the luck of the draw. Invisible to natural selection because they make no difference to the organism, these variations impose no cost in replacement.

In twenty years of copious writing, Kimura has always carefully emphasized that his neutral theory does not disprove Darwinism or deny the power of natural selection to shape the adaptations of organisms. He writes, for example, at the beginning of his epochal book *The Neutral Theory of Molecular Evolution* (1983):

> The neutral theory is not antagonistic to the cherished view that evolution of form and function is guided by Darwinian selection, but it brings out another facet of the evolutionary process by emphasizing the much greater role of mutation pressure and random drift at the molecular level.

The issue, as so often in natural history (and as I emphasize so frequently in these essays), centers upon the relative importance of the two processes. Kimura has never denied adaptation and natural selection, but he has tended to view these processes as quantitatively insignificant in the total picture—a superficial and minor ripple upon the ocean of neutral molecular change, imposed every now and again when selection casts a stone upon the waters of evolution. Darwinians, on the other hand, at least before Kimura and his colleagues advanced their potent challenge and reeled in the supporting evidence, tended to argue that neutral change occupied a tiny and insignificant corner of evolution—an odd process occasionally operating in small populations at the brink of extinction anyway.

This argument about relative frequency has raged for twenty years and has been, at least in the judgment of this bystander with no particular stake in the issue, basically a draw. More influence has been measured for selection than Kimura's original words had anticipated; Darwin's process is no mere pockmark on a sea of steady tranquility. But neutral change has been established at a comfortably high relative frequency. The molecular clock is neither as consistent nor as regular as Kimura once hoped, but even an imperfect molecular timepiece makes little sense in Darwin's world. The ticking seems best interpreted as a pervasive and underlying neutralism, the considerable perturbations as a substantial input from natural selection (and other causes).

Nonetheless, if forced to award the laurels in a struggle with no clear winners, I would give the nod to Kimura. After all, when innovation fights orthodoxy to a draw, then novelty has seized a good chunk of space from convention. But I bow to Kimura for another and more important reason than the empirical adequacy of neutralism at high relative frequency, for his theory so beautifully illustrates the theme that served as an introduction to this essay: the virtue of randomness in the technical as opposed to the vernacular sense.

Kimura's neutralist theory has the great advantage of simplicity in mathematical expression and specification of outcome. Deterministic natural selection yields no firm predictions for the histories of lineages—for you would have to know the exact and particular sequences of biotic and environmental changes, and the sizes and prior genetic states of populations, in order to forecast an outcome. This knowledge is not attainable in a world of imperfect historical information. Even if obtainable, such data would only provide a prediction for a particular lineage, not a general theory. But neutralism, as a random model treating all items and times in the same manner, yields a set of simple, general equations serving as precise predictions for the results of evolutionary change. These equations give us, for the first time, a base-level criterion for assessing any kind of genetic change. If neutralism holds, then actual outcomes will fit the equations. If selection predominates, then results will depart from predictions—and in a way that permits identification of Darwinian control. Thus, Kimura's equations have been as useful for selectionists as for neutralists themselves; the formulas provide a criterion for everyone, and debate can center upon whether or not the equations fit actual outcomes. Kimura has often emphasized this point about his equations, and about random models in general. He wrote, for example, in 1982:

> The neutral theory is accompanied by a well-developed mathematical theory based on the stochastic theory of population genetics. The latter enables us to treat evolution and variation quantitatively, and therefore to check the theory by observations and experiments.

The most important and useful of these predictions involves a paradox under older Darwinian views. If selection controls evolutionary rate, one might think that the fastest tempos of alteration would be associated with the strongest selective pressures for change. Speed of change should vary directly with intensity of selection. Neutral theory predicts precisely the opposite—for an obvious reason once you start thinking about it. The most rapid change should be associated with unconstrained randomness—following the old thermodynamic imperative that things will invariably go to hell unless you struggle actively to maintain them as they are. After all, stability is far more common than change at any moment in the history of life. In its ordinary everyday mode, natural selection must struggle to preserve working combinations against a constant input of deleterious mutations. In other words, natural selection, in our technical parlance, must usually be "purifying" or "stabilizing." Positive selection for change must be a much rarer event than watchdog selection for tossing out harmful variants and preserving what works.

Now, if mutations are neutral, then the watchdog sees nothing and evolutionary change can proceed at its maximal tempo—the neutral rate of substitution. But if a molecule is being preserved by selection, then the watchdog inhibits evolutionary change. This originally counterintuitive proposal may be regarded as the key statement of neutral theory. Kimura emphasizes the point with italics in most of his general papers, writing for example (in 1982): *"Those molecular changes that are less likely to be subjected to natural selection occur more rapidly in evolution."*

Both the greatest success, and the greatest modification, of Kimura's original theory have occurred by applying this principle that selection slows the maximal rate of neutral molecular change. For modification of the original theory, thousands of empirical studies have now shown that watchdog selection, measured by diminished tempo of change relative to predictions of randomness, operates at a far higher relative frequency than Kimura's initial version of neutralist theory had anticipated. For success, the firm establishment of the principle itself must rank as the greatest triumph of neutralism—for the tie of maximal rate to randomness (rather than to the opposite expectation of intense selection) does show that neutralism exerts a kind of base-level control over evolution as a whole.

The most impressive evidence for neutralism as a maximal rate has been provided by forms of DNA that make nothing of potential selective value (or detriment) to an organism. In all these cases, measured tempos of molecular change are maximal, thus affirming the major prediction of neutralism.

1. *Synonymous substitutions.* The genetic code is redundant in the third position. A sequence of three nucleotides in DNA codes for an amino acid. Change in either of the first two nucleotides alters the amino acid produced, but most changes in the third nucleotide—so-called synonymous substitutions—do not alter the resulting amino acid. Since natural selection works on features of organisms, in this case proteins built by DNA and not directly on the DNA itself, synonymous substitutions should be invisible to selection, and therefore neutral. Rates of change at the third position are usually five or more times as rapid as changes at the functional first and second positions—a striking confirmation of neutralism.

2. *Introns.* Genes come in pieces, with functional regions (called exons) interrupted by DNA sequences (called introns) that are

snipped out and not translated into proteins. Introns change at a much higher evolutionary rate than exons.

3. *Pseudogenes.* Certain kinds of mutations can extinguish the function of a gene—for example, by preventing its eventual translation into protein. These so-called pseudogenes begin with nearly the same DNA sequence as the functional version of the gene in closely related species. Yet, being entirely free from function, these pseudogenes should exert no resistance against the maximal accumulation of changes by random drift. Pseudogenes become a kind of ultimate test for the proposition that absence of selection promotes maximal change at the neutral rate—and the test has, so far, been passed with distinction. In pseudogenes, rates of change are equal, and maximal, at all three positions of the triplet code, not only at the third site, as in functional genes.

I was inspired to write about neutral theory by a fascinating example of the value of this framework in assessing the causes of evolutionary rates. This example neither supports nor denies neutralism but forms a case in the middle, enlightened by the more important principle that random models provide simple and explicit criteria for judgment.

While supposedly more intelligent mammals are screwing up royally above ground, Near Eastern mole rats of the species *Spalax ehrenbergi* are prospering underneath. Subterranean mammals usually evolve reduced or weakened eyes, but *Spalax* has reached an extreme state of true blindness. Rudimentary eyes are still generated in embryology, but they are covered by thick skins and hair. When exposed to powerful flashes of light, *Spalax* shows no neurological response at all, as measured by electrodes implanted in the brain. The animal is completely blind.

What then shall we make of the invisible and rudimentary eye? Is this buried eye now completely without function, a true vestige on a path of further reduction to final disappearance? Or does the eye perform some other service not related to vision? Or perhaps the eye has no direct use, but must still be generated as a prerequisite in an embryological pathway leading to other functional features. How can we decide among these and other alternatives? The random models of neutral theory provide our most powerful method. If the rudimentary eye is a true vestige, then its proteins should be changing at the maximal neutral rate. If selection has not been relaxed, and the eye still functions in full force (though not for vision), then rates of change should be comparable to those for other rodents with conventional eyes. If selection has been relaxed due to blindness, but the eye still functions in some less constrained way, then an intermediate rate of change might be observed.

The eye of *S. ehrenbergi* still builds a lens (though the shape is irregular and cannot focus an image), and the lens includes a protein, called αA-crystallin. The gene for this protein has recently been sequenced and compared with the corresponding gene in nine other rodents with normal vision (see article by W. Hendriks, J. Leunissen, E. Nevo, H. Bloemendal, and W. W. de Jong in bibliography).

Hendriks and colleagues obtained the most interesting of possible results from their study. The αA-crystallin gene is changing much faster in blind *Spalax* than in other rodents with vision, as relaxation of selection due to loss of primary function would suggest. The protein coded by the *Spalax* gene, for example, has undergone nine amino acid replacements (of 173 possible changes), compared with the ancestral state for its group (the murine rodents, including rats, mice, and hamsters). All other murines in the study (rat, mouse, hamster, and gerbil) have identical sequences with no change at all from the ancestral state. The average tempo of change in αA-crystallin among vertebrates as a whole has been measured at about 3 amino acid replacements per 100 positions per 100 million years. *Spalax* is changing more than four times as fast, at about 13 percent per 100 million years. (Nine changes in 173 positions is 5.2 percent; but the *Spalax* lineage is only 40 million years old—and 5.2 percent in 40 million corresponds to 13 percent in 100 million years.) Moreover, *Spalax* has changed four amino acids at positions that are absolutely constant in all other vertebrates studied—seventy-two species ranging from dogfish sharks to humans.

"These findings," Hendriks and colleagues conclude, "all clearly indicate an increased tolerance for change in the primary structure of αA-crystallin in this blind animal." So far so good. But the increased tempo of change in *Spalax*, though marked, still reaches only about 20 percent of the characteristic rate for pseudogenes, our best standard for the maximal, truly neutral pace of evolution. Thus, *Spalax* must still be doing enough with its eyes to damp the rate of change below the maximum for neutrality. Simple models of randomness have taught us something interesting and important by setting a testable standard, approached but not met in this case, and acting as a primary criterion for judgment.

What then is αA-crystallin doing for *Spalax?* What can a rudimentary and irregular lens, buried under skin and hair, accomplish? We do not know, but the established intermediate rate of change leads us to ask the right questions in our search for resolution.

Spalax is blind, but this rodent still responds to changes in photoperiod (differing lengths of daylight and darkness)—and apparently through direct influence of light regimes themselves, not by an indirect consequence that a blind animal might easily recognize (increase in temperature due to more daylight hours, for example). A. Haim, G. Heth, H. Pratt, and E. Nevo (see bibliography) showed that *Spalax* would increase its tolerance for cold weather when exposed to a winterlike light regime of eight light followed by sixteen dark hours. These mole rats were kept at the relatively warm temperature of 22°C, and were therefore not adjusting to winter based on clues provided by temperature. Animals exposed to twelve light and twelve dark hours at the same temperature did not improve their thermoregulation as well. Interestingly, animals exposed to summerlike light regimes (sixteen light and eight dark), but at colder temperatures of 17°C, actually *decreased* their cold-weather tolerance. Thus, even though blind, *Spalax* is apparently using light, not temperature, as a guide for adjusting physiology to the cycle of seasons.

Hendriks and colleagues suggest a possible explanation, not yet tested. We know that many vertebrates respond to changes in photoperiod by secreting a hormone called melatonin in the pineal gland. The pineal responds to light on the basis of photic information transmitted via the retina. *Spalax* forms a retina in its rudimentary eye, yet how can the retina, which perceives no light in this blind mammal, act in concert with the pineal gland? But Hendriks and colleagues note that the retina can also secrete melatonin itself—and that the retina of *Spalax* includes the secreting layer. Perhaps the retina of *Spalax* is still functional as a source of melatonin or as a trigger of the pineal by some mechanism still unknown. (I leave aside the fascinating, and completely unresolved, issue of how a blind animal can respond, as *Spalax* clearly does, to seasonal changes in photoperiod.)

If we accept the possibility that *Spalax* may need and use its retina (in some nonvisual way) for adaptation to changing seasons, then a potential function for the lens, and for the αA-crystallin protein, may be sought in developmental pathways, not in direct utility. The lens cannot work in vision, and αA-crystallin focuses no image, but the retina does not form in isolation and can only be generated as part of a normal embryological pathway that includes the prior differentiation of other structures. The formation of a lens vesicle may be a prerequisite to the construction of a retina—and a functioning retina may therefore require a lens, even if the lens will be used for nothing on its own.

Evolution is strongly constrained by the conservative nature of embryological programs. Nothing in biology is more wondrously complex than the production of an adult vertebrate from a single fertilized ovum. Nothing much can be changed very radically without discombobulating the embryo. The intermediate rate of change in lens proteins of a blind rodent—a tempo so neatly between the maximal pace for neutral change and the much slower alteration of functioning parts—may point to a feature that has lost its own direct utility but must still form as a prerequisite to later, and functional, features in embryology.

Our world works on different levels, but we are conceptually chained to our own surroundings, however parochial the view. We are organisms and tend to see the world of selection and adaptation as expressed in the good design of wings, legs, and brains. But randomness may predominate in the world of genes—and we might interpret the universe very differently if our primary vantage point resided at this lower level. We might then see a world of largely independent items, drifting in and out by the luck of the draw—but with little islands dotted about here and there, where selection reins in tempo and embryology ties things together. What, then, is the different order of a world still larger than ourselves? If we missed the different world of genic neutrality because we are too big, then what are we not seeing because we are too small? We are like genes in some larger world of change among species in the vastness of geological time. What are we missing in trying to read this world by the inappropriate scale of our small bodies and minuscule lifetimes?

8 | Reversals— Fragments of a Book Not Written

29 | Shields of Expectation— and Actuality

DISCOVERY, like its soul mate love, is a many-splendored thing. Stumbling serendipity surrounds some great finds—like *Archaeopteryx*, the first bird, unearthed by a quarryman at Solnhofen. Others are the product of dogged purpose. Consider Eugène Dubois who, as a Dutch army surgeon, posted himself to Indonesia because he felt sure that human ancestors must have inhabited East Asia (see Essay 8). There he found, in 1893, the first human fossils of a species older than our own—the Trinil femur and skull cap of *Homo erectus* ("Java man" of the old texts).

The most beautiful specimens in my office, which I happily share with about 50,000 fossil arthropods, rest in the last cabinet of the farthest corner. They are head shields of *Eurypterus fischeri*, a large, extinct freshwater arthropod related to horseshoe crabs. These exquisite fossils are preserved as brown films of chitin, set off like an old rotogravure against a surrounding sediment so fine in grain that the background becomes a uniform sheet of gray (see figure). They were collected in Estonia by William Patten, a professor of biology at Dartmouth.

When I first came to Harvard twenty years ago, I made a reconnaissance of all our 15,000 drawers of fossils—an adventure surely surpassing anything ever achieved by the smallest boy in the largest candy store. I found some of the great specimens of my profession—Agassiz's echinoderms, Raymond's collection from the Burgess Shale. But I got a particular thrill from Patten's eurypterids because I knew exactly why he had gathered them. Patten, like Dubois, had collected with a singular purpose. I had

A head shield of *Eurypterus fischeri* collected by William
Patten in Estonia. *Photograph by Rosamond W. Purcell.*

read his 1912 book—*The Evolution of the Vertebrates and Their Kin*—
one of the curiosities of my profession. Patten's book represents
the last serious defense of the classic, though incorrect, theory
for vertebrate origins—the attempt to link the two great phyla of
complex animals by arguing that vertebrates arose from arthropods.

Patten identified eurypterids as the arthropod ancestors of
vertebrates—hence his strong desire to collect them. But Patten
was even more interested in a group that occurred with the eurypterids in some localities—jawless fishes of the genus *Cephalaspis* (meaning head shield). We now recognize these jawless fishes

(class Agnatha) as the oldest vertebrates and precursors of all later forms, ourselves included. The Agnatha survive today as a small remnant of naked eel-shaped forms—the lampreys (genus *Petromyzon*) and the distantly related hagfishes (genus *Myxine*). But the original armored agnathans, popularly called ostracoderms (shell skinned), dominated vertebrate life for its first hundred million years and included a large array of diverse forms. Patten's fascination with ostracoderms arose from his misinterpretation of their anatomy. Patten viewed *Cephalaspis* and its relatives as intermediary forms between arthropods and true fishes.

We usually tell the history of a profession as a pageant of changing ideas and their proponents. But we can also render a different and equally interesting account from the standpoint of objects studied. One could provide a fascinating history of astronomy from the moon's point of view, and genetics receives a different, multifaceted account through the eyes of a fruit fly. *Cephalaspis* may be our best standard bearer for evolution.

The history of ideas about *Cephalaspis*—from its original misinterpretation as the head of a trilobite in the early 1800s to its present status as the archetypal ostracoderm for all aficionados of the group—provides more than a synopsis of evolutionary thinking. It also illustrates, in an unusually forceful way, the fundamental process of scientific discovery itself.

Popular misunderstanding of science and its history centers upon the vexatious notion of scientific progress—a concept embraced by all practitioners and boosters, but assailed, or at least mistrusted, by those suspicious of science and its power to improve our lives. The enemy of resolution, here as nearly always, is that old devil Dichotomy. We take a subtle and interesting issue, with a real resolution embracing aspects of all basic positions, and we divide ourselves into two holy armies, each with a brightly colored cardboard mythology as a flag of struggle.

The cardboard banner of scientific boosterism is an extreme form of realism, the notion that science progresses because it discovers more and more about an objective, material reality out there in the universe. The extreme version holds that science is an utterly objective enterprise (and therefore superior to other human activities); that scientists read reality directly by invoking the scientific method to free their minds of cultural superstition;

and that the history of science is a march toward Truth, mediated by increasing knowledge of the external world.

The cardboard banner of the opposition is an equally extreme form of relativism, the idea that truth has no objective meaning and can only be assessed by the variable standards of different communities and cultures. The extreme version holds that scientific consensus is no different from any other arbitrary set of social conventions, say the rules for Chinese handball set by my old crowd on 63d Avenue. Science is ideology, and scientific "progress" is no improving map of external reality, but only a derivative expression of cultural change.

These positions are so sharply defined that they can only elicit howls of disbelief from the opposition. How can relativists deny that science discovers external truth? say the realists. Cro-Magnon people could draw a horse as beautifully as any artist now alive, but they could not resolve the structure of DNA or photograph the moons of Uranus. How, reply the relativists, can anyone deny the social character of science when Darwin needed Adam Smith more than Galápagos tortoises and when Linnaeus matched his taxonomy to prevailing views of divine order?

These extreme positions, of course, are embraced by very few thinkers. They are caricatures constructed by the opposition to enhance the rhetorical advantages of dichotomy. They are not really held by anyone, but partisans *think* that their opponents are this foolish, thus fanning the zealousness of their own advocacy. The possibility for consensus drowns in a sea of charges.

The central claim of each side is correct, and no inconsistency attends the marriage once we drop the peripheral extremities of each attitude. Science is, and must be, culturally embedded; what else could the product of human passion be? Science is also progressive because it discovers and masters more and more (yet ever so little *in toto*) of a complex external reality. Culture is not the enemy of objectivity but a matrix that can either aid or retard advancing knowledge. Science is not a linear march to truth but a tortuous road with blind alleys and a rubbernecking delay every mile or two. Our road map is not objective reality but the patterns of human thoughts and theories.

My position, as a variety of apple pie, is easy to state. It is also empty and tendentious as an abstract generality. This middle way, this golden mean, can only permeate our understanding by example. *Cephalaspis* provides one of the best demonstrations I know because this fish played a central role in three important and sequential views of nature's order. Each view embodied its cultural context, but each also provided a framework for new and genuine objective knowledge about *Cephalaspis*. The new knowledge then helped to establish a revised view of natural order. Speaking of rhetoric in the best American tradition, culture and knowledge are rather like liberty and union—one and forever, now and inseparable.

Cephalaspis, as its name implies, enclosed its head in a thick, bony shield. Much thinner scales covered everything behind, from front fins to tail. Since the scales usually disarticulate at death and are rarely preserved at all, most fossils of *Cephalaspis* include only the head shield. By itself, the shield is a peculiar and decidedly unfishlike object. It looks much like the head end of many trilobites (fossil arthropods), and was so classified until Louis Agassiz established the true affinity of *Cephalaspis* in his great monograph *Les poissons fossiles* (*Fossil Fishes*), published in five large volumes between 1833 and 1843.

Agassiz confessed his wonder and puzzlement in his first paragraph on *Cephalaspis:*

> These are the most curious animals that I have ever observed; their features are so extraordinary that I had to make the most careful and scrupulous examination . . . in order to convince myself that these mysterious creatures are really fish.

Agassiz reached the correct solution to his puzzle because his collection included some unusually well-preserved specimens, with the characteristic head shield indubitably attached to an undeniably fishy posterior (see figure). Yet while Agassiz began the modern history of *Cephalaspis* by placing this genus properly among the vertebrates, he could never resolve its relationship with other fishes for lack of crucial evidence. Agassiz particularly bewailed his failure to find any specimen exposing the lower surface of the head shield, where, he surmised (correctly), the mouth would be located. Thus Agassiz could never recognize the chief feature of jawlessness in *Cephalaspis* and could not identify the ostracoderms as structural precursors of all later vertebrates (jaws evolved from bones that supported gill arches behind the mouth of these jawless fishes). *Cephalaspis,* to Agassiz, remained an unplaceable oddball among fishes.

Although Agassiz could not fully resolve the status of *Cephalaspis,* he used this most peculiar of fishes as a linchpin for his theory of biological order. *Les poissons fossiles* is no simple list of old fishes; it is, perhaps most of all, a closely reasoned brief for Agassiz's creationist world view—a theory that embodied the cultural consensus of 1830, but that Agassiz maintained doggedly to his death in 1873, long after its scientific demise in Darwin's favor.

Agassiz rooted his version of creationism in a complex analogy with his favorite subject, comparative embryology. Agassiz viewed embryonic growth as a tale of differentiation—more complex and specialized forms develop from simpler and more generalized precursors. These later specializations may proceed in several directions from a common initial form. Thus, a single (and simple) early embryo, representing a vertebrate prototype,

A figure from Agassiz's *Les poissons fossiles* proving the vertebrate affinities of the head shield of *Cephalaspis*. *Photograph by Rosamond W. Purcell.*

might differentiate along several pathways into advanced fishes, reptiles, or mammals.

Agassiz then argued that the geological history of a group should match the embryological development of its latest and most advanced members. Early (geologically oldest) forms should be few, simple, and generalized; later relatives should be specialized and differentiated versions of these primordial archetypes. This scheme might sound evolutionary, but Agassiz explicitly rejected such a heresy. The geological sequence of separate creations paralleled embryological growth within each group because God's orderly and benevolent plan permeated all developmental processes in nature.

Agassiz remained loyal to the classification of his mentor, the great French zoologist Georges Cuvier. He arranged all animals in four great groups: radiates (a hodgepodge by modern standards, but including such radially symmetrical forms as corals and echinoderms); mollusks; articulates (segmented worms and arthropods); and vertebrates. The four trunks are coequal and do not coalesce at life's dawn, for they represent separately created plans for anatomy, not ancestors and descendants. But since geological history mimics embryological differentiation, prototypes of the four trunks from the oldest strata should be more similar than their modern representatives—for embryology is a tale of divergence from generalized roots.

Agassiz used *Cephalaspis* as a primary illustration of his embryological vision for geological history. As a representative of the oldest fishes, *Cephalaspis* fulfilled all expectations for a primordial creature in Agassiz's vision of differentiation as the guiding principle of history. Agassiz located two different supports for his theory of differentiation in "the bizarre characters of this genus" (*les caractères bizarres de ce genre*). First, he viewed the single solid head shield (not divided into separate cranial bones linked by sutures) and the few, simple scales covering the body as marks of a primitive generality—a source for later differentiation of separate bones and more complex scales. Second, he twisted to his advantage the old bugbear of superficial resemblance between the shield of *Cephalaspis* and the head of trilobites—for this similarity indicated that the major trunks of animal life did draw closer to a common simplicity at life's source.

Finally, Agassiz delighted in the great age of *Cephalaspis*, for its antiquity proved that all four trunks lived simultaneously at the dawn of life. The most complex group of vertebrates did not arise later as a possible evolutionary descendant (heaven forfend) of a simpler trunk.

Agassiz's theory of a God ordering his creation by embryological rules of differentiation was clearly not an interpretation logically entailed by objective facts of nature. It was a vision rooted in a cultural context still unable to embrace evolution, and in the personal psychology of Agassiz's own interests and training. Agassiz imposed his theory upon *Cephalaspis* and highlighted only those facts most congenial with his preferred views. Yet his use of *Cephalaspis* cannot be read as a vindication of relativism. Agassiz may have exploited *Cephalaspis* in the interests of his vision, but he also unearthed the primary fact that fueled all later discussion. He proved that *Cephalaspis* was a vertebrate by discovering the body of a fish behind a head shield that had confused all earlier observers.

When William Patten used *Cephalaspis* as the centerpiece of an important theory eighty years later, the context of science had changed irrevocably. Evolution had triumphed, and *Cephalaspis* would now be invoked in the interest of genealogical claims. Our cardboard relativist might argue that since *Cephalaspis* had played no notable part in fomenting this great revolution in thought, any evolutionary interpretation must be viewed as a new convention impressed upon old information—a new set of rules like the annual revision of Mah-Jongg hands imposed upon the same old tiles. But a realist would rightly reply that the tiles had changed as well. Agassiz had not resolved the anatomical status of *Cephalaspis* among the fishes; he could only affirm that the genus was both old and aberrant. Several of the greatest nineteenth-century evolutionists then studied *Cephalaspis*—including T. H. Huxley, E. R. Lankester, and E. D. Cope. From all their arguments and disagreements, one strong theme emerged: *Cephalaspis* and the ostracoderms were not just a grab bag of peculiar fishes. They formed a coherent group, with a large and consistent set of features all pointing to an anatomically primitive status among fossil vertebrates. *Cephalaspis* therefore became a prime candidate for theories about the ancestry of higher vertebrates. Evolution set the context, but new information about *Cephalaspis* fueled the debate.

Patten presented the most sophisticated case for the oldest theory of vertebrate origins—the attempt, dating to Geoffroy Saint-Hilaire in the early nineteenth century, to derive vertebrates from an inverted annelid or arthropod, a "worm that turned," so to speak. Arthropods run their main nerve cords along their ventral (bottom) surface. The gut lies above, and the esophagus must therefore pierce through nervous tissue to end in a ventral mouth. In vertebrates, on the other hand, the main nerve tract, the spinal cord, is dorsal (on top), and the gut lies below. Turn an arthropod upside down, and you get the right order for vertebrates—nerves above guts. You also obtain a set of additional correspondences that some scientists have read as superficial and analogical, and others as deeply meaningful signs of evolutionary affinity.

But this act of inversion also produces some horrendous problems for the theory of arthropod ancestry. In particular, the vertebrate mouth does not pierce the brain and open on top of the head—though the old arthropod mouth would take this path in its supposedly inverted position. Proponents of the arthropod theory must therefore argue that this original mouth atrophied, and that vertebrates opened a new ventral version below the brain. No one has ever provided a good explanation for how such a topological transformation might plausibly occur.

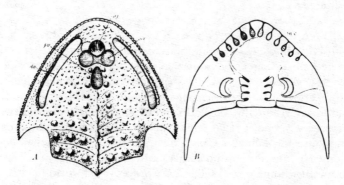

A drawing from Patten's 1912 book showing the superficial resemblance of a head shield of an ostracoderm fish (left) to a trilobite (right). *Courtesy of Department of Library Services, American Museum of Natural History.*

The arthropod theory, though venerable, suffered another major impediment that Patten tried to remedy. This theory was little more than an abstract argument based on a theoretical transformation without hard evidence in the form of intermediary creatures from the fossil record. Patten therefore returned to the oldest folk wisdom about *Cephalaspis*— the basic observation that had been judged false and treacherous ever since Agassiz. Maybe that first idea about a relationship with trilobites had some validity after all. Maybe *Cephalaspis* looked like an arthropod because it wasn't all fish, as its back end seemed to proclaim. Maybe the head shield truly possessed some arthropod characters. Maybe ostracoderms did represent that long-sought intermediary group between arthropods and vertebrates.

Patten eventually argued himself into this position. He identified the marine arachnids (eurypterids and horseshoe crabs) as arthropod ancestors, and he classified ostracoderms, not as primitive jawless fishes, but as a transitional group between the two great phyla. He wrote in 1912:

> We may now confidently affirm that the ostracoderms belong neither to the arthropods nor to the vertebrates, but constitute a new class standing midway between them, the ancestors of the one and the descendants of the other, the long sought missing link between the vertebrates and the invertebrates.

Patten was more than a vertebrate anatomist; he also fancied himself a philosopher and moralist. As such, he used his theory of vertebrate origins as a centerpiece for one of the widest (and wildest) claims ever made for the sweep of evolutionary theory. In a series of works, including published class notes for Dartmouth courses in the late 1920s and in his general book *The Grand Strategy of Evolution* (1920), Patten tried to establish evolution as the source of all morality, proper conduct, and good human relations. He therefore becomes a convenient foil for our cardboard relativist who wishes to see little of the external world (if such a concept be intelligible at all) and much of social context in the claims of science.

I could not be more out of sympathy with Patten's wider effort. I have never read a more tendentious or vainglorious attempt to establish a preferred social morality as the pathway and dictate of nature. I have no particular quarrel with Patten's beliefs—a compendium of unassailable apple-pie virtues, featuring the value of service to others and the wisdom of self-restraint in a world of temptation. But I'll be damned if nature can validate, or even address, such cultural hopes and preferences.

Patten argued that nature could instruct us if we learned her patterns and followed them in all our beliefs and dealings. He wrote in 1920:

> The universal end, or purpose in life, and in nature, is to construct, to create, or grow. The ways and means of accomplishing that end are mutual service, or cooperative action, and rightness.

This universal growth occurs along three cosmic axes—time, space, and rightness. The three-dimensional result is linear and necessary progress:

> There is an abiding compulsion to the action of all these factors which is cumulative, or progressive, producing that increasing architectural organization that we call naturegrowth, or evolution.

The direction of evolution also sets a moral imperative, for it "compels man to accept nature's constructive rightness as his ethical standard, and to adopt her constructive methods as his moral code."

But if nature's progress is the source of our morality, then we had better be able to find an unambiguous direction in evolution. Patten bravely surveyed the entire tree of life, with its complex and ramifying branches shooting forth in a thousand separate directions, and managed to realign this intricate meshwork, with bold upper case, as "The Great Highway of Organic Evolution which leads from the lower forms of animal life up to man."

To produce this remarkable change, Patten had to extract one lineage from life's tree and depict it as a straight, central highway. He then had to view all other groups as side roads leading to the dirt of nowhere. But how can a central artery be discovered (with humans on top) if, as generally held, vertebrates go back into the mists of time and do not arise directly from any other complex phylum (but share closest ties with the lowly echinoderms)? How can we specify a Great Highway if arthropods, representing some 80 percent of animal species and including the most structurally complex invertebrates, do not lie firmly upon this main route?

Obviously, then, the concept of a Great Highway demands a direct linkage of lower arthropod to higher vertebrate. And if no highway can be found, then we have no natural basis for morality and no primacy of evolution among the disciplines. Patten therefore called upon *Cephalaspis* and the ostracoderms to perform the greatest of all services—to form the link that would secure both the direction of life and the laws of moral conduct. Patten explicitly cited his arthropod theory of vertebrate origins as the key to his entire system:

> It shows that the great vertebrate-ostracoderm-arthropod phylum forms the main trunk of the genealogical tree of the animal kingdom; that, emerging from unsegmented, coelenterate-like animals, as though driven by some mysterious internal power, moves with astonishing precision, through broad, predetermined channels—from which neither habit, nor environment, nor heredity, can cause it to diverge— towards its goal.

Patten even thought that he had finally found a mechanism, in his theory of necessary progress, for that most improbable claim of the arthropod theory—the closure of the old mouth above the brain and the opening of a new version below. He argued that progress must be marked by increasing size of the brain, and that expanding nervous tissue would choke off the old brain-piercing esophagus, forcing construction of a new mouth:

> The progressive constriction of the esophagus, by the growth of the surrounding brain, ultimately compels all those with relatively larger brains to suck their food in liquid form through the narrowest possible opening, or give up eating altogether. . . . Without this closing up of what had come to be a very inconvenient gateway to the gut, the growth of the brain, as we see it in the higher vertebrates and in man, would have been a physical impossibility.

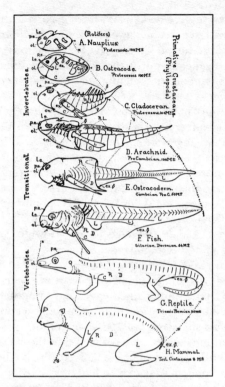

An illustration of Patten's Great Highway of animal evolution from his 1920 book. Note that he places ostracoderms between arthropods and vertebrates. *Courtesy of Department of Library Services, American Museum of Natural History.*

Our cardboard relativist may now exult. Patten's personal need to find moral answers in evolution, and the early twentieth-century vogue among paleontologists for reading life's history as a tale of linear progress, surely fueled his improbable interpretation of ostracoderms as transitional between arthropods and vertebrates. But our realist shouts "wait!" A factual question must still be resolved. People may believe correct things for the damnedest and weirdest of flawed reasons. We still have to know the zoological status of *Cephalaspis*—for this genus ranks somewhere in life's genealogy no matter what cultural blinders we may wear at any moment. We still must find out whether Patten was right or wrong.

This question can be answered definitively, for *Cephalaspis* then had the good fortune to become the subject of this century's greatest work in observational paleontology—a treatise so stunning in care and detail that I thrill every time I pick it up, even though its unremitting technical detail scarcely forms the usual stuff of inspirational literature. In 1927, Erik Andersson Stensiö, professor of paleontology at Stockholm, published his monograph on "The Downtonian and Devonian Vertebrates of Spitsbergen, Part I, Family Cephalaspidae" (Downtonian is an old name for strata now termed Upper Silurian).

One hardly expects revolutionary work after such a humdrum title in the conventional form of a taxon from a time and a place. But Stensiö chose understatement as an antidote to Patten's quest for ultimates. Stensiö had found some exceptionally well-preserved head shields of *Cephalaspis* and other ostracoderms on the island of Spitsbergen. He realized that the unusually heavy ossification of the shield suggested an exciting possibility for research—for bone permeated and tightly surrounded all soft anatomy of the head, including delicate blood vessels and cranial nerves, not to mention the more prominent brain and eyes. The soft parts had decayed after death, and had been replaced by matrix of a much lighter color than the surrounding bone. By distinguishing bone from matrix, Stensiö could reconstruct the soft anatomy of *Cephalaspis* in astounding detail.

Stensiö used two basic methods for resolving the anatomy of ostracoderms. First, he dissected head shields enlarged thirty to fifty times under a binocular microscope. He worked with fine needles on specimens immersed in alcohol or Canada balsam, for these liquids enhanced the contrast between bone and matrix. Each specimen required up to two months of work, but Stensiö managed to remove bone and leave the matrix behind as a perfect cast of soft anatomy. Second, Stensiö ground serial sections at intervals of one-fifteenth of a millimeter through the head. By lining up this long series of parallel cuts, and tracing the path-

ways of matrix and bone, Stensiö could reconstruct the soft parts.
He then made wax models of this internal anatomy. By coordinating these methods, Stensiö was able to trace all the cranial nerves,
identify all major arteries and veins, and provide a detailed reconstruction of the brain.

(If I may be excused one short tangent on the subject of nar-

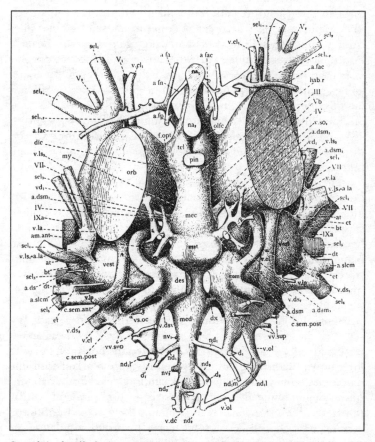

Stensiö's detailed reconstruction of the soft anatomy of the brain and
nervous system in *Cephalaspis*. This figure appeared in his 1927
monograph. *Courtesy of Department of Library Services, American Museum of Natural History.*

row-mindedness in science, we taxonomists and comparative morphologists are often derided as second-class citizens, not quite true scientists, by colleagues who work with more familiar accouterments of *the* scientific method—numbers and experiments. A study without formulae or controls seems to lack the necessary rigor of the stereotype. Let those mired in such myopia try to duplicate the work of Erik Andersson Stensiö. Let them spend months with fine needles, separating matrix from bone, grain by grain. Let them try their hand at serial sections, not through the usual wax and tissue, but through bone and stone. And let them try to interpret the resulting mosaic of holes and connections. Stensiö's work is the most elegant, the most beautiful example ever produced of care and rigor in another dimension. And his results are as firm as anything rooted in numbers and experiment. He was not right about everything; who can be? He misidentified as electric organs, for example, two areas that probably form part of the sensory system for responses to pressure. But the intricate details of his reconstructions for parts of the brain and cranial nerves have been upheld again and again in repeated studies.)

Stensiö's work proved that Patten had been entirely wrong. *Cephalaspis* was all fish, and included not a whiff of arthropod. Moreover, the cranial anatomy of *Cephalaspis* showed detailed similarity, part after part, with the living lampreys—jawless fishes beyond a doubt. After 380 pages of text, Stensiö wrote as his last paragraph and final conclusion: "It is clear now that the Ostracodermi, though very lowly organized, are true craniate vertebrates which have nothing whatever to do either with the Arthropoda or with the Annelida."

In establishing the position of ostracoderms, Stensiö had also resolved the order of early fishes. Lampreys and hagfishes had no jaws, and embryological evidence indicated the homology of gill-arch bones with later jaws. But before Stensiö's treatise, ichthyologists didn't know whether the lampreys and hags were curiously degenerate lines of jawed fishes or remnants of a primordial jawless group. By proving that the ostracoderms were a genealogically coherent group of jawless fishes, predating all jawed vertebrates by tens of millions of years—and by demonstrating the anatomical relationship between lampreys and *Cephalaspis*—Stensiö both established the pathway of early vertebrate evolution and proved that two lineages of the primordial group had survived.

What then can our realists and relativists make of this tale? The relativist correctly identifies three sequential and mutually incompatible worldviews behind the history of change—Agassiz's creationism, Patten's linear progressivism, and Stensiö's branching tree. Yes indeed, each man read *Cephalaspis* in the light of his worldview. Yes again, *Cephalaspis* did not fashion the worldview, but found its inevitable slot in a preconceived structure. Yes once more, the worldviews were products of surrounding culture and personal psychology: Agassiz's accident of birth in a pre-Darwinian world; Patten's need for moral answers in nature.

But this history is not only a tale of social fashion—a story of varying dress lengths, tie widths, or degrees of abstraction in painting. Each worldview was a cultural product, but evolution is true and separate creation is not. *Cephalaspis* may have been buffeted from one social construction to another, but paleontologists also learned important facts about its anatomy at each step—and this accretion of genuine information about the external world must be identified as scientific progress. Agassiz proved that *Cephalaspis* was a fish, but knew nothing of its internal structure. Patten had resolved enough anatomy to know that ostracoderms were a coherent group of primordial fishes, not a hodgepodge of unplaceable oddballs. Stensiö mapped the brain, the cranial nerves, the blood vessels—while Agassiz could not even find the mouth.

Worldviews are social constructions, and they channel the search for facts. But facts are found and knowledge progresses, however fitfully. Fact and theory are intertwined, and all great scientists understand the interaction.

The debate of realists and relativists, when expressed as ends of a dichotomy vying for victory, is silly and tendentious. Science progresses by establishing facts about the world out there—and science is, and must be, socially embedded. The history of interaction between paleontologists and *Cephalaspis* is both a pageant of ideas and a growing compendium of information. I exult in the ideas, but I confess to a special love for the rock-hard primacy of Stensiö's dissections and for Patten's beautiful eurypterids in the corner of my office.

Isaac Newton mused on the interaction of fact and theory in his most famous passage:

> I do not know what I may appear to the world; but to myself I seem to have been only like a boy playing on the seashore, and diverting myself in now and then finding a smoother pebble or a prettier shell than ordinary, whilst the great ocean of truth lay all undiscovered before me.

We would love to fathom that distant ocean, but it is no shabby thing to fondle those pretty pebbles on the shore.

30 | A Tale of Three Pictures

GOETHE, who coined the word *morphology* and therefore ought to know, once proclaimed that "we should talk less and draw more. I personally would like to renounce speech altogether and, like organic nature, communicate everything I have to say in sketches." As a card-carrying member of the guild of essayists, I should resist this heresy tooth and nail. I might also argue that the world is a better place because Goethe did not take his own advice. We would all be a little poorer without *Faust*, while Goethe's sketches, although no disgrace, are not depriving us of quintessential insight by their general oblivion.

Primates are visual animals. No other group of mammals relies so strongly on sight. Our attraction to images as a source of understanding is both primal and pervasive. Writing, with its linear sequencing of ideas, is a historical afterthought in the history of human cognition.

Yet traditional scholarship has lost this root to our past. Most research is reported by text alone, particularly in the humanities and social sciences. Pictures, if included at all, are poorly reproduced, gathered in a center section divorced from relevant text, and treated as little more than decoration. (Natural scientists, although not noted for insights about communication, have better intuitions on this subject. Most scientific papers are illustrated, and slide projectors are automatically provided for scientific talks throughout the world. By contrast, I have, three or four times, suffered the acute embarrassment of arriving before a large audience in the humanities or social sciences, slides in hand, to deliver a talk that would be utterly senseless without

427

pictures: no slide projector, no screen, not even a way to darken the room. My fault: I had forgotten to request the projector because, in my own scientific culture, slides are as automatic as words. And so, all you budding scientists who may read these essays, if I have taught you nothing in twenty years of these monthly efforts, at least remember this and thank me some day for a small boon of advice: If you are ever asked to talk before a department in the humanities, remember that you have to request the slide projector. Call this Gould's law and let it be my immortality—long after everyone has forgotten those upside-down flamingos and pandas' thumbs.)

Pictures are not peripheral or decorative; iconography offers precious insight into modes of thinking that words often mask or ignore—precisely because we tailor our words so carefully but reveal our secrets unconsciously in those "mere" illustrations. (My thanks to M. J. S. Rudwick, great historian of geology, who first taught me this lesson and who supplied the initial quote from Goethe.)

Pictures are revealing enough when they simply claim to represent an object "as it is." Shading, emphasis, context, and surroundings all provide an artistic leeway for expressing (often unconsciously) a social or ideological framework. Have you ever seen a dodo pictured as anything other than alone and forlorn, although they once abounded on Mauritius? The classic dodo reconstruction shows a single bird dominating the foreground of a desolate terrain. For the dodo is both a large flightless pigeon and our conventional metaphor for extinction.

Iconography becomes even more revealing when processes or concepts, rather than objects, must be depicted—for the constraint of a definite "thing" cedes directly to the imagination. How can we draw "evolution" or "social organization," not to mention the more mundane "digestion" or "self interest," without portraying more of a mental structure than a physical reality? If we wish to trace the history of ideas, iconography becomes a candid camera trained upon the scholar's mind.

This essay is a tale of three pictures. It tries to illustrate something crucial in the history of evolutionary thought by analyzing three sequential snapshots of "relationships among animals." These pictures present two favorable features that may promote their expansion from anecdote to illumination. First, all three pictures tell the changing story of the same animal, thereby imparting coherence to a sequence that would otherwise have no anchor. Second, the pictures embody, in a visual epitome that I (at least) found stunning, what may be the most important general issue in our struggle to understand the distinctive character and history of scientific thought.

These pictures were all presented as simple sketches of "objective" relationships among animals; they are also (and primarily, I would argue) iconographies of three strikingly different and incommensurable worldviews. They were drawn by the three men discussed in the previous essay: by Louis Agassiz in the 1830s, by William Patten early in our century, and by Erik Andersson Stensiö in 1927. They all include, as a prominent feature, an attempt to fix the biological position of *Cephalaspis,* prototype of the jawless fishes that gave rise to all later vertebrates, ourselves included of course.

To summarize briefly the sequence of opinions about *Cephalaspis* (see previous essay for the details), Agassiz discovered that *Cephalaspis* was a fish, not a trilobite (the bony head shield looks like the external armor of an arthropod, but Agassiz found heads attached to indubitably fishy bodies). Agassiz denied evolution altogether, but Patten tried to interpret *Cephalaspis* as an intermediary form between arthropods and vertebrates, a key waystation on "the Great Highway of Animal Evolution." Stensiö proved, by meticulous dissection of exceptionally preserved fossils from Spitsbergen, that *Cephalaspis* was "all fish," without a whiff of arthropod. He also demonstrated that modern lampreys and hagfishes are close relatives of the great primordial group of jawless fishes represented by *Cephalaspis* (class Agnatha, meaning, quite appropriately, "jawless").

The previous essay also set the story of *Cephalaspis* in the context of an old debate about progress in the history of scientific thought. I argued that this subject is often obscured by a false dichotomy drawn between equally untenable extremes: realists, who argue that science, with its timeless and universal methods, learns progressively more and more about an objective external reality; and relativists, who hold that the history of theories approximates the vagaries of fashion, a series of equally workable solutions altered by whim or social circumstances.

I think that each side of this controversy possesses a central insight, and that their marriage provides a workable solution sensitive to the fundamental concept of each camp. Notwithstanding a long history of arguments, ranging from the playful to the tendentious to the sophistic, there is a world out there full of stars, amoebas, and quartz crystals. (We must, in any case, behave as if this claim were true in order to negotiate life's numerous difficulties with any success—and this behavior has brought consistent results, at least in the form of technological achievement.) Science does construct better and better maps of this outer reality, so we must assume that change in the history of scientific theories often records more adequate knowledge of the external world, and may therefore be called progress.

On the other hand, we must also admit that the history of scientific theories on any subject is no simple tale of good information driving out bad. Successive theories often display the interesting property of incommensurability. They do not speak the same language; they do not parse the world into the same categories; they embody fundamentally different views about the nature of causality. The new is not simply more and better information heaped upon the explanatory structure of the old. In this sense, the history of theories is a successive replacement of mutually incompatible worldviews, not a stroll up the pathway of objective knowledge.

The three pictures of relationships among early vertebrates demonstrate, with bold literality, this principle of scientific change as a series of incommensurable worldviews, each replacing rather than just building upon the last. Yet the sequence also records increasing objective knowledge about *Cephalaspis;* it is not a passive mirror of social change.

Louis Agassiz (1807–73), the great Swiss zoologist who became America's premier naturalist, was the last great scientific creationist (I am writing this essay in the museum and laboratory that he opened in 1859). He built his career upon two fundamental achievements: the development of the theory of ice ages, and a monumental work on the classification and relationships of all fossil fishes. Agassiz summarized his fifteen-year project on fossil fishes with the first major example of an iconography that paleontologists have since adopted as canonical—the so-called spindle diagram (see figure). In these geological charts of relationships among organisms, the vertical axis represents time, as though the diagram portrayed a sequence of strata in the field. Each group of organisms is drawn as a spindle, with varying widths through time representing a history of fluctuating diversity, and the ends of the spindle marking origin and extinction. The ordering of spindles records degree of relationship, with physical closeness representing biological affinity. We have all seen so many of these diagrams that we read them automatically, rarely stopping to acknowledge that all these features are iconographic conventions, not necessary realities.

These conventions leave great latitude for portraying a theoretical worldview in the guise of objective knowledge—and Agassiz's famous chart is a striking example of concept as iconography. Note two features of Agassiz's fishes. First, of the various geometries that might be used to portray relationships among organisms—circles, chains, ladders, parallel lines like teeth on a comb—he chooses a topology of branching from a central stem in each of his four groups. This iconography embodies his bio-

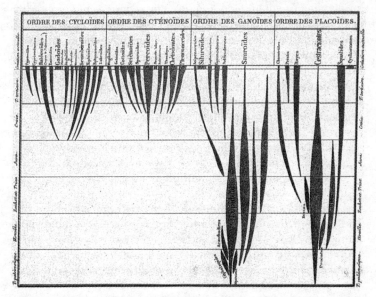

Agassiz's creationist version of the evolution of fishes. Note *Cephalaspis* at the base of the ganoid radiation. *From Louis Agassiz's* Les poissons fossiles, *Vol. 1, p. 170, 1833.*

logical theory of life's history as a tale of differentiation through time from simple and highly generalized archetypes. Life diversifies on an embryological model. Just as all mammalian fetuses begin with a simple and similar form and differentiate later to bat, whale, and camel, so too does the geological history of a group generate more diverse and specialized creatures through time.

This view sounds so evolutionary that we wonder why Agassiz continued his lone holdout against Darwin to the death. But such a feeling only represents the chauvinism of later knowledge imposed upon a fundamentally different worldview. Differentiation from a common archetype need not imply a physical, evolutionary connection among successive forms. Suppose that differentiation is God's grand design for all developmental processes in nature. Embryology proceeds in physical continuity, but geological succession may feature a series of independently fashioned forms, linked together as incarnations of an ordered pattern of thought in their creator's mind.

Agassiz depicted his creationist interpretation in the second striking feature of his iconography. The separate spindles in each of his four groups may converge lovingly towards each other, and towards the central or archetypal line, but they never join! And Agassiz knew exactly what he was doing, and why:

> Nevertheless, I have not joined the lateral branches to the central trunks because I am convinced that they do not descend, one from the other, by pathways of direct procreation or successive transformation, but that they are materially independent, although forming in their ensemble . . . a systematic whole, whose connections must be sought in the creative intelligence of its author.

Agassiz placed *Cephalaspis* as the first side branch from his central stock of the most "primitive" group—the ganoids (sharks and their relatives). He had not been able to excavate the mouth and did not recognize the jawless character of *Cephalaspis* and its relatives. For Agassiz, *Cephalaspis* was both primitive and peculiar—a short-lived side branch of God's early efforts.

Eighty years later, William Patten gave *Cephalaspis* a more central role in the order of life's history. Patten recognized *Cephalaspis* as more than a curiosity, and classified this genus as the proto-

type of a group ancestral to all later vertebrates—the ostracoderms of his terminology. But Patten, firmly committed both to a
general theory about life's progressive advance and to a specific
claim that vertebrates had descended from arthropods, misinterpreted the structure of *Cephalaspis*. The fishlike body he could not
and did not deny, but he also thought (quite incorrectly) that he
had found jaws of arthropod design—and he therefore interpreted the ostracoderms as chimeras of arthropod and vertebrate
characters and as intermediary forms in an evolutionary sequence from horseshoe crab to fish.

This convenient casting of *Cephalaspis* allowed Patten to fulfill
his dream of mapping evolution upon his hopes for morality and
good conduct. Patten yearned to find one true path through the
labyrinthine branching of phylogeny. That path must, of course,
ascend to *Homo sapiens,* thereby making our distinctive features
the goal of life's entire history. This dream would have died if
Patten had not been able to link arthropods with vertebrates—
for arthropods (mostly insects) represent some 80 percent of all
animal species, and their exclusion from our lineage would have
converted Patten's Great Highway of Animal Evolution into a
dinky little road less traveled.

Patten's iconography (see figure) portrays this claim for linear
progress as a single grand highway. This figure is unconventional
in its dubious attempt to compress three dimensions into two.
The vertical axis represents "progress in organization, brain size,
parental care, and adaptability," rather than the usual geological
time. Instead, time runs along the horizontal axis, symmetrically
in both directions, from the origin of life at the center, to younger
and younger strata both right and left. But time correlates with
progress (the chart would be unintelligible otherwise)—so farther from the beginning point (bottom, center) means younger
and better. Thus, Patten draws four bubbles, outlined in black
and shaped like light bulbs, centered on the starting point. Each
bubble adds geologically younger and biologically more advanced forms. The four bubbles represent, in order, Precam-
brian, Paleozoic, Mesozoic, and Cenozoic eras. The third dimension of organic form can't be plotted into this scheme, but Patten
fudges and simply draws the conventional spindle for each
group, radiating from the interior of a bubble toward its edge.

Note the major feature of this iconography—and the obvious

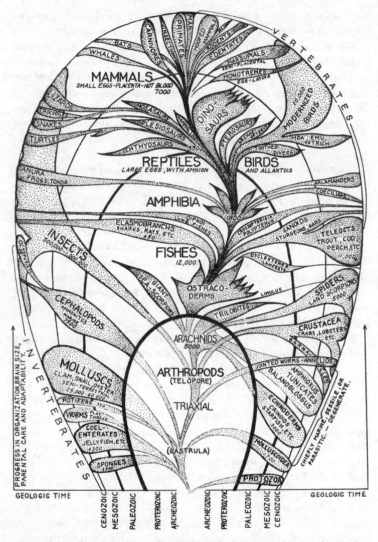

Patten's progressivist tree of animal evolution. Note how arthropods progressed directly to vertebrates through ostracoderm intermediates. *William Patten's picture from his* Evolution. *Plate 1, Dartmouth Press, 1925.*

rationale behind Patten's chosen form: The Great Highway of Animal Evolution ascends right through all the bubbles in the center of the chart. It rises from arthropods in the first bubble; to ostracoderms (the crucial link, including *Cephalaspis*), fishes, and amphibians in the second; to dinosaurs and early mammals in the third bubble; and finally to man, flanked by primates and hoofed mammals, in the fourth bubble. The great highway does become dangerously constricted now and then, but it always perseveres and always moves like the motto of New York State and the sky-scrapers of its metropolis—Excelsior, "ever upward."

How shall we compare Patten's iconography with Agassiz's ear-lier chart? Is it truer? Better? Does it represent a simple accretion of knowledge—scientific progress—in the intervening eighty years? I suppose so, in some sense. Patten does connect the spin-dles to acknowledge the discovery of evolution. He also recog-nizes the coherence of ostracoderms as an ancestral group of vertebrates, whereas Agassiz had regarded *Cephalaspis* alone as a confusing oddity. Yet in other ways, Patten's chart has lost accu-racy. Agassiz correctly classified *Cephalaspis* as a fish, whereas Pat-ten, impelled by his theory of necessary progress, managed to reconstruct the ostracoderms as part arthropod.

But more fundamental than these backings-and-forthings must be the basic incommensurability of these two charts. They are not linearly related by progress or regress in knowledge. In fact, you cannot transform one into the other. They represent two incom-patible worldviews, not a filling-in of new facts on the objective background of history. Each scheme for relationships among or-ganisms depicts a personal theory, not a hatrack stamped out of universal logic. And the main difference isn't even the watershed that we usually identify as the chief contrast between Agassiz's and Patten's worlds—the intervening discovery of evolution. Agassiz's topology is easily converted to an evolutionary scheme by connecting the spindles. But Agassiz's theory cannot be trans-formed into Patten's worldview, for Agassiz based his vision on differentiation (radiation of numerous lineages from common points of origin), while Patten embraced linear progress. You can't turn a hand into an upraised forearm.

When Stensiö resolved the debate about *Cephalaspis* by proving both its jawlessness and its relationship to modern lampreys and hagfishes, he summarized his discoveries in a third iconography.

This more modest chart (see figure) does not show all life (like Patten's) or even all fishes (like Agassiz's); it portrays jawless fishes only—both the ostracoderms and their modern descendants. It argues that a root stock of ostracoderms split into two basic groups: the first containing two groups of fossils (including *Cephalaspis* among the Osteostraci) and the modern lampreys (Petromyzontia); the second including two other fossil groups and the modern hagfishes (Myxinoidea).

Again, Stensiö's iconography does not emerge by accretion of information onto Patten's version. It represents yet another worldview, incommensurate with Patten's and therefore not derivable by any transformation from Patten's grand highway. Stensiö's organizing scheme is diversification, not progress. His iconography is closer to Agassiz's preevolutionary version than to Patten's supposed improvement. Connect Agassiz's spindles and you obtain something more like Stensiö's evolutionary branching than Patten's linear progress.

But Stensiö is not simply Agassiz after a game of connect the dots. We still note a basic incommensurability. On Agassiz's chart, all subsidiary groups radiate from a central axis within each

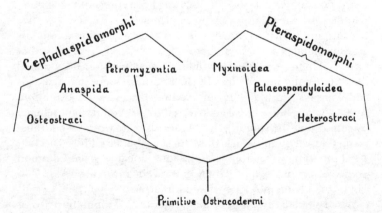

Stensiö's evolutionary tree of ostracoderm fishes. *Erik Stensiö's sketch from his monograph* Downtonian and Devonian Vertebrates of Spitsbergen *Pt., 1, 279, 1927.*

of the four divisions. No geologically younger group splits from a previous side branch; all point toward the central stem. This choice records Agassiz's belief that creation occurs on an embryological plan. Mammals did not evolve directly from reptiles (themselves a side branch of the vertebrate trunk). Mammals were created after the death of ruling reptiles—as a more highly differentiated incarnation of the vertebrate ideal. The central stem of each group is, for Agassiz, the archetype that must serve as a model for each new and independent side branch of created complexity. But Stensiö's iconography is fully evolutionary, with hierarchical diversification. Stems produce side branches, and side branches then bud off other twigs. Agassiz's iconography is like a human hand; on Stensiö's chart, fingers generate fingerlets, which generate fingerletchens, which generate . . .

These three successive iconographies lead me to conclude that scientific change cannot, in this case at least, be viewed as a simple accumulation of knowledge within the unchanging framework of a universal, objective method. We did learn more and more about *Cephalaspis* in particular and about the evolution of vertebrates in general. But the theories of Agassiz, Patten, and Stensiö are three incompatible worldviews—three visions imposed upon our greatly imperfect knowledge—not a progressive fleshing out of nature's bare bones. And ironically, for those wedded to linear progress in ideas (if not in life's history), Stensiö's "final" version shares more of its iconography with Agassiz's creationist vision of differentiation than with Patten's evolutionary dream of linear advance.

Nature does not tolerate chimeras among its more complex animals. You cannot put a man's head and chest upon a horse's torso, and you certainly can't meld an arthropod's head with a vertebrate's trunk (as Patten proposed in arguing that *Cephalaspis* had the jaw of a sea scorpion). Organic designs have integrity as working wholes constructed from coherent patterns of development. Nature is not an interior decorator or a postmodern builder recycling the entire history of architecture in an eclectic display of façades.

I believe that ideas have a similar integrity. Great thinkers build their edifices with subtle consistency. We do our intellectual forebears an enormous disservice when we dismember their visions and scan their systems in order to extract a few disembodied "gems"—thoughts or claims still accepted as true. These disarticulated pieces then become the entire legacy of our ancestors, and we lose the beauty and coherence of older systems that might enlighten us by their unfamiliarity—and their consequent challenge—in our fallible (and complacent) modern world.

This integrity of systems also precludes smooth transitions in intellectual life. Some systems will not transform smoothly into others, and discontinuous change must occur from time to time in the history of ideas, of human social institutions, and in the form of organisms. D'Arcy Thompson, great morphologist and admirer of Goethe (see Essay 10), made his idiosyncratic argument for rapid transitions in his classic treatise *On Growth and Form* (1917). He probably exaggerated the case for organisms with his geometric analogy, but I recommend his words as a stimulating, if overstated, model for the history of great ideas:

> An algebraic curve has its fundamental formula, which defines the family to which it belongs. . . . We never think of "transforming" a helicoid into an ellipsoid, or a circle into a frequency-curve. So it is with the forms of animals. We cannot transform an invertebrate into a vertebrate, nor a coelenterate into a worm, by any simple and legitimate deformation. . . . Nature proceeds from one type to another . . . and these types vary according to their own parameters, and are defined by physico-mathematical conditions of possibility. . . . To seek for stepping-stones across the gaps between is to seek in vain, forever.

31 | A Foot Soldier for Evolution

FRANCIS GALTON, Darwin's cousin and England's most eccentric scientific genius, took all knowledge (and much speculation) for his province. Many of his studies were cranky or dubious. He performed a statistical test on the efficacy of prayer and, as inventor of the word *eugenics,* led a movement for selective marriage among the élite. But he could also be outstandingly right in his originality, as in his successful campaign for using fingerprints to identify criminals.

In his autobiography, Galton tells a story of Herbert Spencer's visit to his fingerprint lab. Galton took Spencer's prints and "spoke of the failure to discover the origin of these patterns, and how the fingers of unborn children had been dissected to ascertain their earliest stages." Spencer, quick to offer certain opinions about almost anything, told Galton that he had been working the wrong way round.

> Spencer remarked . . . that I ought to consider the purpose the ridges had to fulfil, and to work backwards. Here, he said, it was obvious that the delicate mouths of the sudorific glands required the protection given to them by the ridges on either side of them, and therefrom he elaborated a consistent and ingenious hypothesis at great length. I replied that his arguments were beautiful and deserved to be true, but it happened that the mouths of the ducts did not run in the valleys between the crests, but along the crests of the ridges themselves.

Galton then ends his anecdote by giving the original source for one of the top ten among scientific quotes. Spencer, dining with T. H. Huxley one night at the Athenaeum, stated that he had once written a tragedy. Huxley replied that he knew all about the work. Spencer rebutted Huxley, arguing that he had never mentioned it to anyone. But Huxley insisted that he knew anyway and identified Spencer's debacle—"a beautiful theory, killed by a nasty, ugly little fact."

Some beliefs may be subject to such instant, brutal, and unambiguous rejection. For example, no left-coiling periwinkle has ever been found among millions of snails examined. If I happen to find one during my walk on Nobska beach tomorrow morning, a century of well-nurtured negative evidence will collapse in an instant.

This Huxleyan vision of clean refutation buttresses one of our worst stereotypes about science. We tend to view science as a truth-seeking machine, driven by two forces that winnow error: the new discovery and the crucial experiment—prime generators of those nasty, ugly little facts. Science does, of course, seek truth, and even succeeds reasonably often, so far as we can tell. But science, like all of life, is filled with rich and complex ambiguity. The path to truth is rarely straight, marked by a gate of entry that sorts applicants by such relatively simple criteria as age and height. (When I was a kid, you could get into Yankee Stadium for half price if your head didn't reach a line prominently drawn on the entrance gate about four feet above the ground. You could scrunch down, but they checked. One nasty, ugly day, I started to pay full price, and that was that.)

Little facts rarely undo big theories all by themselves—the myth of David and Goliath notwithstanding. Such facts can refute little, highly specific theories, like my conjecture about lefty periwinkles, but they rarely slay grand and comprehensive views of nature. No single, pristine fact taught us that the earth revolves around the sun or that evolution produced the similarities among organisms. Overarching theories are much bigger than single facts, just as the army of Grenada really didn't have much chance against the combined forces of the United States.

Instead, little facts are assimilated into large theories. They may reside there uncomfortably, bothering the honorable proponents. Large numbers of little facts may eventually combine with other social and intellectual forces to topple a grand theory. The history of ideas is a play of complex human passions interacting with an external reality only slightly less intricate. We debase the richness of both nature and our own minds if we view the great pageant of our intellectual history as a compendium of new information leading from primal superstition to final exactitude. We know that the sun is hub to our little corner of the universe, and that ties of genealogy connect all living things on our planet, because these theories assemble and explain so much otherwise disparate and unrelated information—not because Galileo trained his telescope on the moons of Jupiter or because Darwin took a ride on a Galápagos tortoise.

This essay tells the story of a pristine, unexpected little fact that should have mattered, but didn't particularly. The fact was widely reported, discussed, and personally studied by the greatest naturalists of Europe, and then assimilated into each of several contradictory systems. Fifty years later, in 1865, a second discovery resolved the paradox generated by the first fact—and should have won, by Huxley's principle, a big and important victory for Darwin and evolution. It was welcomed, to be sure, but largely ignored. One foot soldier could not decide a battle waged on so many fronts.

Trigonia is a distinctive clam, thick shelled and triangular in shape. It flourished with dinosaurs during the Mesozoic era and then became extinct in the same debacle that wiped out the ruling reptiles—one of the five greatest mass dyings in our geologic record. No trigonian had ever been found in the overlying Cenozoic strata—the entire age of mammals (about sixty million years, as we now know). *Trigonia* had therefore become a valued "guide fossil"; when you found one, you knew you had rocks of the earth's middle age. Everyone (who was anybody) understood this useful restriction in time.

Then, the nasty, ugly little—and quite undeniable—fact: In 1802, P. Péron, a French naturalist, found the shell of a living trigonian washed up on the beaches of southern Australia. Twenty-five years later, and following several failures, J. Quoy and J. Gaimard, naturalists aboard the *Astrolabe*, finally found a live trigonian. They had dredged for several days with definite purpose, but without success. Becalmed one night in Bass Strait and with little else to do, they tried again and brought up their single prize, a molluscan life soon snuffed and preserved in the medium of the collector's trade—a bottle of alcohol. Quoy and Gaimard treasured their booty and wrote later:

> We were so anxious to bring back this shell with its animal that when we were, for three days, stranded on the reefs of Tonga-Tabu, it was the only object that we took from our collection. Doesn't this recall the ardent shell collector who, during seven years' war, carried constantly in his pocket an extraordinary *Phasianella,* which he had bought for twenty-five louis?

A simple story. A fact and a puzzle. *Trigonia* had not disappeared in the great Cretaceous debacle, but lived still in Australia. Yet no fossil trigonians had been found in all the strata in between—throughout the long and well-recorded history of the age of mammals (now called the Cenozoic era). Where were they? Had they ever existed? Could such a distinctive animal die and be reborn (or re-created) later? The "Cenozoic gap" became as puzzling and portentous as the one later associated with Mr. Nixon and Ms. Woods.

Trigonia occupies a specially interesting place in the history of biology because its unexpected fact and consequent puzzle arose and prevailed at such an important time—at the dawn and through the greatest conceptual transition ever experienced by the profession: from creationist to evolutionary views of life. *Trigonia* also (or rather therefore) attracted the attention and commentary of most leaders in nineteenth-century natural history. J. B. Lamarck, most famous of pre-Darwinian evolutionists, formally described the first living trigonian. Darwin himself thought and commented about *Trigonia* for thirty years. Louis Agassiz, most able and cogent of Darwin's opponents, wrote the major technical monograph of his generation on the genus *Trigonia.*

The lesson of the living *Trigonia* can be distilled in a sentence: Everyone made the best of it, incorporating favorable aspects of this new fact into his system and either ignoring or explaining away the difficulties. *Trigonia* became an illustration for everyone, not a crucial test of rival theories. Evolutionists celebrated the differences in form and distribution between ancient and modern trigonians—and ignored the Cenozoic gap. Creationists high-lighted the gap and made light of the differences.

Today, we remember Lamarck best as the author of a rejected evolutionary theory based on the inheritance of acquired charac-ters (quite an unfair designation since so-called Lamarckian in-heritance represents a minor part of Lamarck's own system). But his day-to-day work in post-revolutionary France focused on the description of living and fossil invertebrates in his role as curator at the *Muséum d'Histoire Naturelle* in Paris. He therefore received Péron's precious shell for formal description, and he named it *Trigonia margaritacea* in 1804 (Lamarck didn't know about modern cocktails; *margarita* is a Latin pearl, and the interior of a trigonian shell shines with a beautiful pearly luster). But since 1804 lay squarely between Lamarck's initial (1802) and definitive (1809) statement of his evolutionary theory, he also used his short paper on *Trigonia* to sharpen and defend his developing transmutation-ist views.

Most fossil trigonians are ornamented with concentric ridges at their anterior ends (enclosing the mouth and digestive appara-tus) and radial ribs on the rear flank. A single strong rib usually separates these two areas. But all modern trigonians cover their shells entirely with radial ribs (although the embryonic shell still bears traces of the ancestral concentrics). Lamarck seized upon these differences to claim that changing environments had pressed their influence upon the shell. The shell had then altered in response and the animal within passed the favorable change to future generations by "Lamarckian" inheritance.

Lamarck's original figure of the shell of living trigonian clams.

Fossil trigonians (left) have both concentric and radial ribs.
Living trigonians (right) have only radial ribs.

They have undergone changes under the influence of circumstances that act upon them and that have themselves
changed; so that fossil remains . . . of the greatest antiquity
may display several differences from animals of the same
type living now but nevertheless derived from them.

(But Lamarck had only demonstrated that the fossils looked different from the modern shells. Any theory could account for this
basic datum in the absence of further information—evolution by
use and disuse, by natural selection, or even re-creation by God
for that matter.)

Lamarck then proceeded to extract more from modern trigonians to buttress other pet themes. He was, for example, a partisan at the wrong end of a great debate resolved a decade later to his disadvantage by Cuvier—does extinction occur in nature? Human rapacity, Lamarck believed, might exterminate some conspicuous beasts, but the ways of nature do not include termination without descent (Lamarck, as a transmutationist, obviously accepted the pseudoextinction that occurs when one form evolves into another). Lamarck gave the old arguments against extinction a novel twist by embedding his justification within his newfangled evolutionary views. How can extinction occur if all organisms respond creatively to changing environments and pass their favorable responses to future generations in the form of altered inheritance?

Yet Lamarck's conviction was sorely challenged by burgeoning data in his own field of marine invertebrate paleontology. So many kinds of fossils are confined to rocks of early periods. Where are their descendants today? Lamarck offered the only plausible argument in a world with few remaining terrae incognitae—they live still in the unexplored depths of the sea. Since Lamarck reveals his own discomfort with such an ad hoc solution by repeating it too often and too zealously—recall Shakespeare's "the lady doth protest too much, me-thinks"—we may take as genuine his delight in *Trigonia* as a real case for a generalization devoutly to be wished: "Small species, especially those that dwell in the depths of the sea, have the means to escape man; truly among these we do not find any that are really extinct." Lamarck then ends his paper by predicting that a large suite of creatures apparently extinct will soon be found at oceanic depths. We are still waiting.

Since Lamarck's argument centers upon an explanation for why creatures still living yield no evidence of their continued vitality, we should not be surprised that he ignored the Cenozoic gap entirely. We must assume that trigonians spent the entire Cenozoic safe in the bosom of Neptune, full fathom five hundred or more, and unrecorded in a fossil archive of shallow-water sediments.

Charles Darwin, leading evolutionist of the next generation, selected yet another feature of living trigonians—their geographic distribution—to bolster a different theme dear to his view of life. Darwin's creationist opponents, as we shall see, rendered the history of life as a series of static faunas and floras separated by episodes of sudden extirpation and renewal. To rebut this catastrophist credo, and to advance his own distinctive and uncompromisingly gradualist view of nature, Darwin argued that the extinction of a group should be as smooth and extended as its origin. A group should peter out, dwindle slowly, decrease steadily in numbers and geographic range—not die in full vigor during an environmental crisis. What better evidence than a family once spread throughout the world in stunning diversity but now confined to one small region and one single species. In his private essay of 1844, precursor to the *Origin of Species* (1859), Darwin wrote: "We have reason to believe that . . . the numbers of the species decrease till finally the group becomes extinct. . . . The *Trigonia* was extinct much sooner in Europe, but now lives in the seas of Australia."

Darwin followed Lamarck in dismissing the Cenozoic gap as an artifact of our imperfect fossil record (I can imagine no other option for an evolutionist committed to genealogical connection). But Darwin was explicit where Lamarck had been silent. Darwin also tried to accentuate the positive by arguing that the rarity of such long gaps strongly implied their artificial status. He wrote in the *Origin of Species:*

> A group does not reappear after it has once disappeared; or its existence, as long as it lasts, is continuous. I am aware that there are some apparent exceptions to this rule, but the exceptions are surprisingly few, so few that . . . the rule strictly accords with my theory.

Creationists, meanwhile, looked at *Trigonia* from the other side. They treasured the Cenozoic gap and found nearly everything else puzzling. The major creationist thinkers tended to agree that life's history had been episodic—a series of stages separated by sudden, worldwide paroxysms that removed the old and set a stage for the new. But they divided into two camps on the issue of progress. Did each new episode improve upon the last? Was God, in other words, learning by doing? Or had life maintained a fairly consistent complexity throughout its episodic history? Progressionists and nonprogressionists found different messages in *Trigonia.*

James Parkinson, England's leading progressionist (though he switched allegiances later on), chose *Trigonia* as a premier example in his *Organic Remains of a Former World* (1811). He read the Cenozoic gap literally, extracting the congenial message that life's history features a series of creations not connected by ties of genealogy and physical continuity.

But *Trigonia* also presented a special problem for Parkinson. He argued that each successive episode of creation had been marked "with increasing excellence in its objects," thus matching in all ways but one the Mosaic progression from chaos to Adam as described in Genesis. "So close indeed is this agreement, that the Mosaic account is thereby confirmed in every respect except as to the age of the world" (a problem then resolved by an allegorical interpretation of God's six creative "days"). Now a *Trigonia,* as some folks say about roses, is a *Trigonia* (subtleties evident to the professional eye aside). Why should a modern shell with radial ribs alone be "better" than a fossil representative with radials and concentrics? Why are the modern versions superior, as Parkinson's theory of progressive creation required? Parkinson was evidently troubled. In the summary statement to his three-volume work, he devoted more space to *Trigonia* than to any other genus. He clutched at the one available straw, but clearly without conviction. At least the modern trigonians are different. We don't know why, but different must be better:

> This shell, although really of this genus, is of a different species from any shell, which has been found in a fossil state. So that none of the species of shells of this genus, which are known in a fossil state, have, in fact, been found in any stratum above the hard chalk [the Cretaceous, or last period of dinosaurs], or in our present seas.

Louis Agassiz, most able of all creationists, followed Parkinson's personal route in reverse. He began as an advocate of progress in each successive creation and ended by defending the earliest of God's creatures as fully up to snuff (largely because he

despised Darwinism with such passion and felt that any admission
of progress would bolster the evolutionary cause). For him,
therefore, the apparent lack of improvement in modern trigonians posed no problem, while the Cenozoic gap brought nothing
but pleasure and confirmation. In the major pre-Darwinian work
on these clams, his *Mémoire sur les trigonies* (1840), Agassiz argued
explicitly that a Cenozoic gap, if conclusively affirmed, would effectively disprove evolution (quite a cogent claim, by the way):

> The absence of *Trigonia* in Tertiary [Cenozoic] strata is a
> very important fact for discussions of the origin and rela-
> tionships of species of different epochs; for if it could one
> day be shown that *Trigonia* never existed throughout the
> entire duration of Tertiary time, it would no longer be pos-
> sible to maintain the principle that species of a genus living
> in successive geological epochs are derived from each
> other.

But Agassiz well understood the discomforting uncertainty of
negative evidence. Find one nasty, ugly little Cenozoic trigonian
tomorrow, and the entire argument collapses. So Agassiz decided
to cover his rear and disclaim: No Cenozoic trigonian is dandy,
but future discovery of a Cenozoic trigonian would prove nothing. God may, after all, ordain temporal continuity among a
group of related, created forms.

Although his passage may be an exercise in special pleading, it
also contains one of the most succinct and eloquent defenses ever
written for the Platonic version of creationism.

> Although I now invoke this fact [the Cenozoic gap] to sup-
> port my conviction that the different species of a genus are
> not variants of a single type . . . the discovery of a Tertiary
> trigonian would still not demonstrate, to my eyes, that the
> relationship among species of a genus is one of direct de-
> scent and successive transformation of original types. . . . I
> certainly do not deny that natural relationships exist among
> different species of a genus; on the contrary, I am convinced
> that species are related to each other by bonds of a higher

nature than those of simple direct procreation, bonds that may be compared to the order of a system of ideas whose elements, developed at different times, form in their union an organic whole—although the elements of each time period also appear, within their limits, to be finished products.

In summary, as Darwin's revolution dawned in 1859, the supposedly pure and simple little fact of modern trigonians stood neither as arbiter nor slayer of theories but as touted support for all major conflicting and contradictory views of life—for evolution by Lamarckian and Darwinian agencies, and for creationism in both progressionist and directionless versions. How can something so important be so undecisive, unless Huxley's heroic vision of raw empiricism triumphant rarely describes the history of ideas or even the progress of science? Percepts may not create and drive concepts, but concepts are not intractable and immune to perceptual nudges either. Thought and observation form a wonderfully complex web of interpenetration and mutual influence—and the interaction often seems to get us somewhere useful.

The *Trigonia* story has a natural ending that should be conventional and happy, but doesn't quite work in the expected way. The resolution is not hard to guess, since Darwin's vision has prevailed. The elusive Cenozoic trigonian was found in Australian rocks—at just the right time, in 1865, when nascent evolutionism needed all the help it could get.

H. M. Jenkins, a minor figure in British geology, explicitly defended Darwin in describing the first Cenozoic trigonians. He interpreted the happy closure of the Cenozoic gap as a clear vindication of Darwin's characteristic attitude toward the fossil record and as direct support for evolution. Darwin viewed the fossil record as riddled with imperfections—"a history of the world imperfectly kept . . . of this history we possess the last volume alone. . . . Of this volume, only here and there a short chapter has been preserved; and of each page, only here and there a few lines" (*Origin of Species,* 1859). Gaps, as the old saying goes, represent absence of evidence, not evidence of absence. Jenkins wrote, linking the newly discovered Cenozoic trigonian to this fundamental Darwinian prediction:

Every paleontologist believes that, when a genus of animals is represented by species occurring in strata of widely different ages, it must have been perpetuated by some one or more species during the whole of the intervening period. . . . The only rational meaning that has ever been attached to this presumed general law . . . is that the perpetuation of the genus . . . has been due to "descent with modification." *Trigonia subundulata* [the formal name for the Cenozoic trigonian] is one of the links hitherto wanting; first, in explanation of the existence of the genus *Trigonia* in the Australian seas of the present day; and secondly, as showing that the great gap which before existed in its life-history was . . . simply a consequence of the imperfection of our knowledge of the geological record.

Finally, a personal confession in closing: This essay has been an exercise in self-indulgence and expiation. I put together the trigonian story at the very beginning of my professional career (when I was barely big enough to pay full price at the stadium). I published a rather poor account in a technical journal in 1968 (frankly, it stunk).

I got part of the story right. I did recognize that everyone managed to slot the living trigonian into his system and that simple, single facts did not (at least in this case) undo general theories. But I got the end all wrong because the traditional, Huxleyan view still beguiled me. I told the happy ending because I read Jenkins's quote and took it at face value—as an evolutionary prediction fulfilled and an empirical vindication provided. I forgot (or hadn't yet learned) a cardinal rule of scholarly detection: Don't only weigh what you have; ask why you don't see what you ought to find. Negative evidence is important—especially when the record is sufficiently complete to indicate that an absence may be genuine.

I now read the Cenozoic discovery quite differently, because I have confronted what should have happened but didn't. If Darwin's vindication required a set of new, clean, pristine, unexpected facts, then why didn't the Cenozoic trigonian inspire a wave of rejoicing? Darwin had predicted the discovery; Agassiz had invested much hope in its nonexistence.

Sure, Jenkins said the right things in his article; I quoted them and regarded my task as complete. But the key to the story lies elsewhere—in the nonevents. Jenkins wrote a two-and-a-half-page note in a minor journal. No one else seemed to notice. Darwin never commented, though the *Origin of Species* still had several editions to run. *Trigonia* did not become a textbook example of evolution triumphant. Most curiously, Jenkins did not find the Cenozoic trigonian. It was unearthed by Frederick McCoy, an eminent leader of Australian science, the founder and head of the Museum of Natural History and Geology in Melbourne. He must have known what he had and what it meant. But he didn't even bother to publish his description. I should have taken my clue from the opening lines of Jenkin's paper, but I passed them by:

> The very interesting discovery of a species of *Trigonia* in the Tertiary deposits of Australia has in England remained entirely in the background, and I have been several times surprised at finding students of Tertiary paleontology, generally *au courant* with the progress of their special branch of science, unacquainted with the circumstance. Its importance, in a theoretical point of view, is beyond all question, hence the deep interest always exhibited by those to whom I have spoken on the subject.

I had, in short, succumbed to the view I was questioning. I had recognized that the original discovery of the living trigonian upended no theory, but I had let the Cenozoic fossil act as a Huxleyan nasty fact because Jenkins had so presented it. But when we consider what the Cenozoic trigonian did *not* provoke, we obtain a more general and consistent account of the entire affair. The living trigonian changed no theory, because it could fit (however uncomfortably) with all major views of life. The Cenozoic trigonian did not prove evolution either, because Agassiz's position of retreat was defensible (however embarrassing) and because evolution was too big a revolution to rely critically on any one datum. *Trigonia* didn't hurt, but a multitude of fish were frying, and one extra clam, however clean and pretty, didn't bring the meal to perfection.

In a case that has become a cliché in our language, Sherlock Holmes solved a mystery by noting no bark and inferring that no dog had been present. Nonevents matter, not only new and nasty facts. Which reminds me: I must have looked at a thousand periwinkles this morning. Still no lefties. Maybe someday.

Postscript

Circumstances demand that this essay receive the best, the most indisputable, of all conceivable epilogues—though it must occur at my expense. The factual correction of error may be the most sublime event in intellectual life, the ultimate sign of our necessary obedience to a larger reality and our inability to construct the world according to our desires. For science, in particular, factual correction holds a specially revered place for two reasons: first, because we define the enterprise as learning more and more about an external reality; second, because we know in our hearts that we can be as stubborn and resistant to change as petty bureaucrats and fundamentalist preachers—and undeniable factual correction therefore becomes a kind of salvation from our own emotional transgressions against a shared ideal. Nothing, therefore, can be quite so joyful, quite so appropriate for a finale to this book, as a "nasty, ugly little fact" that, Samsonlike, brings down a conceptual edifice.

As my example in the foregoing essay, I wrote: "Some beliefs may be subject to such instant, brutal, and unambiguous rejection. For example, no left-coiling periwinkle has even been found among millions of snails examined. If I happen to find one during my walk on Nobska beach tomorrow morning, a century of well-nurtured negative evidence will collapse in an instant." I then ended the essay by reporting my walk in a world of righties: "I must have looked at a thousand periwinkles this morning. Still no lefties. Maybe someday."

I expected no overturn. Occasional left-coiling specimens are common enough in right-coiling species, but their frequency varies, and many species have never yielded a single left-coiling individual. Since the periwinkle, *Littorina littorea*, is among the most common of all snails within our purview (as the standard shoreline mollusk of both western European and eastern American coasts), this form has become the classic example of an all right-coiled species (literally *dextrous* and *righteous*, as opposed to *gauche* and *sinister* left-handers). So the textbooks always say, and so did I report.

On the auspicious day of February 14, Solene Morris, then curator of mollusks at the British Museum, decided to check in her museum's incomparable collections. (Solene now holds the neatest job available in our bailiwick—for she is, as curator of Darwin's home at Downe, the closest thing to a grail keeper that our profession can muster.) She sent me a letter, circumscribed with a huge blood red heart, and labeled as the new St. Valentine's Day Massacre.

Museum drawers are the greatest sequesterers of unpublished and unacknowledged treasures. (I have written several essays on great discoveries made not directly in the field, but among forgotten and misclassified material in museum drawers—see Essay 16, page 240, in *The Flamingo's Smile,* or my entire book *Wonderful Life.*) Sure enough, one lefty periwinkle lay in a vial—and had so resided since 1937—in the back of drawer 3 among the collection of Littorinidae. Solene wrote to me (partly in verse I later realized, though she masked the poetry in a prose paragraph that I now disentangle):

> However, in the best tradition of my (incidentally left-handed) father, whenever it has been stated that such a thing cannot be found, it is our duty to find it. So, on the morning of the 14th of February, I searched . . . amongst the monstrosities and littorinids until . . . Eureka!
>
> What did I see
> Lodged in a vial in the back of drawer three,
> amongst the dextrally coiled Littorinidae . . .
> Purchased of a Mr. E. F. Smith of Acton, in 1937, for twenty-five shillings (old currency).
> I can almost hear you cry:
> "You can't be serious!"
> When the sinistral face of the fact you spy—
> *Littorina littorea* var. *johnmacenroei!*

I include Solene's picture of the specimen, lest any zealots, recalcitrants, or other species of doubters remain. There the specimen had resided, since 1937, unknown to all as textbooks continued to propagate their little falsehood.

A nasty, ugly little fact: A left-coiling periwinkle exists. *Courtesy of The Natural History Museum, London.*

How very lovely that my own point should be proved—at my expense to be sure—by the quick and unambiguous destruction of my own example.

One final aspect of this tale should give us further hope. Solene also sent me the correspondence between Mr. Smith of Acton and the British Museum. He knew what he had (I guess he had read the texts too), and he wasn't parting cheaply. He didn't quite demand recompense at the scale of Van Gogh's swirling flowers or that Honus Wagner baseball card, but twenty-five bob, in those days, could at least get you a fancy meal or two. He wrote making his offer. The Museum responded on March 22 stating that they needed permission from the trustees (talk about bureaucracies saddled with small items because they don't properly delegate authority!), but allowing that such should be forthcoming if Mr. Smith still wanted to sell. Mr. Smith responded— now get this, for here's the point of the tale—*on March 22,* stating that he would be happy to accept. This means, of course, that the London mails were then so efficient that a letter posted one morning could reach its destination and elicit a response *on the same afternoon!* (Of course, all readers of mystery stories, and anyone who lived in Britain—as I did for a year—before 1970, knows perfectly well that their postal service was once so quick and impeccable. In one of Dorothy Sayers's wonderful crime novels, Lord Peter Wimsey struggles for months to close a case and then absolutely must get his solution to London by the next morning, or all is lost. He calmly writes his brief as a letter and simply drops it in the nearest public postal box, absolutely confident that it will reach the right hands on time! Can you imagine doing such a thing in America today? But then, do not despair. Our postal service may not so deliver (without the added expense of ten bucks for Express Mail). Yet we now have those demons, those horrid ultimate invaders of all privacy—e-mail and the FAX machine. Who says that life and culture do not progress!

Bibliography

Adams, H. 1909. *The education of Henry Adams; an autobiography.* Boston: Houghton Mifflin Co.

Agassiz, L. 1833–1843. *Recherches sur les poissons fossiles.* Neuchâtel: Petitpierre.

Agassiz, L. 1840. *Mémoire sur les trigonies.* In *Etudes critiques sur les mollusques fossiles.* Neuchâtel: Petitpierre.

Alexander, R. McN. 1966. Physical aspects of swim bladder function. *Biological Reviews of the Cambridge Philosophical Society* 41: 141–76.

Allin, E. F. 1975. Evolution of the mammalian middle ear. *Journal of Morphology* 147:403–38.

Baier, J. J. 1708. *Oryktographia norica, sive rerum fossilium et ad minerale regnum pertinentium in territorio Norimbergensi.* Nuremburg: Wolfgang Michanel.

Balouet, J-C., and A. Alibert. 1990. *Extinct species of the world.* Translated by K. J. Hollyman. New York: Barron's.

Barlow, N. (ed.) 1933. *Charles Darwin's diary of the voyage of H.M.S. Beagle.* Cambridge, Eng.: Cambridge University Press.

Barr, J. 1985. Why the world was created in 4004 B.C.: Archbishop Ussher and biblical chronology. *Bulletin of the John Rylands University Library* 67:575–608.

Barrett, P. H. (ed.) 1974. *Darwin's early and unpublished notebooks.* New York: E. P. Dutton & Co. (with *Darwin on Man* by H. E. Gruber).

Barrington, D. 1770. Account of a very remarkable young musician. *Philosophical Transactions of the Royal Society of London* 60:4–64.

457

Basolo, A. 1990. Female preference predates the evolution of the sword in swordtail fish. *Science* 250:808–10.

Bateson, P. 1982. Preference for cousins in Japanese quail. *Nature* 295:236–37.

Bengtson, S. 1991. Oddballs from the Cambrian start to get even. *Nature* 351:184.

Borges, J. L. 1977. *The book of sand.* Translated by N. T. di Giovanni. New York: E. P. Dutton.

Buckland, W. 1836. *Geology and mineralogy considered with reference to natural theology.* London: Pickering.

Burnet, T. 1691. *Sacred theory of the earth.* London: R. Norton.

Burns, J. McL. 1975. *Biograffiti: A natural selection.* New York: Demeter Press.

Carlton, J. T., G. J. Vermeij, D. R. Lindberg, D. A. Carlton, and E. C. Dudley. 1991. The first historical extinction of a marine invertebrate in an ocean basin: The demise of the eelgrass limpet *Lottia alveus. Biological Bulletin* 180:72–80.

Chen J., Hou X., and Lu H. 1985. Early Cambrian netted scale-bearing worm-like sea animal. *Acta Paleontologica Sinica* 28:1–16.

Chomsky, N. 1969. *Syntactic structures.* The Hague: Mouton.

Clack, J. A. 1989. Discovery of the earliest known tetrapod stapes. *Nature* 342:425–27.

Clark, T. H., and C. W. Stearn. 1960. *The geological evolution of North America.* New York: Ronald Press.

Clarke, B., J. Murray, and M. Johnson. 1984. The extinction of endemic species by a program of biological control. *Pacific Science.* 38:97–104.

Clarke, B., J. Murray, and M. Johnson. 1988. The extinction of *Partula* on Moorea. *Pacific Science* 42:150–53.

Coates, M. I., and J. A. Clack. 1990. Polydactyly in the earliest known tetrapod limbs. *Nature* 347:66–69.

Colp, R. 1979. Charles Darwin's vision of organic nature. *New York State Journal of Medicine* 79:1622–29.

Conway Morris, S. 1977. A new metazoan from the Cambrian Burgess Shale, British Columbia. *Palaeontology* 20:623–40.

Conway Morris, S., and J. S. Peel. 1990. Articulated halkieriids from the lower Cambrian of north Greenland. *Nature* 345:-802–5.

Cooke, J. 1990. Proper names for early fingers. *Nature* 347:14–15.

Coppinger, R., J. Glendinning, E. Torop, C. Matthay, M. Sutherland, and C. Smith. 1987. Degree of behavioral neoteny differentiates canid polymorphs. *Ethology* 75:89–108.

Crampton, H. E. 1917. Studies on the variation, distribution and evolution of the genus *Partula*. The species inhabiting Tahiti. *Carnegie Institution of Washington Publications* 228:1–311.

Crampton, H. E. 1925. Studies on the variation, distribution and evolution of the genus *Partula*. The species of the Mariana Islands, Guam and Saipan. *Carnegie Institution of Washington Publications* 228a:1–116.

Crampton, H. E. 1932. Studies on the variation, distribution and evolution of the genus *Partula*. The species inhabiting Moorea. *Carnegie Institution of Washington Publications* 410:1–335.

Cubelli, R. 1991. A selective deficit for writing vowels in acquired dysgraphia. *Nature* 353:258–60.

Cuvier, G. 1812. *Discours préliminaire*, vol. 1 of *Recherches sur les ossemens fossiles des quadupèdes*. Paris: Deterville.

Darwin, C. 1840. On the formation of mould. *Transactions of the Geological Society of London* 5:505–9.

Darwin, C. 1845. *A naturalist's voyage. Journal of researches into the natural history and geology of the countries visited during the voyage of H.M.S. Beagle round the world, under the command of Capt. FitzRoy, R.N.* London: John Murray. (Voyage of the Beagle)

Darwin, C. 1859. *On the origin of species by means of natural selection.* London: John Murray.

Darwin, C. 1868. *The variation of animals and plants under domestication.* London: John Murray.

Darwin, C. 1871. *The descent of man and selection in relation to sex.* London: John Murray.

Darwin, C. 1872. *The expression of the emotions in man and animals.* London: John Murray.

Darwin, C. 1881. *The formation of vegetable mould, through the action of worms.* London: John Murray.

Dollo, L. 1892. Sur l'origine de la nageoire caudale des ichthyosaures. *Procès-Verbaux de la Société Belge de Géologie* 26 July.

Dollo, L. 1893. Les lois de l'évolution. *Bulletin de la Société Belge de Géologie, de Paléontologie et d'Hydrologie,* vol. 7.

Dubois, E. 1897. Sur la rapport du poids de l'encéphale avec la grandeur du corps chez les mammifères. *Bulletins de la Société d'Anthropologie de Paris* 8:337–76.

Dubois, E. 1928. The law of the necessary phylogenetic perfection of the psychoencephalon. *Proceedings of the Section of Sciences of the Koninklijke Akademie van Wetenschappen* 31:304–14.

Dubois, E. 1932. Early Man in Java. *Nature* 130:20.

Dubois, E. 1935. On the gibbon-like appearance of *Pithecanthropus erectus. Proceedings of the Section of Sciences of the Koninklijke Akademie van Wetenschappen* 38:578–85.

Eco, U. 1985. *The name of the rose.* Translated by William Weaver. San Diego: Harcourt, Brace, Jovanovich.

Eldredge, N., and S. J. Gould. 1972. Punctuated equilibria: An alternative to phyletic gradualism. In *Models of paleobiology,* ed. T. J. M. Schopf, 82–115. San Francisco: Freeman, Cooper and Co.

FitzRoy, R., and Darwin, C. 1836. A letter, containing remarks on the moral state of Tahiti, New Zealand &c. *South African Christian Recorder* 2(4):221–38.

Galton, F. 1869. *Hereditary genius.* London: MacMillan.

Galton, F. 1889. *Natural inheritance.* London: MacMillan.

Galton, F. 1909. *Memories of my life.* London: Methuen & Co.

Geoffroy Saint-Hilaire, E. 1831. *Principes de philosophie zoologique, discutés en mars 1830, au sein de l'Académie Royale des Sciences.* Paris.

Geoffroy Saint-Hilaire, I. 1838. Sur les travaux zoologiques et anatomiques de Goethe. *Comptes rendus des séances de l'Académie des Sciences de Paris* 6:320–21.

Goethe, J. W. von. 1790. *Versuch die Metamorphose der Pflanzen zu erklären.* Gotha: Ettingersch.

Goethe, J. W. von. 1831. Reflexions de Goethe sur les débats scientifiques de mars 1830 dans le sein de l'Académie des Sciences. *Annales des Sciences Naturelles* 22:179–88.

Goethe, J. W. von. 1832. Derniers pages de Goethe expliquant à l'Allemagne les sujets de philosophie naturelle controversées au sein de l'Academie des Sciences de Paris. *Revue Encyclopédique* 53:563–73 and 54:54–68.

Gould, S. J. 1969. An evolutionary microcosm: Pleistocene and Recent history of the land snail *P. (Poecilozonites)* in Bermuda. *Bulletin of the Museum of Comparative Zoology* 138(7):407–532. (Ph.D. diss., Columbia University).

Gould, S. J. 1977. *Ontogeny and phylogeny.* Cambridge, Mass.: The Belknap Press of Harvard University Press.

Gould, S. J. 1985. *The flamingo's smile.* New York: W. W. Norton & Company.

Gould, S. J. 1987. *Time's arrow, time's cycle.* Cambridge, Mass.: Harvard University Press.

Gould, S. J. 1989. *Wonderful life.* New York: W. W. Norton & Company.

Gould, S. J. 1991. *Bully for Brontosaurus.* New York: W. W. Norton & Company.

Gould, S. J., and N. Eldredge. 1977. Punctuated equilibria: The tempo and mode of evolution reconsidered. *Paleobiology* 3:115–51.

Griffis, K., and D. J. Chapman. 1988. Survival of phytoplankton under prolonged darkness: Implications for the Cretaceous-Tertiary boundary darkness hypothesis. *Palaeogeography, Palaeoclimatology, Palaeoecology* 67:305–14.

Gulick, J. T. 1905. Evolution, racial and habitudinal. *Carnegie Institution of Washington Publications* 25:1–269.

Haim, A., G. Heth, H. Pratt, and E. Nevo. 1983. Photoperiodic effects on the thermoregulation in a 'blind' subterranean mammal. *Journal of Experimental Biology* 107:59–64.

Halley, E. 1714–1716. A short account of the cause of the saltiness of the ocean, and of the several lakes that emit on rivers; with a proposal, by help thereof, to discover the age of the world. *Philosophical Transactions of the Royal Society of London* 29: 296–300.

Hawkins, T. 1834. *Memoirs of the Ichthyosauri and Plesiosauri.* London: William Pickering.

Hawkins, T. 1840. *Book of the great sea-dragons.* London: William Pickering.

Heaton, T. H. 1988. Patterns of evolution in *Ischyromys* and *Titanotheriomys* (Rodentia: Ischyromiidae) from Oligocene deposits of western North America. Ph.D. diss., Harvard University.

Hendriks, W., J. Leunissen, E. Nevo, H. Bloemendal, W. W. de-Jong. 1987. The lens protein α-A-crystallin of the blind mole rat *Spalax ehrenbergi*: Evolutionary change and functional constraints. *Proceedings of the National Academy of Sciences* 84:5320–24.

Hinchliffe, J. R. 1989. Reconstructing the archetype: Innovation and conservatism in the evolution and development of the pentadactyl limb. In *Complex organismal functions: Integration and evolution in vertebrates,* ed. D. B. Wake and G. Roth, 171–90. New York: John Wiley & Sons.

Hutchinson, G. E. 1931. Restudy of some Burgess Shale fossils. *Proceedings of the United States National Museum* 78(11):565–87.

Jarvik, E. 1980. *Basic structure and evolution in vertebrates.* New York: Academic Press.

Jastrow, R. 1981. *The enchanted loom: The mind in the universe.* New York: Simon and Schuster.

Jenkins, H. M. 1865. On the occurrence of a Tertiary species of *Trigonia* in Australia. *Quarterly Journal of Science* 2:363–64.

Kimura, M. 1968. Evolutionary rate at the molecular level. *Nature* 217:624–26.

Kimura, M. 1982. The neutral theory as a basis for understanding the mechanism of evolution and variation at the molecular level. In *Molecular evolution, protein polymorphism and the neutral theory,* ed. M. Kimura, 3–56. Berlin: Springer-Verlag.

Kimura, M. 1983. *The neutral theory of molecular evolution.* Cambridge, Eng.: Cambridge University Press.

Kitchell, J. A., D. L. Clark, and A. M. Gombos, Jr. 1986. Biological selectivity of extinction: A link between background and mass-extinction. *Palaios* 1:504–11.

Krafft-Ebing, R. von. 1984. *Psychopathia sexualis.* Munich: Matthes & Seitz.

Lamarck, J. P. B. 1802. *Recherches sur l'organisation des corps vivans.* Paris: Maillard.

Lamarck, J. P. B. 1804. Sur un nouvelle espèce de Trigonie et sur une nouvelle d'Huître découvertes dans le voyage du capitaine Baudin. *Annales du Muséum d'Histoire Naturelle* 4:353–54.

Lamarck, J. P. B. 1809. *Philosophie zoologique.* Paris: Dentu.

Leakey, R., and M. Leakey. 1986. A new Miocene hominoid from Kenya. *Nature* 324:143–46.

Leakey, R., and M. Leakey. 1986. A second new Miocene hominoid from Kenya. *Nature* 324:146–48.

Liem, K. F. 1988. Form and function of lungs: The evolution of air breathing mechanisms. *American Zoologist* 28:739–59.

Loftus, E. F. 1979. *Eyewitness testimony.* Cambridge, Mass.: Harvard University Press.

Lyell, C. 1832. *Principles of geology, being an attempt to explain the former changes of the earth's surface by reference to causes now in operation.* vol. 2. London: John Murray.

Malthus, T. R. 1803. *An essay on the principles of population.* London: J. Johnson.

McGowan, C. 1989. The ichthyosaurian tailbend: A verification problem facilitated by computer tomography. *Paleobiology* 15: 429–36.

McGowan, C. 1989. *Leptopterygius tenuirostris* and other long-snouted ichthyosaurs from the English Lower Lias. *Palaeontology* 32:409–27.

Mivart, St. G. 1871. *On the genesis of species.* London: MacMillan.

Monod, J. 1971. *Chance and necessity; an essay on the natural philosophy of modern biology.* New York: Knopf.

Murray, J., and B. Clarke. 1980. The genus *Partula* on Moorea: Speciation in progress. *Proceedings of the Royal Society of London* 211:83–117.

Owen, R. 1840. Note on the dislocation of the tail at a certain point observable in the skeleton of many ichthyosauri. *Transactions of the Geological Society of London* 5:511–14.

Owen, R. 1849. *On the nature of limbs.* London: J. Van Voorst.

Owen, R. 1861–1881. *Monograph on the fossil Reptilia of the Liassic Formation.* London: Paleontographical Society.

Paley, W. 1802. *Natural theology: or, evidences of the existence and attributes of the deity, collected from the appearances of nature.* London: E. Paulder.

Parkinson, J. 1811. *Organic remains of a former world, examination of the mineralized remains of the vegetables and animals of the antediluvian world generally termed extraneous fossils.* London: Sherwood, Neely and Jones.

Patten, W. 1912. *The evolution of vertebrates and their kin.* Philadelphia: P. Blakiston's Son.

Patten, W. 1920. *The grand strategy of evolution.* Boston: Gorham Press.

Prothero, D. R., and N. Shubin. 1983. Tempo and mode of speciation in Oligocene mammals. *Geological Society of America Abstracts with Programs* 15(6):665.

Quoy, J., and J. Gaimard. 1834. *Voyage de découvertes de l'Astrolabe exécuté par ordre du Roi pendant les années 1826–29; Zoologie,* vol. 3. Paris: J. Tatsu.

Ramsköld, L., and Hou X. 1991. New early Cambrian animal and onychophoran affinities of enigmatic metazoans. *Nature* 351: 225–28.

Raup, D. M. 1979. The size of the Permo-Triassic bottleneck and its evolutionary implications. *Science* 206:217–18.

Ray, J. 1691. *The wisdom of God manifested in works of the creation.* London: William Innys.

Reichert, C. B. 1837. Entwicklungsgeschichte der Gehörknöchelchen der sogenannte Meckelsche Fortsatz des Hammers. *Müller's Archiv für Anatomie, Physiologie und wissenschaftliche Medicin,* 178–88.

Ryan, M. J., and A. S. Rand. 1990. The sensory basis of sexual selection for complex calls in the Tungara frog *Physalaemus pustulosus* (sexual selection for sensory exploitation). *Evolution* 44:305–14.

Ryan, M. J., J. H. Fox, W. Wilczynski, and A. S. Rand. 1990. Sexual selection for sensory exploitation in the frog *Physalaemus pustulosus. Nature* 343:66–67.

Sacks, O. 1985. *The man who mistook his wife for a hat.* New York: Summit Books.

Sagan, C., and A. Druyan. 1985. *Comet.* New York: Random House.

Scheuchzer, J. 1708. *Piscium querelae et vindiciae.* Zurich: Gessner.

Shubin, N. H., and P. Alberch. 1986. A morphogenetic approach to the origin and basic organization of the tetrapod limb. In *Evolutionary Biology,* vol. 20, ed. M. K. Hecht, B. Wallace, and G. T. Prance, 319–87. New York: Plenum Publications.

Smith, A. 1776. *The wealth of nations.* London: Strahan and Cadell.

Snell, O. 1891. Die Abhängigkeit des Hirngewichtes von dem Körpergewicht und den geistigen Fähigkeiten. *Archiv für Psychiatrie und NervKrankheiten* 23:436–46.

Stauffer, R. C. (ed.) 1975. *Charles Darwin's Natural Selection; being the second part of his big species book written from 1856 to 1858.* London: Cambridge University Press.

Stensiö, E. A. 1927. *Downtonian and Devonian vertebrates of Spitsbergen, part 1, family Cephalaspidae.* Oslo.

Theunissen, B. 1989. *Eugène Dubois and the ape-man from Java; the history of the first 'missing link' and its discoverer.* Boston: Kluwer Academic Publishers.

Thompson, D. W. 1917. *On growth and form.* Cambridge, Eng.: Cambridge University Press.

Tinbergen, N. 1960. *The herring gull's world.* New York: Lyons & Burford.

Ussher, J. 1645. *A body of divinity: or, the sum and substance of Christian religion.* Dublin: John Downham.

Ussher, J. 1650. *Annales veteris testamenti, a prima mundi origine deducti.* Vols. 8–11 in the *Whole works* of Ussher, ed. C. R. Elrington and J. H. Todd. Dublin. 17 vols. (1847–64).

Ussher, J. 1653. *The annals of the world. Deduced from the origin of time, and continued to the beginning of the emperour Vespasians reign, and the totall destruction and abolition of the temple and commonwealth of the Jews.* London: E. Tyler.

Vermeij, G. J. 1986. The biology of human-caused extinction. In *The Preservation of Species,* ed. B. G. Norton, 28–49. Princeton: Princeton University Press.

Vermeij, G. J. 1987. *Evolution and escalation.* Princeton: Princeton University Press.

Walcott, C. D. 1911. Middle Cambrian Merostomata. Cambrian Geology and Paleontology, II. *Smithsonian Miscellaneous Collections* 57:17–40.

Walcott, C. D. 1911. Middle Cambrian holothurians and medusae. Cambrian Geology and Paleontology, II. *Smithsonian Miscellaneous Collections* 57:41–68.

Walcott, C. D. 1911. Middle Cambrian annelids. Cambrian Geology and Paleontology, II. *Smithsonian Miscellaneous Collections* 57:109–44.

Walcott, C. D. 1912. Middle Cambrian Branchiopoda, Malacostaca, Trilobita and Merostomata. Cambrian Geology and Paleontology, II. *Smithsonian Miscellaneous Collections* 57:145–228.

Wallace, A. R. 1889. *Darwinism, an exposition of the theory of natural selection with some of its applications.* London: MacMillan.

Wayne, R. K. 1986. Cranial morphology of domestic and wild canids: The influence of development on morphological change. *Evolution* 40:243–61.

Whitman, C. O. 1904. The problem of the origin of species. *Congress of Arts and Science, Universal Exposition, St. Louis,* Vol. 5.

Whitman, C. O. 1919. *Orthogenetic evolution in pigeons.* Edited by Oscar Riddle. *Carnegie Institution of Washington Publications* 257.

Index

Acanthostega
  digits on, 66, 67, *68, 75,* 76
  stapes bones of, 101
Achatina (tree snail), 36–39
Adams, Henry, 56
adaptation, 144
  Darwin on, 315
  in evolution of land snails, 30
  by ichthyosaurs, 91
  Lamarckian and Darwinian models
    of, 27
  in number of vertebrate digits, 75
  variations favoring, 362
adaptationism, 369, 373
Afropithecus, 294
Agassiz, Louis, 359, 409, 429
  on *Cephalaspis*, 413–16, 425, 435
  on early vertebrates, 430–32, *431,*
    436–37
  on *Trigonia*, 442, 447–49, 451
agnathans (jawless vertebrates), 99,
  411
Alberch, Pere, 69–74
Albert (prince consort, England), 371
Alexander (the Great; king, Macedon),
  188*n*, 340
Alexander, R. McNeill, 118
Alexandria (Egypt), 340–41
Allin, Edgar F., 106, 107
allometry, 387
Alvarez, Luis, 306–7
Alvarez, Walter, 306–7

Amana Colonies (Iowa), 207–10
Amana Refrigeration, Inc., 208–9
Amana Society, 208
ammonites, 304, 312
Amniota, 68, 69
amphibians
  digits on, 67–69, 75
  frog calls, 376–79
Amphioxus, 336
Amrou Ibn el-Ass, 340
angiosperms (flowering plants), *Zostera
  marina* (eelgrass), 57–60
annelids, 344, 345, 351
Anning, Mary, 83
anti-Semitism, 269
anvil (incus; bone), 102, 103, 106
apes
  error of humans having evolved
    from, 284–87
  evolution of monkeys and, 287–95
archaeocyathids, 330
Archaeopteryx (first bird), 276, 409
archeology, 276
archetypes, 63–64
Aristotle, 193, 304, 342, 352
arthropods, 344, 345, 410, 417–18,
  420, 421, 433
articulates, 415
Asaro, Frank, 306
Atdabanian fauna, 329
Attucks, Crispus, 60
Audubon, John James, 53–54

Australopithecus, 321
authenticity, 239–40
Aysheaia, 346, 347, 350

Bacon, Francis, 153
bacteria, 98, 323
  oldest fossil evidence of, 328
Baier, J.J., 83
Balistes (triggerfish), 119
Barr, James, 188, 190, 192
Barratier, John, 260
Barrington, Daines, 19, 249–53,
    260–61
Basket, Fuegia, 271
Basolo, Alexandra, 375–76, 380
Bateson, Patrick, 379–80
Beagle (ship), 267
  in Tahiti, 263–65
  in Tierra del Fuego, 270–72
Bede (the Venerable), 187, 190
behavior of domesticated animals,
    385–86
  of herding and guarding dogs, 394
  neoteny in, 392
Bengtson, Stefan, 350
Berlioz, Louis Hector, 95–96
Bermuda, 37
Berra, Yogi, 238
Bible
  assumption of inerrancy of, 186–87
  Ussher's chronology based on,
    187–92
  Ussher's chronology printed in, 184
Big Badlands (South Dakota), 277,
    278
birds
  Archaeopteryx (first bird), 276, 409
  cuckoos, 258–59
  dodos, 428
  English fantail pigeon, *361*
  evolution of wings of, 372–73
  Galápagos finches, 25–26
  herring gulls, 257–58
  passenger pigeons, 52–54
  peacocks, 374
  pigeons, 382
  pigeons, Whitman on, 356–68

  quail, 379–80
  reproduction among, Paley on,
    138–40
  spotted owls, 44
Birker, Ingrid, 229–36
blacks, Darwin on, 266–67, 272–74
Bligh, William, 24
Bloemendal, H., 404
Bochte, Bruce, 11
Borges, Jorge Luis, 177
Boston Massacre (1770), 60
brain
  of *Cephalaspis*, 420, *423*
  Dubois's study of size of, 130–36
  human, evolution of, 255, 321
  human, modularity in, 259–60
Brazil, Darwin on slavery in, 272–74
Brule Formation (South Dakota), 277
Buckland, William, 83, 90, 94
Buckner, Bill, 78
Bulimulidae (land snail), 27
Burch, Jack, 38*n*
Burgess Shale, 224–29, 232–36, 329,
    334, 338, 345–46, 351
  Hallucigenia of, 347–50, *348*, 352
Burnet, Thomas, 176
Burns, John, 108
butterflies, 373
Button, Jemmy, 271

calendars, Julian and Gregorian, 191
Cambrian explosion, 110, 224, 329,
    338
canids, 387–88, 391
Canis lupus (wolf), 386
Carlton, James T., 56, 58
Carnegie, Andrew, 224
cats, 389
  skulls of, *390*
Cenobita diogenes (crab), 55–56
Cenozoic gap, 442, 443, 445–48
Cephalaspis (jawless fish), 17, 18,
    410–11, *414*
  Agassiz on, 413–16
  brain and nervous system of, *423*
  Patten on, 416–22
  portrayals of, 429–38

Stensiö's reconstruction of, 422–25,
    423
cercopithecoid (Old World) monkeys,
    287–88
Cerion (land snail), 27
Chapman, David J., 309, 311
Chaucer, Geoffrey, 41
Chautauqua (New York), 206
Chengjiang fauna, 333, 346–47, 350
Chen Jun-yuan, 333
chickens, 383
Childs, family of Abijah and Sarah,
    210–13, *213, 214*
chimpanzees, common genes of
    humans and, 290
Chomsky, Noam, 321
Christian, Fletcher, 24, 265
Christianity
    spread by missionaries, 263–65
    see also religion
Cittarium pica (whelk), 56
Civil War (U.S.), 77–78
Clack, J.A., 67, 72, 75, 101–2
clams, *Trigonia*, 441–51, *443, 444*
Clark, David L., 309–10
Clark, T.H., 227–36, *230, 231, 233,
    237*
Clarke, Bryan, 35, 37, 38
Coates, M.I., 67, 72, 75
Cobb, Ty, 357
coelacanths, 114
Colbert, Ned, 223, 262
Coleridge, Samuel Taylor, 326
color, of pigeons, 362–65, *363, 365*
Colp, Ralph, 301
Columba livia (rock pigeon), 358, 362
comets
    Halley's, 168, 179–80
    impacts with earth of, 308
consciousness, evolution of, 320–24
conservation movement, 50
contingency
    in human history, 77–78
    in number of vertebrate digits,
        75–77
convergence, of ichthyosaurs and fish,
    81–83, 90–94

Conway Morris, Simon, 334, 336,
    347–49
Conybeare, W., 83
Cook, James, 23, 265
Cooke, Jonathan, 76–77
Coolidge, Calvin, 224
Cope, E.D., 416
Copeland, Michael D., 45–47, 49
Coppinger, Raymond, 394
coral, 304
cotyledons (of plants), 161–63
crabs
    Cenobita diogenes, 55–56
    evolution of stronger claws in,
        304
Crampton, Henry Edward, 13–14,
    29–35, 39–40
creationism, 176, 201
    Agassiz's, 414–16, 430, 432
    on bones of inner ear, 96–97
    functionality of organs in, 117–18
    Trigonia and, 446–49
    Ussher's setting date of creation in,
        181–93
creativity, 97–98
Cretaceous mass extinction, 49,
    305–11, 316
crinoids, 304, 312
Cro-Magnons, 126
Cromwell, Oliver, 183
cuckoos, 258–59
Cuvier, Georges, 131, 155, 415,
    445
    on evolution, 254–55
cynodont therapsids, 106
cypriniforms (fish), 119

Darwin, Charles, *141*, 216–17,
    222–23, 277
    on archetypes, 63–64
    calling card of, 221, *222*
    death of daughter of, 209
    on domestication of animals,
        382–83
    Dubois and, 133
    on earthworms, 79
    errors in work of, 109–11

Darwin (continued)
  on evolution of partial wings,
    372–73
  on facial gestures, 255–57
  on false ideas, 164
  first published work of, 262–63
  on Fuegians, 269–72
  on Galápagos finches, 26
  on Galápagos tortoises, 52
  on limits to inheritance, 384
  on lungs and swim bladders,
    111–17
  on mass extinctions, 305–6
  metaphors used by, 300–302
  on modularity, 259
  on "moral state" of Tahiti, 263–65
  on multiple functions of organs,
    119
  on natural selection, 146–47, 303–4,
    315
  on non-Western peoples, 265–66
  on origins of features, 373–74
  Owen and, 80
  Paley and, 140
  on pigeons, 357–59
  on progress, 302–3
  on race, 266–67
  on race and sex, 19
  on race and slavery, 272–74
  on redundancy in evolution, 120
  on scope of time, 179
  on sexual differences, 267–68
  on source of evolutionary change,
    397–98
  sources of inspiration for, 148–50
  on struggle in nature, 323–24
  on Tahitian women, 23–24
  on Trigonia, 445–46
  on vox populi, 15
  Whitman's attacks on, 359–69
Darwinism, 146
  Lamarckian evolution versus,
    216
  randomness in, 397–98
death, human consciousness of,
  322
Deere, John, 209
de Jong, W.W., 404

de la Beche, Henry, 83, 84
Delhi (India), 313, 315
De Smet (South Dakota), 220, 221
Devil's Tower (Wyoming), 203–4
DeVries, Hugo, 133
Diarthrognathus, 106
diatoms, 308–11, 316
digits, on pentadactyl limbs, 64–77
digs (archeological), 276
DiMaggio, Joe, 11
diners, 244–45
dinosaurs, 304
  mass extinction of, 308
disease, germ theory of, 215
dissociability, 255
DNA (deoxyribonucleic acid)
  molecular clock of evolution in,
    398–99
  neutral theory and, 402–3
dodos, 428
dogs
  domestication and breeding of,
    385–95
  skulls of, 390, 393
Dollo, Louis, 91–94
Dollo's Law, 91–92
domestication of animals, 382–83,
    385–86
  of dogs, 386–95
  of pigeons, 356–68
  skulls of animals, 393
Donne, John, 262
Druyan, Ann, 169
Dubois, Eugène, 19, 126–37, 409

earth
  age of life on, 327–28
  extinctions caused by impact with,
    306–8
  Halley on age of, 168–79
  Ussher on age of, 181–93
earthquakes, San Francisco (1989),
    245–46
earthworms, 79
Eco, Umberto, 42
Ediacara fauna, 329
Edison, Thomas, 170
eelgrass (Zostera marina), 57–60

Eldredge, Niles, 277
Elijah (biblical), 245
embryological development
    Agassiz on, 415
    of bones of inner ear, 103–5
    of mammalian brains, 134
    ontogeny in, 367
    of vertebrate limbs, 69–73
emotions, evolution of, 255–57
English fantail pigeon, *361*
environment
    in evolution of land snails, 28–30
    impact of humans on, 13
    in Lamarckian and Darwinian
        models of evolution, 27
    during mass extinctions, 306
    in natural selection, 362
environmentalism, 50
    role of humans in, 48–59
    species supported by, 44
Eric the Red, 326
Ericson, Leif, 326–27
eternity, 176–77, 193
Euglandina (land snail), 37–39
eukaryotes, 319–20
Eurypterus fischeri, 409–10, *410*
evolution
    analogies of human cultural change
        with, 241–42
    of bones of inner ear of mammals,
        98–108
    competing theories of, 356–57
    of complex cells, 319–20
    Cuvier on, 254–55
    Dubois on, 133
    of hearing, 96–108
    of human consciousness, 320–24
    of humans, 294–95
    impact of environment on, 27
    impact of mass extinction on,
        311–12
    as interaction of selection and
        constraints, 385
    irreversibility in, 91–92
    islands as laboratories of, 25–26
    issue of "progress" in, 302–5
    lack of direct human observation of,
        201

ladder and bush metaphors for,
    284–87
Lamarckian, 216, 443–45
of mammalian brains, 134–36
mass extinctions and, 316–17
molecular clock of, 398–99
of monkeys and apes, 287–95
multiple copies of genes in, 318–19
neutral theory in, 399–403
by oddities and imperfections, 383
origins of features in, 373–74
Paley on, 144–47
Patten on, 418–21
of pentadactyl limbs, 65, 72, 74–76
of pigeons, 357–68
punctuated equilibrium in, 277–79
randomness in, 397–98
redundancy in, 120
of vertebrate lungs, 111–17
wedge metaphor for, 301–2
of wings, 372–73
exons, 402–3
extinctions
    inevitability of, 46
    Lamarck on, 55, 445
    of *Lottia alveus* (limpet), 56–60, *57*
    mass extinctions, 46–49, 305–12,
        316–18
    of *Partula,* 38–39
    of passenger pigeon, 54
    present trend of, 13
    of subspecies, 43
eyes, of mole rats *(Spalax ehrenbergi),*
    403–6
eyewitness accounts, 201–2

females (non-human)
    sexual selection by, 374–79
    see also women
finches, Galápagos, 25–26
fingerprints, 439
Fischer-Dieskau, D., 95
fish
    Agassiz on evolution of, *431*
    Cephalaspis (jawless fish), 410–11,
        413–25, *414, 423,* 429–38
    convergence of ichthyosaurs and,
        81–83, 90–94

fish *(continued)*
  evolution of lungs and swim
    bladders of, 111–19
  sense of hearing in, 96, 99
  swordfish *(Xiphias gladius)*, 374
  swordtail *(Xiphophorus helleri)*,
    375–76
FitzRoy, Robert, 19
  Fuegians and, 270–72
  on "moral state" of Tahiti, 263–65
Foerstner, George C., 208
foot (unit), 138
fossil record
  Cenozoic gap in, 442, 443, 445–48
  punctuated equilibrium in, 277–79
fossils
  abundance of, 275–77
  amphibians and reptiles, 67–69
  of Burgess Shale, 224–29, 232–36
  of *Cephalaspis* (jawless fish), 18, 413,
    414
  Cuvier's "correlation of parts" and,
    254
  of *Eurypterus fischeri*, 409–10, *410*
  evolution of ear bones
    demonstrated by, 105–6
  of *Halkieria*, 334–36, *335*
  of *Hallucigenia*, 347–50, *348*, 352
  of humans, earliest, 126–37
  of ichthyosaurs, 83–94, *85*, *87*
  of *Ischyromys*, 278–79
  of *Microdictyon*, 332–34, *332*, *333*,
    346–47
  of Miocene apes, 289–94
  oldest, 328
  of onychophorans, 345
  of SSF (small, shelly fauna), 330–32
  of *Trigonia* (clam), 441, 443–45, *444*
fraud, in science, 109
Freud, Sigmund, 322
frogs
  calls of, 376–79
  digits on, 74
Fuegians, 267, 269–72

Gadsden Purchase, 42
Gaimard, J., 441–42
Galápagos finches, 25–26

Galápagos Islands, 25–27, 52
Galápagos tortoises, 52, 53
Galton, Sir Francis
  on inheritance, 382
  polyhedron metaphor used by, 368,
    384–87
  on Spencer, 439–40
Garland, William, 213
genes
  amount of variation in, 399
  exons and introns, 402–3
  pseudogenes, 403
genetic engineering, 383
genetics
  flexibility in, 318–19
  Galton on, 368
Geoffroy Saint-Hilaire, Etienne, 155,
  158, 159, 417
Geoffroy Saint-Hilaire, Isidore,
  155–56
geology
  Halley's work in, 168–79
  oldest rocks in, 328
  Ussher's chronology criticized in
    texts in, 184–88
germ theory, 215
gestures, human, 255–57
Gettysburg, battle of, 77–78
giant pandas, 44
Gideon Society, 184
Gilbert, Sir William Schwenck, 94,
  206
God
  creation of universe by, 191–93
  invisible hand versus, 149, 151
  Paley's "watchmaker" argument for,
    138–44
  Walcott on, 227
Goethe, Johann Wolfgang von, 19,
  154–65, *154*, 427
Goethe's bone (premaxillary), 157–58
Gombos, Andrew M., Jr., 309–10
Gosse, Philip Henry, 18
Gould, Ethan, 25
Gould, Leonard, 124–25
gradualism, 110
Greenwood, J.M., 198
Gregorian calendar, 191

Gregory XIII (pope), 191
Gregory, W.K., 71, 223
Griffis, Kathy, 309, 311
Guilding, Lansdown, 342
Gulick, John T., 27–30
gulls, 257–58

hagfish *(Myxine)*, 411, 424, 436
Haim, A., 405
Halkieria, 334–41, *335*
Halley, Edmund, 19, 168–79
Halley's comet, 168, 179–80
Hallucigenia, 347–50, *348*, 352
hammer (malleus; bone), 102, 103,
   106
Hammerstein, Oscar, 24
Handel, George Frederick, 260
harmonics, 377
Hawkins, Thomas, 83, 84
hearing
   evolution of, 96–108
   by frogs, 377–79
   swim bladders as organ of, 119
Heaton, Tim, 278–79
Hendriks, J., 404, 405
Henry V (king, England), 201
Herder, J.G., 160
Herod (king, Judea), 189–90
herring gulls, 257–58
Heth, G., 405
Hinchliffe, J.R., 66–67, 73
history (human), generalizing from,
   279–80
Holmes, Oliver Wendell, 304
Holzmaden deposits, 86, 87, *87*, 93
Homo erectus, 19, 127–30, 132, 409
   brain size of, 134–36
Hooker, Joseph, 146
Hoover, Herbert, 226
horses, 286, 390
Hou Xian-guang, 333, 347, 349–50
humans
   arms and hands of, 65
   common genes of chimpanzees and,
      290
   evolution of, 294–95
   evolution of consciousness in, 318,
      320–24

as evolved from apes, 284–86
facial gestures of, 255–57
first fossils of, 126–37
impact on environment of, 13
"nature" of, 280–83
races among, Darwin on, 266–67
role in environment of, 48–59
toes of, 75
hummingbirds, 64
Hutchinson, G. Evelyn, 227, 346
Huxley, Thomas Henry, 152, 223,
   416, 440
   death of child of, 209
   on swim bladders, 119
   on vertebrate digits, 71

ichthyosaurs, 80–94, *82*, *87*
   tail bends of, *85*, *89*
Ichthyostega, *66*, 101
   digits on, 66, 67, *68*, 75, 76, 78
inbreeding depression, 379
incest, 379–80
inch (unit), 138
India, 313
"indicator" (umbrella) species, 44
infinity, 177
inheritance, Galton on, 368, 382, 384
insects, butterflies, 373
instinct, Paley on, 139
interspecific scaling, 387, 388
introns, 402–3
invisible hand metaphor, 149–51
Ischyromys, 278–79
islands, as laboratories of evolution,
   25–26
isolation, in evolution of land snails,
   30
Ives, Charles, 168

James, William, 18
Jarvik, Erik, 64–66
Jastrow, Robert, 287
Java Man *(Homo erectus)*, 126–30, 132,
   134–36, 409
jaws, evolution of, 99–100, *100*
   of ear bones from, 100–105
Jefferson, Thomas, 168, 272
Jenkins, H.M., 449–51

Jesus Christ, 190, 236
Johnson, Mike, 35, 37, 38
Joly, John, 173, 178
Joplin, Scott, 355
Julian calendar, 191
jumping genes (transposons), 319

Kant, Immanuel, 308
Keith, Arthur, 137
Kemp, T.S., 98
Kepler, Johannes, 190
Khayyam, Omar, 49
Kimeu, Kamoya, 291–94, *292*
Kimura, Motoo, 398–402
Kingsley, Charles, 152
Kitchell, Jennifer A., 309–11
Kotzebue, Otto von, 263
Krafft-Ebing, Richard von, 145
Kubla Khan (khan of Mongols), 326
Kuehneotherium (early mammal), 106

Labarbera, Mike, 347
Labyrinthula (slime mold), 59, 60
"ladder of life" model, 67
La Guardia, Fiorello, 109
Lamarck, Jean Baptiste, 55, 56, 144,
    215, 216
    on *Trigonia*, 442–45, *443*
Lamarckian evolution, 356, 443–45
    Darwinian versus, 216
    Paley on, 144–45, 147
lampreys *(Petromyzon)*, 411, 424
land snails, 26–29
    Achatina, 36–39
    Euglandina, 37–39
    Partula, 13–14, 29–40, *33*
    Poecilozonites, 37
language, 321, 328
    Romance languages, 337–38
Lankester, E. Ray, 223, 416
Larsen, Don, 240
La Shana brothers, 213
Laughton, Charles, 24
Leakey, Louis, 291
Leakey, Meave, 293, 294
Leakey, Richard, 290–94
leaves (on plant), 159–64
Lebedev, O.A., 66

Lee, Robert E., 77–88
Leonardo da Vinci, 153
Leunissen, J., 404
Liem, Karel F., 114
life on earth
    age of, 327–28
    first multicellular, 329
Lillie, F.R., 359
limpets, 56–60
Lincoln, Abraham, 272
Linnaeus, Carolus, 183
litotes, 249–50
Littorina littorea (periwinkle), 452–53,
    454
Loftus, Elizabeth, 201–2
Lorenz, Konrad, 395
Lottia alveus (limpet), 56–60, *57*
Louis IX (saint; king, France), 355
Lubbock, John, 140
Lu Hao-zhi, 333
lungfishes, 114, 117
lungs
    evolution of, 111–17
Luolishania, 347, 350
Lyell, Sir Charles, 386

McCoy, Frederick, 451
McGowan, Chris, 87–90
Macroperipatus, 343
Mahler, Gustav, 209
males (non-human)
    fish, swords on tails of, 375–76
    frogs, calls of, 376–79
    "showy" organs of, 374
Malthus, Thomas, 148, 150, 301–2
mammals, 304
    bones of inner ear of, 98–108, *100*,
        104
    brain size in, 131–36
    fossil record of, 278, 290–94
    mole rats *(Spalax ehrenbergi)*, 403–6
    pentadactyl limbs of, 65
    see also vertebrates
marine invertebrates
    extinctions of, 55–56
    Lottia alveus (limpet), 56–60, *57*
    Trigonia (clam), 441–51, *443*, *444*
Marshall, Thurgood, 272

Martin, Mary, 23, 38, 40
mass extinctions, 46–49, 305–12,
    316–18
McDonald's (restaurants), 244
Meckel, J.F., 103
medicine (discipline), 215
memory, errors of, 202–4
Mendel, Gregor, 30
Mesozoic era, 441
meter (unit), 138
Micah (biblical), 295
Michel, Helen, 306
Michelangelo, 19, 153, 257
Microdictyon, 332–34, *332*, 336–41,
    346–47, 350
Microdictyon sinicum, *333*
Milton, John, 184, 242
mimicry, 373
Minster, York, 271
Miocene era, 287–94
mitochondria, 320
Mivart, St. George, 385
modularity, principle of, 254–55
    in animals, 257–59
    in humans, 259–60
mole rats (*Spalax ehrenbergi*), 403–6
mollusks, 415
monkeys, Old World (cercopithecoid),
    287–88
Monod, Jacques, 398
Moorea (French Polynesia), 13, 14,
    24–25, *25*
    land snails of, 27, *28*, 29, 31–39
Morganucodon (early mammal), 106
Morris, Solene, 453, 454
mosaic evolution, 255
Mount Graham Red Squirrel
    (*Tamiasciurus hudsonicus
    grahamensis*), 43–48
Mozart, Leopold, 251
Mozart, Wolfgang Amadeus, 19,
    20, 250–53, *251*, 260–61,
    262
multicellular animals, first, 329
Murray, Jim, 35, 37, 38
mutationism, 356
mutations
    in evolution of land snails, 30

polydactylous, 73–74
pseudogenes resulting from, 403
rates of, 398–99
Myxine (hagfish), 411

Nairobi (Kenya), 313, *314*
natural selection, 146–47
    Darwin's inspirations for, 148–49
    in evolution of human brain, 321
    in evolution of land snails, 27–30
    in evolution of pigeons, 360–62
    limits on, 384–85
    mass extinctions and, 317
    neutral theory versus, 401–2
    possible modes of, 304
    progress not implied in, 302–3
    as source of evolutionary change,
        397–98
    theories competing with, 356–57
    Walcott on, 226
    wedge metaphor for, 302
nautiloids, 304, 312
Neanderthals, first fossils of, 126
Nebuchadnezzar (king, Bablyon),
    189
Neckwinder, Emil, 210, *211*, *212*
Neckwinder, Emma, 210, *212*
Neckwinder, Eva, 210
Neckwinder, Evaline, 210
neoteny (paedomorphosis), 391–95
    in faces of domesticated animals,
        393
neutral theory, 398–403
Nevo, E., 404, 405
Newton, Sir Isaac, 176, 426

observation, 200–202
oceans, salt in, 170–74, 178
Old World (cercopithecoid) monkeys,
    287–88
Omar I (caliph), 340, 341
O'Neill, Eugene, 304
ontogeny, 368, 387
    of dogs, 388
    of pigeons, 367
onychophorans, 337, 343–47, 350–51
    Hallucigenia as, 349–50
organelles, 320

orthogenesis, 357
  in evolution of pigeons, 363–68
Osborn, Henry Fairfield, 223, 226
ostracoderms, 411, 413–14, 418, 420,
    421, 424, 436, *436*
Otala (land snail), 37
Owen, Richard
  on archetypes, 63
    Darwin on, 64
  on ichthyosaurs, 80–81, 83–86, 91,
    94
  on number of digits, 72
oxen, *393*

paedomorphosis (neoteny), 391–92
pain, 142–43
paleontology
  age of life on earth determined by,
    327–28
  CT scans used in, 89
  inevitability of extinction in, 46
  regulation of, 276
  time scale in, 48
Paley, William, 138–52
pandas, 44, 74
Pareto, Vilfredo, 164
Parkinson, James, 447
Partula (land snail), 13–14, 29–40, *33*
passenger pigeons, 52–54, 364
Patten, William, 409–11, *410*, 429
  on animal evolution, *434*
  on *Cephalaspis*, 416–22, 425, 432–36
  Great Highway of evolution of, *431*
peacocks, tail feathers of, 374
Peel, J.S., 334, 336, 350
Peking Man *(Homo erectus)*, 127
pelycosaurs, 107
pentadactyl limbs, 64–77, *70*
Pentreath, Dolly, 249
Peripatus, 342–46, 352
periwinkles, left-coiling, 440, 452–55,
    454
Permian mass extinction, 305, 308
Péron, P., 441
Petrified Forest National Park
    (Arizona), 275
Petromyzon (lampreys), 411
Physalaemus coloradorum (frog), 378–79

Physalaemus pustulosus (Tungara frog),
    377–80
pigeons, 355, 382
  color patterns of, *363, 365*
  English fantail, *361*
  passenger pigeons, 52–54
  Whitman on, 356–68
pigs, 390, 391, *393*
Pikaia, 225
Pilbeam, David, 289
Pinaleno Mountains (Arizona), 44–45
Pinza, Ezio, 14, 23, 40
Pithecanthropus erectus (Homo erectus),
    127–30, 132, 134–36
plankton, Cretaceous mass extinction
    of, 308–10
plankton line, 308
plants
  genetic engineering of, 383
  Goethe on forms of, 158–65
  plankton, 308–10
  Zostera marina (eelgrass), 57–60
Plato, 63
Platyhelminthes, 336
Poecilozonites (land snail), 26–27, 37
Polo, Marco, 326
Polypterus (fish), 114
population, Malthus on, 148
Pratt, H., 405
prehallux digits, 74
premaxillary (Goethe's bone), 157–58
prepollux digits, 74
primates, 183–84
  evolution of, 284–95
  sense of sight in, 427
progress
  Darwin on, 302–3
  as delusion, in evolution, 323
  in evolution, Patten on, 419,
    433–35, *434*
  through extinctions, 304–5
  mass extinctions and, 307, 316–18
  scientific, theories of, 411–12
prokaryotes, 319–20, 323
Prothero, Donald, 278
pseudogenes, 403
Ptolemaic dynasty (Egypt), 188*n*,
    340

punctuated equilibrium, theory of, 277–79

quail, 379–80
Quoy, J., 441–42

race, Darwin on, 266–67, 269, 272–74
radiates, 415
Ramapithecus, 290
Ramsköld, L., 347, 349–50
randomness
    different meanings of, 396–97
    in evolution, 397–98
    in mass extinctions, 307
    in neutral theory, 401
    in variation, 361–62
Raphael, 153
Raup, David, 46, 308
Ravel, Maurice Joseph, 95
Ray, John, 140
Raymond, Percy, 227–29, *230,* 409
redundancy, of organs, 117–18
Reichert, C.B., 102–5
religion
    in Amana colony, 207
    Freud on origin of, 322
    in orthogenesis, 366–67
    science versus, 183
    spread by missionaries, 263–65
    Walcott on science and, 226
reproduction
    avoiding incest in, 379–80
    by land snails, 26
    Paley on, 138–40
    sexual selection in, 374–78
reptiles, 68–69
    ichthyosaurs, 80–94
    jaws of, 106
restaurants, 244–45
rock pigeon *(Columba livia),* 358, 362
rocks, oldest, 328
rodents
    fossil record of, 278
    mole rats *(Spalax ehrenbergi),* 403–6
Roman Catholicism, Ussher on, 183–84
Romance languages, 337–38
Roosevelt, Theodore, 224

roots (on plant), 159–60
Rosenberg, Joseph A., 198–200, *199,* 204
Rückert, Friedrich, 209
Rudwick, M.J.S., 428
Russell, Joseph B., 250*n*
Ruth, Babe (George Herman), 357
Ryan, Michael J., 377

Saarinen, Eero, 355
Sacks, Oliver, 260
Sagan, Carl, 169
Saint Louis (Missouri), 355–56
Saint Louis World's Fair (1904), 355–56
salt, in oceans, 170–74, 178
San Francisco (California), 238–41
    earthquake in (1989), 245–46
sap (in plants), 162–64
Säve-Soderbergh, Gunnar, 66
Sayers, Dorothy, 455
Scaliger, J.J., 187
Schevill, Bill, 227
Scheuchzer, J.J., 83
science
    ambiguities in, 440
    changing theories in, 430
    comparative morphology as, 424
    false ideas in, 164
    fraud versus error in, 109
    historical, 77
    illustrations used in, 427–28
    observation in, 200–202
    parochialism and sexism in, 156–57
    religion versus, 183
    theories of progress in, 411–12
    Walcott on religion and, 226
Scottsbluff (Nebraska), 204
Sears (restaurant), 238–41
Sepkoski, Jack, 47
Septuagint (Greek Bible), 187, 188
sex
    human, Darwin on differences between, 267–68
    as pleasurable, Paley on, 139–40
sexism
    Darwin's, 268
    in science, 156

sexual selection
  by females, 374–79
  incest and, 379–80
Shaler, N.S., 18
sharks, 114, 118, 144
Shubin, Neil H., 69–74
Simmonds, R.T., 337, 338
skulls
  of cats and dogs, *390*
  of dogs and other canids, 387–88
  of domesticated animals, *393*
slavery, 267
  Darwin on, 272–74
Smith, Adam, 148–51
Smith, E.F., 453–55
snails
  Achatina, 36–39
  Euglandina, 37–39
  evolution of defenses against crabs
    in, 304
  found at West Turkana, 293
  land snails, 26–29
  Littorina littorea (periwinkle), 452–53
  Lottia alveus (limpet), 56–60, *57*
  Partula, 13–14, 29–40, *33*
  periwinkles, left-coiling, 440,
    452–55, *454*
snakes, 107
Snell, Otto, 131–32
Society of True Inspirationists
    (Amanaites), 207, 209
Sontag, Susan, 162
Spalax ehrenbergi (mole rats), 403–6
species
  diversity of, 242–43
  extinction as inevitable for, 46
  importance of preserving, 241
  mass extinctions of, 305–12
  punctuated equilibrium in evolution
    of, 277–79
  subspecies and, 43
  "umbrella" (indicator), 44
Spencer, Herbert, 439–40
Spinoza, Baruch, 370
spotted owls, 44
squirrels, Mount Graham Red Squirrel
    (*Tamiasciurus hudsonicus
    grahamensis*), 43–48

SSF (small, shelly fauna), 330–32,
    334–36
stapes (stirrup; bone), 100–103, 107
Steno, Nicolaus, 18
Stensiö, Erik Andersson, 422–25, 429
  Cephalaspis brain and nervous
    system reconstructed by, *423*
  on ostracoderms, 435–37, *436*
Stetson, Henry, 228
subspecies, 43
Sullivan, Sir Arthur Seymour, 206
super-normal stimuli, 258–59
swim bladders (of fish), 111–19
swordfish *(Xiphias gladius)*, 374
swordtail fish *(Xiphophorus helleri)*,
    375–76
synonymous substitutions, in DNA,
    402

Tahiti, 23–24
  land snails of, 27, 30–33, 39
  "moral state" of, 263–65
  tree snails of, 36
Tamiasciurus hudsonicus grahamensis
    (Mount Graham Red Squirrel),
    43–48
taxonomy, 343–45
teleosts (bony fish), 114, 118–19
Tennyson, Alfred, 371
tetrapods
  pentadactyl limbs of, 64–77, *70*
  stapes bones of, 101–2
textbooks, Ussher's chronology
    criticized in, 184–88
Theunissen, Bert, 129–30
Thomas, Lewis, 215
Thompson, D'Arcy W., 438
thumbs, 74
Tierra del Fuego (Argentina-Chile),
    270–72
time
  geological, 277
  Halley on limits of, 175–79
  physical and biological, 216
Tinbergen, Niko, 19, 257–58
Titanotheriomys, 279
Tommotian fauna, 329–30
  SSF (small, shelly fauna), 330–32

Toxodon, 80
transposons (jumping genes), 319
tree snails, 36–39
Trigonia (clam), 441–51, *443*, *444*
Trigonia margaritacea (clam), 443–45
trilobites, 329
  Cephalaspis mistaken for, 411, 415, 418
  ostracoderms and, *417*
Truman, Harry, 308
Tulerpeton, 66
Tungara frog *(Physalaemus pustulosus)*, 377–80
Turkanapithecus, 294
Twain, Mark (Samuel Clemens), 168

"umbrella" (indicator) species, 44
Uniramia, 344
urodeles, 72
Ussher, James, 18–19, 181–93, *182*

variation
  at gene level, 399
  randomness in, 361–62
Vermeij, G., 56, 304
vertebrates
  Cephalaspis (jawless fish) as, 413–25, 429–30
  evolution of lungs and swim bladders of, 111–19
  fossils of, 277
  Goethe's bone (premaxillary) in, 157–58
  ichthyosaurs, 80–94, *82, 85, 87*
  incorrect theory of origin of, 410–11
  Owen's archetype of, 63–64
  pentadactyl limbs of, 64–77, *70*
  terrestrial, bones of inner ear of, 98–108
  theories of relationships among, 430–38, *431, 434, 436*
Victoria (queen, England), 371
Virchow, Rudolf, 135
Voltaire, 142

Wagner, Richard, 41
Walcott, Charles Doolittle, 223–24, 229–32, *230*
  on *Aysheaia*, 346
  Burgess Shale discovered by, 224–26
  calling card of, 221, *222*
  Clark on, 232–36
  on religion and science, 226–27
Walcott, Mary Vaux (Mrs. Charles Doolittle Walcott), 221–22*n*, 229
Wallace, Alfred Russel, 29, 30, 302
Wambsganss, Bill, 129
Washington, George, 191
watchmaker metaphor, 142
Watson, D.M.S., 71
Wayne, Robert K., 387–91
whales, 64
Whitman, Charles Otis, 18, 356–69
Whittington, Harry, 225, 229, 235
Wilde, Oscar, 396
Wilder, Laura Ingalls, 220–21
Wilder, Rose, 221
Wilford, John Noble, 169
William IV (king, England), 270
Wilson, Alexander, 53
Wilson, Mookie, 78
Wilson, Woodrow, 224
Wiwaxia, 337
wolf *(Canis lupus)*, 386
women
  Darwin on, 267–68
  as scientists, 156
  of Tahiti, 23–24
World Series, 240, 246
Wright, Sewall, 168

Xenusion, 346, 350
Xiphias gladius (swordfish), 374
Xiphophorus helleri (swordtail; fish), 375–76, 380
Xiphophorus maculatus (fish), 375–76

yard (unit), 138
Yeats, William Butler, 98

Zeno, 372
Zostera marina (eelgrass), 57–60